Limit state theory for reinforced concrete

PLATE I **The Bristol Road Bridge, Birmingham**
(Reproduced by kind permission of the Birmingham City Council and the Bournville Village Trust. Consulting Engineers, JEC Farebrother & Partners)

PLATE II Examples of the blemishes that can occur if the limit state of crack width is not considered

Limit state theory for reinforced concrete
SI units

B P Hughes
B Sc, Ph D, C Eng, M I C E, M I Struct E
Senior Lecturer in Civil Engineering and
Reader in Concrete Technology
University of Birmingham

PITMAN PUBLISHING

First published 1971

SIR ISAAC PITMAN AND SONS LTD
Pitman House, Parker Street, Kingsway, London WC2B 5PB
PO Box 6038, Portal Street, Nairobi, Kenya

SIR ISAAC PITMAN (AUST) PTY LTD
Pitman House, 158 Bouverie Street, Carlton, Victoria 3053, Australia

PITMAN PUBLISHING COMPANY SA LTD
PO Box 11231, Johannesburg, South Africa

PITMAN PUBLISHING CORPORATION
6 East 43rd Street, New York, NY 10017, USA

SIR ISAAC PITMAN (CANADA) LTD
495 Wellington Street, Toronto 135, Canada

THE COPP CLARK PUBLISHING COMPANY
517 Wellington Street, Toronto 135, Canada

ISBN 0 273 36002 7 (Cased edition)
0 273 36037 X (Paperback edition)

Made in Great Britain at the Pitman Press, Bath
G1 (T231/1343 : 73)

Preface

This book aims to give an understanding of reinforced concrete and the rational concepts implicit in the new Unified Code. Limit state theory and its applications in both analysis and design are presented as clearly and concisely as possible for both students and practising engineers. The text is based on lecture notes which have been developed over many years in the Civil Engineering Department of the University of Birmingham. It has been suitably extended where necessary to give adequate coverage of the new Unified Code as well as to include topics which are known to be required by other universities. It is hoped to publish companion volumes on plain structural concrete and on prestressed and precast concrete shells and models.

Ease of reading and learning has been facilitated by developing the subject matter in rational stages and providing suitable section and paragraph headings throughout. References to important research papers which would be difficult for the reader to understand in the present context have been omitted. These can usually be found in the few carefully selected references which have been included. For example, a vast research effort has been devoted to shear effects in reinforced concrete beams and numerous research papers on this topic deserve mention. However, one reference only has been given, chosen because it attempts to simplify a complex problem by a process of elimination which is similar to the approach adopted by the Code.

A sign convention should (i) conform with the natural use of the materials, (ii) be consistent throughout the text, and (iii) conform as far as possible with existing practice for steel and other structures, in that order of importance. This has been achieved with the sign conventions which have been used throughout and described in Appendix Y. SI units, as described in appendix Z, have also been used. The general notation and symbols are in accordance wherever possible with BS 1991:1961 *Recommendations for letter symbols, signs and abbreviations; part 4: Structures, materials and soil mechanics.*

It is suggested that most of chapters 1 to 5, section 6.1, chapter 8 and sections 9.1 to 9.4 could be included in a second-year undergraduate course and the remainder reserved for the final year, but this can vary considerably between universities. At Birmingham the concrete structures and technology course is complementary to the general structures course. However, problems which are concerned entirely with general structural analysis, and which have been set in the concrete section of the examination papers of, for example, King's College, London, have been included in the exercises. One or two examples of such questions are to be found in the exercises at the ends of the chapters. Questions from recent examination papers are reproduced by permission of the Universities of Birmingham and London (King's College). The university course and year, and the time allotted, are shown in brackets where appropriate, e.g. "(UL2 1969, 36 min)" indicates a 36-minute question from the 1969 second-year paper for internal students of King's College.

Any publications which I have read, such as books by Military Engineering, Reynolds, Faber, Rao, Baker, Jones, Evans & Wilby, Davies, and Johnson, must have had some influence on the present text quite apart from that of the Unified Code itself. My thanks are, however, especially due to the following friends and colleagues for their comments and assistance: Dr M. R. Prince (general and appendix E), Dr J. E. Ash and Dr W. B. Cranston (general), Mr J. E. C. Farebrother, Dr J. M. Illston, Professor L. L. Jones and Dr P. Swannell (chapter 22) and Mr J Guest (appendix F); and to my secretary Mrs S. Palin who typed the manuscript throughout.

Extracts from *The structural use of concrete: draft code for comment* (known as the Unified Code); CP3 *Codes of basic data for the design of buildings*, chapter V part 1: 1967 *Dead and imposed loads;* BS4449:1969 *Specification for hot rolled steel bars for the reinforcement of concrete;* and BS4466:1969 *Bending dimensions and scheduling of bars for the reinforcement of concrete*, are reproduced by permission of the British Standards Institution, 2 Park Street, London W1Y 4AA, from whom copies of the complete documents may be obtained. Extracts from the reports on *The detailing of reinforced concrete* and *A standardized procedure for the presentation of calculations for reinforced and prestressed concrete structures: draft for comment* are reproduced by courtesy of the Councils of the Concrete Society and the Institution of Structural Engineers. Where material has been extracted from a draft document it should be understood that the extract is from the latest version available to the author but it is subject to modification and only the published document is authoritative.

Finally I wish to thank my wife and family for their patience and encouragement during the writing of this book.

Contents

To Anne

Introduction

The introduction of *ultimate load theory* for the design of reinforced concrete at first tended to discredit the traditional *elastic theory* approach. Steadily increasing knowledge, however, has brought each theory into perspective and has shown that both have their application within the wider concepts provided by *limit state theory*. The limit states of deflection and local damage both utilize elastic theory while the limit state of collapse incorporates ultimate load theory.

The rational approach and philosophy of limit state theory does more than simply adopt existing elastic and ultimate load theories. It introduces the concepts of partial safety factors, characteristic loads and characteristic strengths based on a semi-probability approach. It provides the framework which will enable design to be more accurate and therefore more economical. The former use of a single safety factor included a large allowance for ignorance which can now be drastically reduced. Limit state theory has enabled design methods to be presented in the Unified Code in such a way that the results of research work can be incorporated as they become available. It provides guide lines for future research by focusing attention on those areas where knowledge is lacking.

The elastic theory as given in CP114:1957 used a (stress) *factor of safety*—i.e. ultimate-strength/working-stress ratio—of 3 for concrete and 2 for steel. Ultimate load theory as originally given in CP114:1957 used a *load factor* (of safety)—i.e. ultimate-load/working-load ratio—of 2, together with a factor of safety for concrete strength only. Limit state theory uses (at least) two partial safety factors, in association with characteristic loads and strengths, to allow for predicted variations in a rational manner. This enables the design to be carried out with greater confidence and greater economy.

List of Symbols

a	lever arm ($= a_1 d_1$) [z]; volume fraction of coarse aggregate
a_1	lever arm factor
A	cross-sectional area (with suffixes as for f); cross-sectional area of one bar of diameter D
A_c	(gross) concrete area ($= bd$) [A_g]
A_{cc}	actual (or net) concrete area in compression
A_e	area of equivalent (cracked) concrete section
A_o	area of equivalent (uncracked) gross concrete section ($= bd + mA_s$)
A_{sa}	cross-sectional area of additional main steel for torsion [A_{sl}]
A_{sh}	cross-sectional area of both legs of hoop steel for torsion
A_{sq}	cross-sectional area of both legs of stirrup [A_{sv}]
b	breadth; volume fraction of fine aggregate
b_r	breadth of rib [b_w]
c	distance from nearest point of zero strain [d_c]; volume fraction of cement
C	constant
${}_N C_n$	number of combinations of n items from N items
C_t	ratio of creep strain to short-term strain
d	depth [h]; greater lateral dimension of column [h]
d_1	effective depth [d]
d_2	depth to compression reinforcement [d']
d_n	neutral axis depth ($= n_1 d_1$) [x]
d_s	depth of flange [h_f]
D	diameter of reinforcing bars [d_b]; deviation ($= x_i - \mu$)
D_i'	deviation of class i from class o in terms of class width w
e	base of natural logarithms
e	eccentricity; apparent longitudinal tensile strain in concrete
E_c	effective modulus of concrete
E_s	modulus of elasticity of steel
f_b	average bond stress
f_c	(compressive) concrete stress (+ve or −ve values)
f_c'	cylinder crushing strength
f_{cb}	concrete stress in bending at extreme fibre (+ve values)
f_{cc}	concrete stress in compression (+ve values) [f_c]
f_{ct}	concrete stress in tension (+ve values) [f_r]
f_{cu}	ultimate compressive stress in concrete
f_i	frequency of values within class i
f_s	(tensile) steel stress (+ve or −ve values)
f_{sc}	steel stress in compression (+ve values) [f_s']
f_{sq}	steel stress (in tension) in shear reinforcement
f_{st}	steel stress in tension (+ve values) [f_s]
f_{st}'	tensile stress in steel at a crack
f_{ya}	yield strength of additional main steel for torsion
f_{yc}	yield strength of steel in compression
f_{yh}	yield strength of hoop steel for torsion

f_{yq}, f'_{yq}	yield strength of shear reinforcement [f_{yv}, f_{yb}]
f_y, f_{yt}	yield strength of steel in tension [f_y]
F	force (with suffixes as for stress, f); cumulative frequency
G	shear modulus
h	constant; effective depth of column (with suffixes d and t) [l_e (with suffixes x and y)]
i	integer
I	second moment of area
I_b	second moment of area of (gross) concrete section of beam
I_c	second moment of area of (gross) concrete section of column
I_e	second moment of area of equivalent (cracked) concrete section
I_o	second moment of area of uncracked section
J	torsion constant [C]
k, K	various constants or coefficients [ϕ, k, K]
K_b, K_l, K_u	stiffnesses of beam, lower column and upper column respectively
K'_r	relative stiffnesses column to beam [β]
l	effective span
L	span; length [l or L]
m	modular ratio ($= E_s/E_c$); positive resistance moment per unit length of yield line
m'	negative resistance moment per unit length of yield line
M	bending moment
M_c	effective cracking moment [M_{cr}]
$M_{cx}(M_{cy})$	total centre moment about major (minor) axis for column [M_{tx}, M_{ty}]
M_e	end moment
$M_{ex}(M_{ey})$	end moment about major (minor) axis for column [M_{ix}, M_{iy}]
M_r	design moment of resistance (lesser of M_{rc} and M_{rt})
M_{rc}	moment of resistance in compression
M_{rct}	moment of resistance of concrete in tension
M_{rt}	moment of resistance in tension (cracked section)
M_u	ultimate moment of resistance by analysis (with suffixes as for M_r)
m_x	resistance moment, per unit length, of x-direction reinforcement
m_y	resistance moment, per unit length, of y-direction reinforcement
n	number in a sample; neutral axis depth factor in terms of d
n_1	neutral axis depth factor (cracked section)
N	population size; neutral axis depth factor for uncracked section
o	perimeter of one bar of diameter D [b_o]
p	permissible stress (with suffixes as for f); probability of success of an event
p''	percentage decrease in moment due to redistribution
P	permissible force
q	shear stress [v, τ]; shear stress due to normal bending
q_b	flexural bond stress
q_c	critical shear stress given by table C3 [v_c]
q_t	shear stress due to torsion [v_{tc}]
Q	shear force (with suffixes $*$ and k as for W) [V]
r	tension-steel ratio ($= A_{st}/bd_1$) [ρ]
r'	tension-steel ratio on gross area ($= A_s/A_c$ or A_{st}/bd); total steel ratio on column gross area ($= (A_{sc} + A_{st})/bd$)
r_c	compression-steel ratio ($= A_{sc}/bd_1$)
r_{crit}	critical steel ratio (always on gross area)
r_e	equal-strength steel ratio in tension (elastic design)
R	radius [r]; restraint; ratio [α, β]

s, s'	spacing of secondary reinforcement [t]
s	crack spacing; estimate of standard deviation
t	thickness; time; smaller lateral dimension of column [b]
T	torsional moment (with suffixes as for E)
u	cube strength of concrete [f_{cu}]
u_k, u_w	characteristic (or works) cube strength of concrete [f_u]
u_t	tensile strength of concrete
V	coefficient of variation ($= \sigma/\bar{x}$) [δ]
w	class width; intensity of distributed load [$g + q$]; volume fraction of water
$\bar{W}$	mean load applied due to particular type of occupancy
W_d	characteristic dead load [G_k]
W_i	characteristic imposed load [Q_k]
W_k	characteristic loads generally [F_k]
W_{sh}	characteristic load for shrinkage, creep and thermal effects if significant
W_v	variable (design) load ($= 0{\cdot}4W_d + 1{\cdot}6W_i$ for collapse)
W_w	characteristic wind load [W_k]
$_*W$	design load ($= \gamma_l W_k$) suffix $_*$ used only if necessary for clarity)
x, y, z	independent variables
x_i	i-th value of an independent variable
x'_i	value of an independent variable for midpoint of class i
$\bar{x}$	estimate of population mean or sample mean
x_1	smaller centre-line dimension of hoop
X, Y, Z	various factors as given by design rules
y	distance from neutral axis, the positive direction being *downward* for beams
y_1	larger centre-line dimension of hoop
z	coefficient ($= D/\sigma$)
Z	section modulus ($= I/y$)
α, β, γ	elastic torsion functions; various coefficients
α_c	coefficient of thermal expansion of concrete
α_s	coefficient of thermal expansion of steel
γ_l	partial safety factor for loads [γ_f]
γ_m	partial safety factor for material resistance
δ	deflection [a]
Δ	crack width [w]
ε	strain (with suffixes as for stress)
ε_{ct}	short-term strain in concrete
ε_{cu}	ultimate compressive strain in concrete
ε_{sh}	free shrinkage strain in plain concrete
ε_{ult}	ultimate tensile strain in concrete
θ	temperature change; angle [α, β]
μ	true mean
$\mu M_{ex} (\mu M_{ey})$	primary centre moment for column about major (minor) axis [kM_{ix}, kM_{iy}]
ν	Poisson's ratio
ρ	average bond strength
σ	true standard deviation
σ_1, σ_2	principal stresses
σ_k	characteristic strength or resistance [f_k]
$\sigma_{\bar{x}}$	standard error of the mean
$\bar{\sigma}$	mean strength or resistance

$_*\sigma$	design strength or resistance
ΣM_x	total end moment about major axis for unrestrained long column
ϕ	angle; curvature $(= 1/R)$ $[1/r_b]$
ϕ_{ct}	increase in curvature with time

Note: Proposed symbols based on the notation agreed internationally at the September 1969 meeting of the Comité Européen du Beton are shown above in square brackets wherever they are expected to differ from those used in this book.

Chapter 1

Introduction to limit state theory

Limit state theory is conveniently introduced by considering the fundamental purpose of design so that *structural* design can be seen in its correct perspective. Rowe *et al.* [1.1] have defined the purpose of design as "the provision of a structure complying with the client's (and users') requirements". The client is often uncertain of his requirements when he first approaches an engineer (or architect), so the latter should assist him to define them more precisely.

Although the client automatically assumes that the design will satisfy such essential requirements as structural safety and durability, many joint decisions must be made before the engineer has a satisfactory design brief. For example, what are the client's relative priorities regarding early completion, low initial cost and low maintenance? What is the proposed design life of the structure and does the client envisage a change of use within this period that could increase the applied loadings? If so, should the structure (or parts of the structure such as the foundations) be designed for the increased loadings or will the client meet the cost of strengthening the structure as and when necessary? Accurate selection of design loadings cannot be expected to include an allowance for changes in use which are not intended at the design stage. Although the large factors of safety (or ignorance!) used in the past would often allow such changes, it is clear that there will be an increasing need in the future for the engineer to help the client to clarify his requirements and intentions.

Once the basic requirements have been defined, the structural engineer becomes a member of a design team. It should be realized, therefore, that it is the greatest overall economy that is important, rather than the economy of the structural design or any other individual part. For example, the successful enclosure of space provided by a building requires the creation of an environment which is pleasant to all the senses of the occupant. It should be adequately illuminated, heated and ventilated, have an attractive appearance and be satisfactory acoustically, as well as being structurally sound. The cost of the mechanical and electrical services in a hospital, for instance, can be as high as 40% of the total.

The object of structural design is to obtain a structural solution which gives the greatest overall economy by providing the maximum assistance in satisfying all the other requirements of the structure. It is clear that structural costs rise rapidly if other requirements become too onerous—if, for example, very large clear spans are specified in a multi-storey garage. It is equally clear, however, that structural costs cannot be considered in isolation. A multi-storey car park with closely spaced columns would not perform its primary function very efficiently.

1.1 Philosophy of structural design

Satisfactory design must ensure the achievement of an acceptable probability that the specified life of a structure is not curtailed prematurely due to the attainment of an unsatisfactory condition, or *limit state*. The acceptable probability should be chosen in such a way that a satisfactory balance is achieved between the cost of construction and the combined costs of insurance premiums (which are related to the probability of structural or functional failure) and of maintenance (the costs for design, normal operation and use, and insurance for flooding etc., are normally independent of structural safety and serviceability). Limit state design is therefore based, where possible, on the application of the methods of statistics to the variations that occur in practice.

Limit states

Limit states are concerned with structural safety and serviceability and cover all forms of failure. Unfitness for use can occur in many ways. It is therefore often convenient to group several limit states together (e.g. collapse, local damage and deflection) or even to consider two major categories only, as follows:

(i) Ultimate limit states: the usual collapse limit states, also collapse due to fire, explosive pressure, etc.

(ii) Serviceability limit states: local damage and deflection limit states, durability, vibration, fire penetration and heat transmission, etc.

Often some of the limit states are inapplicable and others can be satisfied by suitable detailing of the structure. The usual approach is to design on the basis of the most critical limit state and then check that the criteria governing the remaining limit states are satisfied.

Limit states of collapse

Collapse of the structure, or part of the structure, may arise from the rupture of one or more critical sections, from buckling due to elastic or plastic instability (including the effects of sway where appropriate), from the loss of static equilibrium (e.g. due to transformation into a mechanism, due to loss of stability by overall tilting etc), or deterioration arising out of fatigue effects.

Limit states of deflection

Deflection or deformation becomes excessive when it impairs the appearance or efficiency of the structure, affects non-load-bearing members or finishes adversely, or causes discomfort or alarm to the users of the structure.

Limit states of local damage

Local damage, such as cracking or spalling of the concrete, is excessive if it impairs the appearance, efficiency or durability of the structure or causes damage to finishes, partitions etc. For example, in water-retaining structures the limit state of crack formation in tension on the water-retaining face should be avoided. In structures exposed to normal weathering, on the other hand, the formation of cracks can be tolerated provided that the width of the individual cracks is insufficient to allow the ingress of moisture, which would corrode the steel reinforcement, hence it is necessary to check the limit state of crack width. Similarly, it may be necessary to check the limit state of crack formation in compression to ensure that no initial microcracking, which could be harmful to the durability of the member, is produced at any stage of construction in zones subject to high compressive stresses.

Limit state of vibration

Excessive vibration causes discomfort, alarm or actual damage, or interferes with the proper function of the structure. Acceptable limits to the level of vibration vary according to usage. For example, pedestrians on a bridge can be very alarmed by vibrations which are not even noticed by the occupants of a vehicle. It may be necessary to isolate the source of vibration or, in certain circumstances—for example, if sensitive measuring instruments are to be installed—to isolate a part or even the whole of the structure.

Limit state of fatigue

The effects of fatigue should be considered if the imposed load on a structure, or part of a structure, is predominantly cyclic in character. Particular attention should be given to the deflections which would occur under repetitions of load, to ensure that these are within acceptable limits. Failure of the materials in fatigue is only likely to govern design when welding of reinforcement is introduced, or when new types of reinforcing or prestressing steels are used.

Limit state of durability

Special precautions may be necessary for exceptionally severe environments. However, compliance with the requirements for concrete cover to reinforcement, cement contents and local damage ensures adequate durability for most structures.

Limit state of fire resistance
There are three conditions to be considered for structural elements which may be subjected to fire: resistance to structural collapse, resistance to penetration of flames and resistance to heat transmission.

Limit states of impact resistance
Structures or structural elements which may be subjected to impact, explosions or earthquakes (inertia forces) must always be considered for structural collapse. This is particularly necessary for precast construction where the small quantities of continuity steel can render such structures very vulnerable, compared with continuously reinforced in-situ construction. If there is a fairly high probability of shock loading, e.g. in earthquake zones, suitable loadings should be included when considering the limit states of collapse, deflection and local damage.

1.2 Avoidance of limit states

An acceptable probability that the structure will not reach a limit state throughout its specified life can only be provided if variations in strengths of the constructional materials, and variations in loads on the structure, can be estimated. Ideally, all the probable variations for each limit state should be predicted from statistical data. Although only some of the relevant data are at present available, it is still possible to implement the main principles of the limit state philosophy. Each limit state is considered individually and the variations (or uncertainties) can be due to one or more of the following causes:

Variations in strength of the structure
Variations which can affect the material strength in the structure are:

(i) Those caused by inherent variability of material strength due to the material's method of manufacture and quality of constituents. These variations can be allowed for by a *characteristic strength* which is calculated using statistical principles.

(ii) Variations due to other causes, such as construction procedure (treatment, curing); state of stress (axial, biaxial, triaxial); speed, duration and repetition of loading; and environmental conditions in the actual structure. Statistical principles cannot be applied, either because the data are inadequate or because the usual assumptions made in statistical analysis are invalid. These variations are therefore allowed for by a *partial safety factor for material strength.*

Variations in loads on the structure

Variations which can increase the loading on a structure are:

(i) Those due to the inherent variability of loads applied to a structure of a particular type throughout its design life. These can be allowed for by a *characteristic load.*

(ii) Variations due to other causes. For example, the load on a member may be excessive due to errors in the design assumptions. These variations are covered by a *partial safety factor for loads.*

Gross defects

Gross defects in design or construction, such as positioning of the reinforcement in the wrong face of a member, must be avoided entirely by checking and inspection. Although the absence of gross defects and exceptional events (see below) are assumed, it may also be assumed that a major catastrophic failure could not, in general, be caused by a single such defect or event, but only by a combination of at least two major errors from either (i) design (ii) construction, and/or (iii) use. A major error by the user, such as gross overloading, or installation of explosive equipment not in accordance with the designer's brief, should not cause complete catastrophic failure unless combined with a major design or construction error.

Exceptional events

Truly exceptional events, such as major national disasters or enemy action, can only be covered completely by insurance, if at all, and one such event alone would be sufficient to cause catastrophic failure. However, events that are somewhat less exceptional, such as explosions, earthquakes or accidents due to fire or internal gas explosions, should be considered in the design even though extensive damage to the structure can occur if such events take place. Damage to the structure can be localized, and possible loss of life reduced, by taking sensible precautions to provide a certain minimum resistance to fire and impact. A particularly disturbing feature of designs based on CP114, CP116, etc., was that impact could be ignored. This disregard of limit states of impact, combined with an exceptional event, could result in a progressive and catastrophic collapse. Exceptional events with an even higher probability of occurrence, such as earthquakes in regions where they are known to occur, should be considered specifically in the calculations for collapse and serviceability.

1.3 Some basic statistical principles

Limit state design is a probabilistic design method in which the values of certain basic design parameters are considered as random. It therefore uses statistical techniques to analyse the variations which occur in practice, either in the strengths of constructional materials (due to their method of manufacture) or in the loads on structures of a given type. A brief outline of some elementary statistical principles (which are covered more fully elsewhere, e.g. references 1.2 and 1.3) is therefore very relevant since it is assumed that the above variations effectively follow a normal (gaussian) distribution for practical purposes.

Errors

There are two main types of error:

(i) *Systematic errors.* These are due to effects which act consistently under the given circumstances (e.g. an incorrect zero on a compression testing machine, or a measuring cylinder calibrated at one temperature and used at another). Such errors should be avoided by suitable experimental techniques.

(ii) *Random errors.* These are due to a large number of small effects such as imprecisions in interpolating between graduations on a scale, or imprecisions in allowing for temperature variations when using equipment which has been calibrated at a standard temperature. Notice that these effects have an equal probability of being positive or negative; hence the probability of a given error being positive is one half and the probability of its being negative is also a half.

Gauss function

If it is assumed for convenience that all the small contributory random errors are of equal absolute magnitude ΔD, and for any particular case there is a surplus (or deficiency) of positive errors equal in number to $2r$, then the resultant positive error D equals $2r\Delta D$. If the total number of errors is $2n$ then the total error can range in individual cases from minus $2n\Delta D$ (all negative) to plus $2n\Delta D$ (all positive).

These are the errors that cause variations between observations even under the most carefully controlled conditions. It can be shown from mathematical considerations that the distribution of these errors is given by the so-called Gauss function:

$$y = Ce^{-h^2D^2} \qquad (1.3.1)$$

where $y\Delta D$ is the frequency with which the error lies in the range D to $D + \Delta D$, C is a constant equal to the maximum value of y, and h is a constant related to the spread of the errors. These constants are related to the total number of observations, n, and the standard deviation, σ (which is defined later), as follows:

$$C = n/\sigma\sqrt{(2\pi)}$$

$$h = 1/\sigma\sqrt{2}$$

We shall reconsider the Gauss function later on.

Normal distribution

The distribution given by the gauss function is commonly known as the *normal* distribution. This distribution is shown in figure 1.3.1

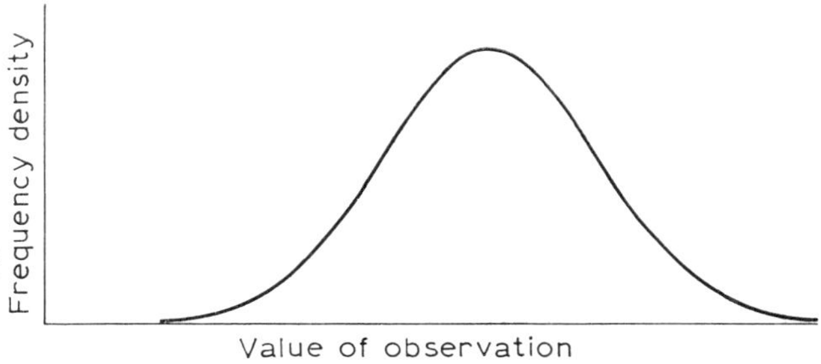

Fig 1.3.1 Frequency density plotted against value of observation for normal probability distribution.

by plotting frequency density against value of observation. Notice that while the individual values are equally distributed about the mean value, the extreme values can be at $\pm\ \infty$. Hence although the probability of the occurrence of an exceptionally high or low value is extremely small, *it is never zero.*

Standard deviation

The *deviation* of an individual observation x_i is given by $(x_i - \mu)$ where μ is the true mean. The *standard deviation* is the root-mean-square deviation and is therefore always positive. The true standard deviation is given by

$$\sigma = \sqrt{\left[\frac{1}{n}\sum_{i=1}^{n}(x_i - \mu)^2\right]} \qquad (1.3.2)$$

It is important to notice that μ is the true mean and not the mean of the actual number, n, of observations which have been taken. The true mean is usually unknown and it is necessary to calculate the standard deviation using the sample mean, $\bar{x}$. The sample mean will generally differ from the true mean, and the direct substitution of $\bar{x}$ for μ in the above expression gives a minimum value of $\sum(x_i - \mu)^2$. It is therefore necessary to apply a correction when using deviations from the sample mean instead of deviations from the actual mean, such that the *estimated* standard deviation is given by

$$s = \sqrt{\left[\frac{1}{n-1}\sum_{i=1}^{n}(x_i - \bar{x})^2\right]} \qquad (1.3.3)$$

Standard deviation calculations

The computation of the standard deviation, s, as given by equation 1.3.3, is tedious for all but the smallest values of n. This computation is expedited by the following alternative expression for $\sum(x - \bar{x})^2$ which can be utilized if all the individual results are grouped together into *classes*, or ranges of values.

Let w = class interval (i.e. range of values for each class)

x'_i = class midpoint (rather than an individual value)

x'_0 = midpoint of lowest class

f_i = frequency of values within class i

D'_i = deviation of class i from class o in terms of w (i.e. deviation from an arbitrary origin in terms of class width w).

Also, let $\sum_{i=1}^{n} x_i$ be written in abbreviated form as $\sum x_i$ for convenience.

It is clear that

$$n = \sum f_i \quad \text{and} \quad \sum x_i = \sum f_i x'_i$$

Also

$$\bar{x} = (1/n)\sum x_i = (1/n)\sum f_i x'_i$$

Therefore

$$\begin{aligned}\sum(x_i - \bar{x})^2 &= \sum f_i(x'_i - \bar{x})^2 \\ &= \sum f_i x_i'^2 - 2\bar{x}\sum f_i x'_i + \bar{x}^2\sum f_i \\ &= \sum f_i x_i'^2 - (2/n)(\sum f_i x'_i)^2 + (\sum f_i x'_i)^2(1/n)\end{aligned}$$

i.e.

$$\sum f_i(x'_i - \bar{x})^2 = \sum f_i x_i'^2 - (1/n)(\sum f_i x'_i)^2 \qquad (1.3.4)$$

Now

$$\bar{x} = (1/n)\sum f_i x'_i = x'_0 + (w/n)\sum f_i D'_i$$

and

$$x'_i = x'_0 + wD'_i$$

Therefore

$$\sum f_i(x'_i - \bar{x})^2 = w^2\sum f_i(D'_i - n^{-1}\sum f_i D'_i)^2$$

since x'_0 cancels out. Thus x'_i can be replaced by D'_i (and w) in r.h.s. of equation 1.3.4 also. Hence, from equation 1.3.3,

$$s = w\sqrt{\frac{\sum f_i D_i'^2 - (1/n)(\sum f_i D'_i)^2}{n - 1}} \tag{1.3.5}$$

Notice that the expression for σ equivalent to equation 1.3.5 is

$$\sigma = \sqrt{\left(\frac{\sum x^2}{n} - \mu^2\right)} \tag{1.3.6}$$

Various "mean" values

The *mean* value should always refer to the *arithmetic* mean. Less common types of average or mean values are as follows. The *median* is the middle observation when the observations are arranged in order of magnitude. The median therefore divides a histogram and a frequency polygon into two equal areas. The *mode* is the value which occurs most frequently if the variable is discrete, or the class interval which has the highest frequency if the distribution is continuous. The mode therefore represents a peak value in a frequency distribution. Consider, for example, the values 6, 6, 7, 8, 8, 10, 18. The mean is 9, median 8, and mode 6 or 8.

Coefficient of variation

The coefficient of variation is the ratio of the standard deviation to the mean, expressed as a percentage, i.e.:

$$V = 100\sigma/\bar{x}$$

Fractional cumulative frequency

The number of results higher or lower than a particular value is given by the *cumulative frequency*. The cumulative frequency divided by the total number of results gives the *fractional* cumulative frequency. A grading curve showing the percentage passing a particular sieve size is a typical example of a fractional cumulative frequency curve.

Histogram

A histogram is one way of representing a grouped frequency distribution diagrammatically. The class intervals are set out on the horizontal axis and the frequency in a given class interval is shown by the rectangular area above the horizontal axis at the position representing the given class interval. The width of all class intervals should generally be made the same, in which case the ordinate is proportional to the area of each rectangle. A typical histogram will be shown in connection with example 1.3.1, below.

Frequency distribution curve

If the class width is decreased indefinitely and the total number of results correspondingly increased, the histogram is transformed, in the limit, into a *frequency distribution curve.*

Frequency polygon

If the frequency within a class interval is plotted as a point whose abscissa is that of the class midpoint, adjacent points can be connected to form a frequency polygon (see example 1.3.1). Notice that the line between the points has no significance, hence intermediate values must not be read off the line.

EXAMPLE 1.3.1

The cube strength results (in MN/m^2) for the first 40 separate batches of concrete produced under actual operating conditions on a construction project were as follows:

30·5	32·0	42·0	39·0	29·0	21·0	24·5	30·5
27·0	25·0	30·5	31·5	30·0	36·5	34·0	33·0
33·0	36·5	31·5	34·0	36·0	33·0	30·5	34·5
43·0	33·0	33·0	32·5	36·0	37·0	34·5	28·0
26·5	23·5	34·0	30·0	35·5	37·5	38·0	31·5

Group the strengths into a frequency distribution table with a class interval of 2·5 MN/m^2 and draw a histogram and a frequency polygon. Hence calculate the mean, median, mode, standard deviation, coefficient of variation and range of the given data.

Solution

Let all stresses be in units of MN/m^2. The results are tabulated as follows:

Class interval	Class mid-point x'_i	Frequency f_i	Cumulative frequency F	Fractional cumulative frequency $F/n(\%)$	Deviation from origin in terms of class width D'_i	$f_iD'_i$	$f_iD'^2_i$
20·0–22·4	21·2	1	1	2·5	0	0	0
22·5–24·9	23·7	2	3	7·5	1	2	2
25·0–27·4	26·2	3	6	15·0	2	6	12
27·5–29·9	28·7	2	8	20·0	3	6	18
30·0–32·4	31·2	10	18	45·0	4	40	160
32·5–34·9	33·7	11	29	72·5	5	55	275
35·0–37·4	36·2	6	35	87·5	6	36	216
37·5–39·9	38·7	3	38	95·0	7	21	147
40·0–42·4	41·2	1	39	97·5	8	8	64
42·5–44·9	43·7	1	40	100·0	9	9	81
		40				**183**	**975**

The histogram and frequency polygon can be drawn using the values in the frequency and class interval columns, as shown in figure 1.3.2.

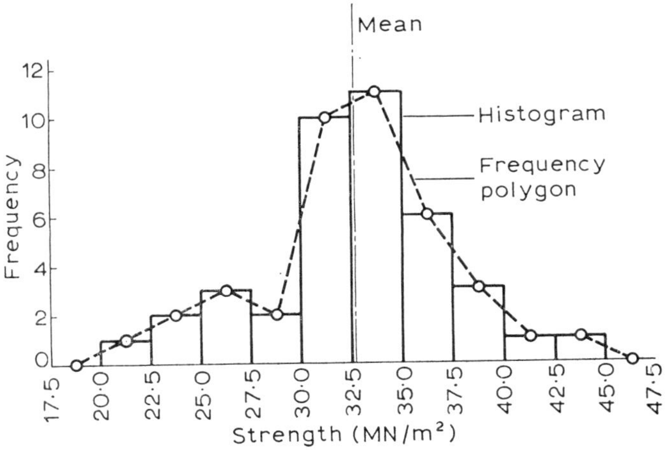

Fig 1.3.2 Example 1.3.1.

The totals for the appropriate columns of the above table give the following values:

$$\sum f_i = 40 \quad \sum f_iD'_i = 183 \quad \text{and} \quad \sum f_iD'^2_i = 975$$

Hence the mean is

$$\bar{x} = x_0 + w\sum f_i D'_i/\sum f_i = 21{\cdot}2 + 183 \times 2{\cdot}5/40$$
$$= 32{\cdot}6\ (\text{MN/m}^2) \qquad \text{(Answer)}$$

The standard deviation is

$$s = w\sqrt{\frac{\sum f_i D_i'^2 - (\sum f_i D_i')^2/\sum f_i}{n-1}}$$
$$= 2{\cdot}5(975 - 183^2/40)^{\frac{1}{2}}(39)^{-\frac{1}{2}}$$
$$= 4{\cdot}70\ (\text{MN/m}^2) \qquad \text{(Answer)}$$

The coefficient of variation is

$$V = 100s/\bar{x} = 100 \times 4{\cdot}70/32{\cdot}64 = 14{\cdot}4\% \qquad \text{(Answer)}$$

The range is $43{\cdot}0 - 21{\cdot}0 = 22{\cdot}0$ (MN/m^2) (Answer)

The median is the centroid of the histogram, i.e. the value for $i = 20$, or

$$32{\cdot}5 + 2w/11 = 32{\cdot}95\ (\text{MN/m}^2) \qquad \text{(Answer)}$$

The mode is the midpoint of the class interval with highest frequency, i.e.:

$$\text{mode} = 33{\cdot}7\ (\text{MN/m}^2) \qquad \text{(Answer)}$$

Note: columns for F and F/n are necessary if a fractional cumulative frequency curve is required.

Probability

For convenience let the occurrence of a certain event be success and its non-occurrence be failure. The probability of success is given by the ratio of successful possibilities to all possibilities. The examination of the possibilities must be exhaustive and all the possibilities must be equally likely and mutually exclusive. If there are n possibilities and success occurs in p cases, the probability of success is p/n and the probability of failure is $1 - (p/n)$. Hence probability is expressed as a number not greater than 1. Unity denotes certainty and zero denotes impossibility. Two important rules of probability should be noted.

(i) The probability of *simultaneous* occurrence of a number of *independent* events is the product of the separate probabilities. For example, if the probability of an excessive load on a structure is 1/20 and the probability of the material in a structure being substandard is also 1/20, the probability of an excessive load being applied to substandard material in a structure is $(1/20) \times (1/20)$, or 1/400.

(ii) The probability of occurrence of any one of two or more events, out of a number of independent and mutually exclusive events, is simply the sum of the probabilities of the separate events. For example, if a box contains 2 red balls, 4 blue balls and 6 green balls, the probability of drawing either a red or a blue ball from the box is $2/12 + 4/12 = 1/2$. Notice that the sum of of the probabilities of *all* the possible events is always unity.

Normal probability density function

The normal *frequency distribution* curve can be plotted from the Gauss function (equation 1.3.1) and the total number of observations equals the area under the curve:

$$\int_{-\infty}^{+\infty} y\, dD = n$$

To obtain the normal *probability* curve from the above it is necessary to make the area under the curve equal unity, since the sum of all probabilities must equal unity.

That is,

$$\int_{-\infty}^{+\infty} y_1\, dD = 1$$

where $y_1 = y/n$. In addition, it is also convenient to express D in terms of standard deviations, i.e. to replace D in equation 1.3.1 by $z\sigma$. where z is a coefficient. The new ordinate y_2 must therefore be given by $\sigma y_1 = \sigma y/n$ if the area under the y_2-z curve is to be unity.

Hence from equation 1.3.1:

$$y_2 = (\sigma/n)C \exp(-h^2\sigma^2z^2)$$

or

$$y = \frac{1}{\sqrt{(2\pi)}} \exp(-\tfrac{1}{2}z^2) \tag{1.3.7}$$

where the suffix 2 for y has been omitted. This is the normal probability density function which is usually referred to in standard statistical tables and which defines the normal distribution. Notice that the mean, μ, is constant for any one distribution, its effect being to determine the position of the normal curve along the x-axis. The shape of the curve (i.e. the horizontal spread) depends on the value of the standard deviation only. The normal probability distribution curve is shown in figure 1.3.3. The areas under the curve between the limits $0 \leqslant z \leqslant \frac{1}{2}$, $\frac{1}{2} \leqslant z \leqslant 1$, etc., are also shown.

Confidence limits

The limits which provide a given probability that any single value will lie between them are known as the *confidence limits.* Thus the 95% confidence limits are given by $\mu \pm 1{\cdot}96\sigma$.

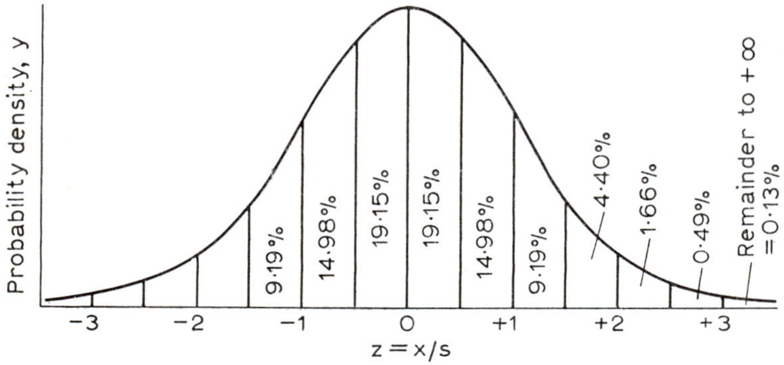

Fig 1.3.3 Areas under the normal probability curve.

Level of significance

It is often convenient to consider the probability that any single value will *not* lie between certain limits, and this probability is known as the *level of significance.* Thus the 5% level of significance, for example, is also given by the limits $\mu \pm 1{\cdot}96\sigma$.

EXAMPLE 1.3.2

What percentage of all results would lie below the value of $\bar{x} - 2{\cdot}5s$ for a normal distribution? What is the stress which provides this number of failures for the data in example 1.3.1? What are the stress limits at the 1·24% level of significance? What are the confidence limits for the band $\bar{x} \pm 2{\cdot}5s$?

Solution From figure 1.3.3, the percentage of all results that lie below $\bar{x} - 2{\cdot}5s$ is

$$(0{\cdot}49 + 0{\cdot}13)\% = 0{\cdot}62\% \qquad \text{(Answer)}$$

Hence for the previous data there are 0·62% failures for a stress (MN/m^2 units) of

$$32{\cdot}6 - 2{\cdot}5 \times 4{\cdot}70 = 20{\cdot}85\ (MN/m^2) \qquad \text{(Answer)}$$

An equal number of results lies above the value $\bar{x} + 2{\cdot}5s$. Hence at the 1·24% (= 2 × 0·62%) level of significance we have

stress limits: 20·85 and 44·35 (MN/m^2) (Answer)

The confidence limits for the band $\bar{x} \pm 2{\cdot}5s$ are

(100 − 1·24)% = 98·76% (Answer)

1.4* Accuracy of the sample mean

Sampling is necessary because it is usually impracticable to test the entire population; this is especially true of destructive tests. It is essential, therefore, that each and every sample is selected in a random manner so that it is representative of the population from which it is drawn. This means that every member of the population has an equal chance of being drawn in every trial. The word "sample" is used in a very broad sense to mean an observation of any type. A value arrived at by sampling is absolutely accurate only for the sample itself. For the population which it represents, a sample can only give an estimate whose accuracy is expressed in terms of probability. The greater the size of the sample (i.e. number of observations), the greater the confidence that can be placed on the estimate for the population.

Population

The term "population" refers to all the observations that could be made. Samples are tested in order to make inferences about the properties of the population and it is vital to be clear which population is of interest. For example, when considering cube results are we interested in a particular batch of concrete, or a complete day's placing, or all the concrete in a particular structure?

Sample means

If we consider a single batch of concrete as the population, each individual cube test can be considered as the sample. However, if we consider the total output of concrete for a particular day as the population, each group of cube tests taken for a single batch could be considered as the sample. Similarly, if we consider all the concrete in a given structure as the population, all the cube tests taken for a particular day's concreting could be considered as the sample. The question which then arises is: how many observations should be made for a single batch of concrete, or a single day's output, or a week's output? Alternatively, consider a unique quantity (e.g. the

* This section concerning further statistical principles should be omitted on first reading.

mean cube strength of a batch of concrete) and assume that the mean of the observations (e.g. the cube test results) is the appropriate single measure. How is the accuracy (e.g. of the assessment of the mean cube strength of the batch of concrete) increased as the number of observations is increased? These questions can be answered by considering the distribution of the sample means, and the relation between the standard deviation of the means and the standard deviation of the original population.

Permutations and combinations

The number of permutations of n from N is the number of ways of filling n different places with N different items, i.e. it is the number of different arrangements by which these places can be filled. This can be determined as follows.

The 1st place can be filled in N ways
,, 2nd ,, ,, ,, ,, ,, $N - 1$ ways
,, n-th ,, ,, ,, ,, ,, $N - n + 1$ ways

Therefore the number of permutations of n from N, written as ${}_NP_n$, is given by

$${}_NP_n = N(N-1)\ldots(N-n+1) = N!/(N-n)!$$

where $N!$ means *factorial* N, i.e.:

$$N(N-1)(N-2)\ldots\times 2\times 1$$

and 0! is one, by definition.

For a combination, the order in which the places are filled (i.e. the order of selection) does not matter, since in this case one is only concerned with the actual items which are selected. Now n items can be arranged in $n!$ ways, hence the number of combinations of n items from N, ${}_NC_n$, is $n!$ times smaller than the number of permutations of n items from N. Hence

$${}_NC_n = \frac{{}_NP_n}{n!} = \frac{N!}{n!(N-n)!} \tag{1.4.1}$$

Standard error of the mean

The number of different samples, each of n observations, that can be drawn from a population of N is the number of combinations of n items from a total of N, that is, ${}_NC_n$. Substituting in equation 1.3.2 gives the standard deviation for the means of samples of size n drawn from a population of N, $\sigma_{\bar{x}}$, as follows:

$$\sigma_{\bar{x}} = \sqrt{\frac{\sum(\bar{x}-\mu)^2}{{}_NC_n}} \tag{1.4.2}$$

where the summation is from 1 to ${}_NC_n$.

Substitution into the alternative expression for σ, equation 1.3.6, gives

$$\sigma_{\bar{x}} = \sqrt{\left(\frac{\sum \bar{x}^2}{{}_N C_n} - \mu^2\right)} \tag{1.4.3}$$

The standard deviation of the mean, $\sigma_{\bar{x}}$, is also called the *standard error* of the mean. Once this expression can be evaluated we can answer the question: how is the accuracy increased as the number of observations in a sample is increased?

Distribution of sample means

To evaluate $\sigma_{\bar{x}}$ in equation 1.4.3, notice that

$$\sum \bar{x}^2 = \sum (n^{-1} \sum x)^2 = n^{-2} \sum (x_1 + \ldots + x_n)^2.$$

This means that it is possible to write:

$$\sum \bar{x}^2 = A \sum x^2 + B \sum x_1 x_2$$
$$= A(x_1{}^2 + x_2{}^2 + \ldots + x_N{}^2) + B(x_1 x_2 + x_1 x_3 \ldots + x_{N-1} x_N)$$

where x_1, x_2, ... are the individual observations and $\bar{x}$ stands for each sample mean (for n observations only). The number of samples (of n observations) containing any single observation, x_1, is given by ${}_{N-1}C_{n-1}$, i.e. the number of combinations of $n - 1$ observations from a total of $N - 1$. Similarly, the number of samples containing both x_1 and x_2 is given by ${}_{N-2}C_{n-2}$. Hence

$$A = \frac{1}{n^2} \times {}_{N-1}C_{n-1} = \frac{1}{n^2} \times \frac{(N-1)!}{(n-1)!(N-n)!}$$

and

$$B = \frac{2}{n^2} \times {}_{N-2}C_{n-2} = \frac{2}{n^2} \times \frac{(N-2)!}{(n-2)!(N-n)!}$$

Therefore

$$A/{}_N C_n = (1/n^2) \times (n/N) = 1/nN$$

and

$$B/{}_N C_n = 2/n^2 \times [n(n-1)/N(N-1)] = [2(n-1)/nN(N-1)]$$

Now

$$\mu^2 = (1/N^2)(\sum x^2 + 2 \sum x_1 x_2)$$

Substituting in equation 1.4.3 and squaring both sides gives

$$\sigma_{\bar{x}}^2 = \left(\frac{1}{nN} - \frac{1}{N^2}\right)\sum x^2 + \left[\frac{2(n-1)}{nN(N-1)} - \frac{2}{N^2}\right]\sum x_1x_2$$

$$= \frac{N-n}{nN^2}\left(\sum x^2 - \frac{2\sum x_1x_2}{N-1}\right)$$

Putting $n = N$ in equation 1.3.6 and squaring both sides gives

$$\sigma^2 = \sum x^2/N - \mu^2 = (1/N - 1/N^2)\sum x^2 - (2/N^2)\sum x_1x_2$$

$$= \frac{N-1}{N^2}\left(\sum x^2 - \frac{2\sum x_1x_2}{N-1}\right)$$

Hence it is now possible to compare the variance of the sample means with the variance of the original population, i.e.:

$$\sigma_{\bar{x}}^2 = \frac{N-n}{nN^2} \times \frac{N^2}{N-1}\sigma^2$$

or

$$\sigma_{\bar{x}} = \left[\frac{N-n}{n(N-1)}\right]^{\frac{1}{2}}\sigma \tag{1.4.4}$$

If N is very large (i.e. $N \gg n$) then

$$\sigma_{\bar{x}} = \sigma/\sqrt{n} \tag{1.4.5}$$

In general, σ is not known and must be estimated from the sample. Substituting s for σ gives the following estimate for the standard deviation of the mean (or standard error of the mean)

$$s_{\bar{x}} = s/\sqrt{n} \tag{1.4.6}$$

Notice that a sample of one gives no information about the accuracy of the estimated mean since s cannot be estimated from a single observation. A sample of 2 can provide useful information but the accuracy is poor. This accuracy is doubled if the sample size is increased to 8 since the standard error of the mean varies inversely as the square root of the sample size. To double the accuracy again it would be necessary to increase the sample size to 32: an additional 24 observations. It is clear that the gain in precision for large samples is relatively small compared with the additional effort required. A sample size of 4 or 5 observations will often provide maximum information for a given effort.

Figure 1.4.1 shows the (probability) distribution of individual observations and the (probability) distribution of sample means drawn from the underlying distribution. Notice the much narrower distribution of the sample means and the fact that the peak for this curve is therefore higher, since the areas under both curves are the same, equal to the sum of all probabilities (i.e. unity).

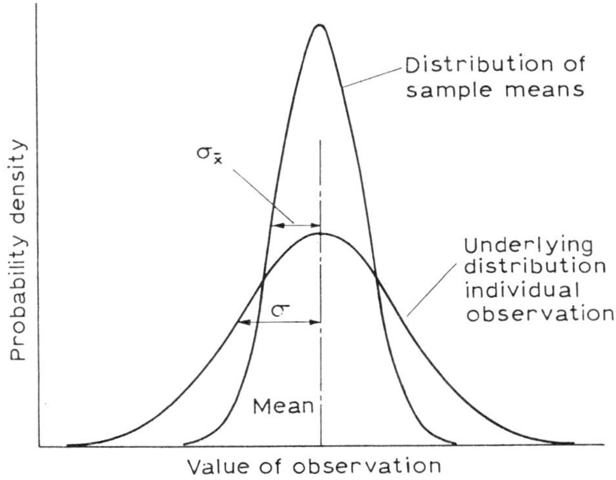

Fig 1.4.1 Probability distribution of sample means compared with the underlying distribution.

EXAMPLE 1.4.1

Calculate the sample mean and the standard error of the mean for the following five results from example 1.3.1: 30·5, 27·0, 33·0, 43·0 and 26·5 MN/m^2. Hence determine the 68·26% confidence limits, as estimated from these five cubes, for the following:

(i) All the concrete represented by the tests
(ii) The 40 actual cube tests.

Solution This example emphasizes the need for distinguishing between different populations. The population of 40 cube results in (ii) is quite distinct from the total population from which the 40 cubes were selected. Let all stresses be in MN/m^2 units. The computation of the appropriate standard deviations is most conveniently done from equations 1.3.2 and 1.3.3. The alternative form given in equation 1.3.5 is only useful if the number of results is sufficiently large to warrant grouping into classes, since it gives a less accurate result, especially when using a slide rule. Tabulation of the values can be made as follows:

(i) Total population

x_i	$x_i - \bar{x}$	$(x_i - \bar{x})^2$
30·5	−1·5	2·25
27·0	−5·0	25·0
33·0	1·0	1·0
43·0	11·0	121·0
26·5	−5·5	30·25
160·0	**0**	**179·5**

(ii) Population of 40

x_i	$x_i - \mu$	$(x_i - \mu)^2$
30·5	−2·1	4·4
27·0	−5·6	31·4
33·0	0·4	0·2
43·0	10·4	108·2
26·5	6·1	37·2
160·0		**181·4**

(i) The sample mean is

$$\bar{x} = 160/5 = 32{\cdot}0 \text{ (MN/m}^2\text{)} \qquad \text{(Answer)}$$

From the standard deviation

$$s = \sqrt{\frac{\sum(x_i - \bar{x}^2)}{n-1}} = \sqrt{\frac{179{\cdot}5}{4}} = 6{\cdot}7$$

the standard error of the mean can be calculated:

$$s_{\bar{x}} = s/\sqrt{n} = 6{\cdot}7/\sqrt{5} = 3{\cdot}0 \text{ (MN/m}^2\text{)} \qquad \text{(Answer)}$$

The 68·26% confidence limits for the mean of all the concrete, estimated from the 5 cube tests only, are

$$\bar{x} \pm s_{\bar{x}} = 32{\cdot}0 \pm 3{\cdot}0 = 29{\cdot}0 \quad \text{and} \quad 35{\cdot}0 \text{ (MN/m}^2\text{)} \qquad \text{(Answer)}$$

(ii) In this case the population is the 40 actual cube tests for which the mean value of 32·6 has been calculated in example 1.3.1. However, assuming that neither $\mu = 32{\cdot}6$ nor $\sigma = 4{\cdot}70$ is known then table (ii) above (which gives $\sigma = 6{\cdot}0$ and $\sigma_{\bar{x}} = 2{\cdot}55$) cannot be determined, and it is necessary to substitute $\mu = \bar{x}$ and $\sigma \simeq s = 6{\cdot}7$ as calculated in (i) above. Hence

$$s_{\bar{x}} = s\sqrt{\frac{N-n}{n(N-1)}} = 6{\cdot}7\sqrt{\frac{40-5}{5 \times 39}} = 2{\cdot}8$$

The 68·26% confidence limits for the mean of 40 results estimated from the 5 cube tests only are

$$\bar{x} \pm s_{\bar{x}} = 32{\cdot}0 \pm 2{\cdot}8 = 29{\cdot}2 \quad \text{and} \quad 34{\cdot}8 \text{ (MN/m}^2\text{)} \quad \text{(Answer)}$$

(*Note:* compare with $\mu \pm \sigma_{\bar{x}} = 32{\cdot}6 \pm 1{\cdot}99$ obtained below.)

EXAMPLE 1.4.2

Calculate the standard error of the mean for the data in example 1.3.1 and explain its significance. Estimate the probability that a mean of a sample of 5 cubes from the population of 40 cubes will be less than 30 MN/m^2. What are the limits for the mean of a sample of 5 cubes at the 5% level of significance? Assume 9·7% and 2·5% failures for $z = 1{\cdot}3$ and 1·96 respectively.

Solution Let all stresses be in MN/m^2 units. The standard error of the mean is

$$s/\sqrt{n} = 4{\cdot}70/\sqrt{40} = 0{\cdot}74$$

This means, for example, that there is a 68·26% probability that the mean of the total population (i.e. all the concrete represented by the test cubes) lies within the range 32·6 ± 0·74.

(*Note:* compare these limits, of 31·9 and 33·3, with the wider limits of 29·0 and 35·0 obtained in example 1.4.1 for the first 5 results only.)

The total population of 40 cubes is known exactly so the exact form of the standard error of the mean can be used, i.e.:

$$\sigma_{\bar{x}} = \sigma\sqrt{\frac{N-n}{n(N-1)}} = 4{\cdot}70\sqrt{\frac{40-5}{5\times 39}} = 1{\cdot}99$$

The deviation is

$$D = \mu - 30 = 2{\cdot}6$$

Hence

$$z = D/\sigma_{\bar{x}} = 2{\cdot}6/1{\cdot}99 = 1{\cdot}3$$

When $z = 1{\cdot}3$ there are 9·7% failures (given). Hence the probability that a mean of 5 cubes is less than 30 MN/m^2 is 0·097 (Answer)

The level of significance is the probability of the value falling outside the range $\mu \pm z\sigma$ on *either* limit, and at 5% level of significance (i.e. $2\frac{1}{2}$% failures), z is 1·96. Hence

$$D = 1{\cdot}96 \times 1{\cdot}99 = 3{\cdot}9$$

and the limits are

$$32{\cdot}6 \pm 3{\cdot}9 = 28{\cdot}7 \quad \text{and} \quad 36{\cdot}5\ (MN/m^2)$$ (Answer)

1.5 Characteristic strengths and loads

In this section the derivation of characteristic strengths and loads, using statistical principles, will be outlined. One important consequence of statistical theory should already be apparent: although the probability of a particular limit state being reached during the working life of a structure can be reduced (by increasing design loads and reducing design stresses), this probability cannot be zero. It is impossible to guarantee that a failure will *never* occur in *any* structure.

Mean strength, or resistance

This is the arithmetic mean of a series of test results made to determine the resistances of samples of a material which are nominally similar, or supposedly conform to the same specification. The term "resistance" commonly implies strength but can also include such properties as resistance to cracking etc.

Characteristic resistance

This is the resistance which takes account of the statistical variation of the resistance of a material. It corresponds to the "specified" strength or "minimum" strength referred to in earlier codes. It is defined by the equation

$$\sigma_k = \bar{\sigma} - zs \qquad (1.5.1)$$

where σ_k = characteristic resistance
$\bar{\sigma}$ = mean resistance
s = standard deviation
z = coefficient, depending on acceptable probability of test results being below the value σ_k.

CEB recommended, and the Unified Code has adopted, a value of 1·64 for z, which implies that 5% of the test results will be below the characteristic value. Steel manufacturers usually give a guaranteed minimum strength (either yield, 0·2% proof stress, elastic limit or ultimate strength) which can often be accepted as the corresponding characteristic strength provided that the probability of having results less than the guaranteed value is not more than that envisaged by the selected value for z.

Characteristic cube strength

The characteristic cube strength, u_k, corresponds to the *minimum* works cube strength formerly given in codes of practice. Hence from equation 1.5.1

$$u_k = \bar{u} - 1{\cdot}64s \qquad (1.5.2)$$

The standard deviation depends upon the degree of control to be exercised during concrete production and should be calculated from recent test results on at least 40 separate batches of concrete produced by the same plant under the same supervision. Equation 1.5.2 applies to either 7- or 28-day specified strengths.

EXAMPLE 1.5.1

Check whether the distribution of the data in example 1.3.1 is approximately normal by plotting F/n against strength on normal probability paper. Assuming a normal distribution, determine the characteristic cube strength for the concrete. How many of the actual results fall below this value? Check your values for mean and characteristic strengths by inspection of your probability plot.

Solution Let all stresses be in MN/m^2 units. Values of the fractional cumulative frequency, F/n%, have been tabulated in example 1.3.1, and F/n is shown plotted against strength on normal probability paper in figure 1.5.1. These values are fairly well distributed about a straight line and indicate an approximately linear plot. The data therefore possess an approximately normal distribution.

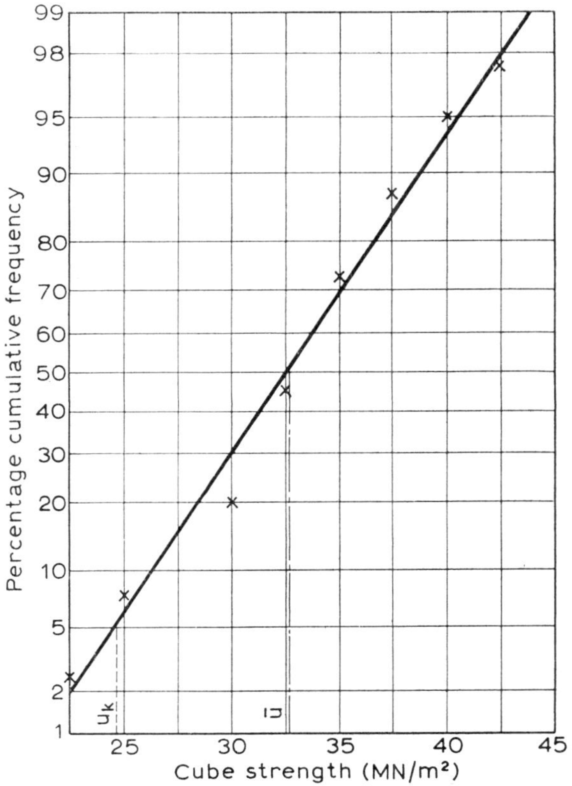

Fig 1.5.1 Example 1.5.1.

We know from example 1.3.1 that $\bar{x}$ is 32·6 and s is 4·70. Hence the characteristic cube strength is

$$32{\cdot}6 - 1{\cdot}64 \times 4{\cdot}70 = 24{\cdot}9 \text{ (MN/m}^2\text{)} \qquad \text{(Answer)}$$

The number of results falling below this value (by inspection) is three. (Answer)

$\bar{u}$ and u_k are given by 50% and 5% failures respectively. Hence from figure 1.5.1 the mean and characteristic cube strengths are approximately 32·6 and 24·9 (MN/m^2) respectively.

Minimum margin for mix design

The Unified Code also requires that the minimum margin (i.e. 1·64s) should not be less than the appropriate values given in table 1.5.1.

Table 1.5.1 Minimum margins allowable based on previous data for at least 40 cube results.

Concrete grade	Minimum margin
3NF, 5NF & 7·5NF; 10 & 15	One-third specified strength
22·5 and above	7·5 MN/m^2
F3, F4 & F5	0·55 MN/m^2
IT2, IT2·5 & IT3	0·35 MN/m^2

The rational method of mix design [1.4] utilizes Erntroy's data in which he put $z = 1{\cdot}75$, which corresponds to 4% failures. This more than satisfies the basic Unified Code requirement which allows 5% failures. Table 1.5.1 however may not be satisfied if very good control is to be used with low-strength concrete, so a slight adjustment in the mix design may sometimes be necessary.

EXAMPLE 1.5.2

Grade 22·5 concrete has been designed assuming a control ratio of 0·82 for materials providing the T-28 curve in figure 1 of reference 1.4. Would this design satisfy the Unified Code requirement of Table 1.5.1 for either temporary works or permanent works when 40 cube results would be available?

Solution The required c/w ratio for a cube strength of 22·5 MN/m^2 is, from figure 1 of reference 1.4, equal to 0·46. Hence the mean c/w ratio is 0·46/0·82 = 0·561, which corresponds to a strength of 30·7 MN/m^2 and a margin of 8·2 MN/m^2. This design therefore satisfies the additional Unified Code requirement. Notice, however, that a grade 15 concrete designed with a control ratio of 0·82 would certainly need to be checked.

Preliminary site trials

The rational method of mix design assumes that the initial concrete produced on the site can be used for temporary works and so enable at least 40 separate batches of concrete to be produced under actual operating conditions. If sufficient data are not available, the Unified Code requires that the margin between mean strength and characteristic strength should be taken as twice the minimum values given in table 1.5.1: e.g., 15 MN/m^2 for grades 22·5 and above. In this case the mix design should be carried out assuming a very poor standard of supervision (i.e. control ratio of 0·70) irrespective of the actual standard which is expected. The initial mix design will then generally conform with the large margin necessary until adequate data becomes available.

EXAMPLE 1.5.3

Some grade 22·5 concrete is required for immediate use for permanent concrete. It has been designed assuming a control ratio of 0·70 with the materials of example 1.5.2. Would this design satisfy the Unified Code requirement of twice the margin given in table 1.5.1?

Solution The mean c/w ratio is 0·46/0·70 = 0·657. This corresponds to a strength of 38·3 MN/m^2 and a margin of 15·8 MN/m^2. This satisfies the Code requirement of at least 15·0 MN/m^2.

Limitations of statistical theory

In connection with mix design it is shown in reference 1.4 that the major variable affecting strength when batching concrete is the c/w ratio. Hence a normal distribution in values of c/w ratio should be considered rather than a normal distribution of strengths. Since strength does not vary linearly with c/w ratio over the whole range it follows that a normal distribution of c/w values must result in a skew distribution of strengths. Although other factors affect variations in strength of the concrete it is clearly most logical to consider the major variable (c/w ratio) directly in mix design, rather than indirectly through strength. Once a mix has been designed and is being used on site, over 68% of the actual c/w ratios, for example, lie within $\pm s$ of the mean value, so that a linear relation between strength and c/w ratio can certainly be assumed over this restricted range. A statistical analysis of the first forty batches assuming a normal distribution of strengths is then perfectly valid. Other factors affecting strength (e.g. variations in cement, aggregate, or impurities) can reasonably be expected to produce a normal distribution also.

The foregoing considerations emphasize the fact that while strengths and loads can be expected to follow a normal distribution in the vicinity of their mean values, they may not do so at extreme values for a variety of reasons. For example, very high strengths and negative strengths are both impossible. Again, the laws of chance may apply predominantly to a variable which is not always linearly related to the one under consideration (as with strength and c/w ratio, above), or the laws of chance may themselves be made inapplicable by human intervention. Such limitations of the application of simple statistical theory to extreme cases should not be overlooked.

Characteristic load

The characteristic load, W_k, takes account of normal variations in load and, ideally, should be determined as a function of the mean load and its standard deviation in the same way as strength. For example, consider the live loading on the floors of a certain type of building such as a block of flats. Let $W_1, W_2, \ldots W_n$ be the maximum values of live loading that occur on the floors of n blocks of flats during their design life. It is assumed that $W_1, W_2, \ldots W_n$ are normally distributed about a mean value $\bar{W}$ having a standard deviation s. The characteristic load is given by

$$W_k = \bar{W} + zs \qquad (1.5.3)$$

Hence if z is taken as 1·64 the probability that the maximum load will exceed W_k is 0·05. (Notice the positive sign in equation 1.5.3, written because it is the *maximum* load in conjunction with *minimum* strength which is usually of interest.) However, until loads can be expressed in statistical terms, the minimum design loads given in CP3; chapter V *Dead and imposed loads* [1.5] or in BS153 [1.6] can be used for the characteristic loads.

Minimum loads

Notice that in certain cases it is the minimum load which can give the worst effect. For example, the maximum midspan moment in a continuous beam is given by maximum loading on that span while the adjacent spans on each side are subjected to the minimum loading. Theoretically, therefore, the characteristic load for the adjacent spans would be given by

$$W_k' = \bar{W} - zs \qquad (1.5.4)$$

However, the dead load can be estimated with much greater precision than the live load, so any differences between the minimum and maximum characteristic dead loads are relatively small and can be allowed for, if necessary, by a slight adjustment in the partial safety factor for load (as explained in the next section). In the case of the live load, the minimum characteristic load is zero so that the possibility of using a lower value does not arise. It may be assumed in this case, therefore, that for all practical purposes equation 1.5.4 need never be used.

EXAMPLE 1.5.4

If the mean unit weight of normal concrete for a particular locality is 23 kN/m³ and the standard deviation is found to be 0·5 kN/m³, what is the difference between the characteristic (upper) load, W_k, and a characteristic lower load, W_k'? Would a value of 0·9 W_k allow for this discrepancy?

Solution 5% failures (and successes) are given by $z = 1{\cdot}64$. Hence

$$W_k' = (23 - 1{\cdot}64 \times 0{\cdot}5)10^3 = 22{\cdot}2 \times 10^3$$

and

$$0{\cdot}9W_k = 0{\cdot}9(23 + 1{\cdot}64 \times 0{\cdot}5)10^3 = 21{\cdot}5 \times 10^3$$

Thus a 10% reduction in W_k could allow for not considering the correct limit for the dead load in this particular instance.

Note that the factor of 0·9 is necessary, for example, when the overall stability of a structure is being considered, since *all* the concrete in the structure could be lighter than normal due to the use of a particular aggregate. In the case of adjacent spans of a continuous beam, however, it is reasonable to assume that the aggregate will not have been changed from one span to the next. Hence any variations in unit weight of the concrete will be mainly due to variations in water content only. This is controlled (via strength) and should be small enough for the factor of 0·9 to be unnecessary. The minimum load effects due to variable loads in this case are given by maximum variable loads on adjacent spans and zero variable load on the given span.

1.6 Partial safety factors and design values

The characteristic values allow for the inherent variations which occur in loads and material strengths. It is now necessary to consider the other factors and causes of variations which can occur and to make allowances (up and down) for these uncertainties by introducing suitable partial safety factors.

Partial safety factor for resistance

The material in the actual structure, under long-term loading, may be weaker than the characteristic value, for a short-term test, for any of the following additional reasons:

(i) The strength of a material in the actual structure may be reduced as a result of constructional faults or difficulties. For example, the concrete strength can be reduced by inadequate compaction or curing, by bad weather, or by impurities. Similarly, bond strength can be reduced if the surface of the steel is contaminated by dirt, oil, etc.

(ii) The strength of a material under sustained or repeated loading is less than that obtained for a short-term test. The long-term strength of a material may be further reduced by corrosion during the life of the structure even though the requirements for the limit state of durability are complied with. If the imposed loads are predominantly cyclic in character the limit state for fatigue should be considered.

Design resistance

The additional errors given above are taken into account by the partial safety factor for material deficiencies, γ_m. The design strength of the material which is assumed in calculating the strength of the structure is given by:

$$\text{Design resistance, } {}_{*}\sigma = \sigma_k/\gamma_m \tag{1.6.1}$$

Partial safety factor for load

The partial safety factors for loads enable the following additional uncertainties to be taken into account for each limit state:

(i) *Modification of loads* (γ_{l1}). Possible unfavourable deviations of load beyond those considered in deriving the characteristic loads. For example, there may be an increase in the general level of loading in structures of a particular type, due to the introduction of modern machinery etc., without any change of use.

(ii) *Load combinations* (γ_{l2}). The reduced probability that various loadings act together simultaneously at their design (maximum) values. Transient loads, such as those due to wind and earthquakes, rarely reach their design values and when they do it is highly unlikely that all the other loads are also at their maximum values.

(iii) *Modification of load effects* (γ_{l3}). Adverse modification of load effects can occur due to errors in design assumptions and calculations or due to constructional discrepancies. These include the possible neglect of thermal, creep and shrinkage stresses, errors in assessing the effects of loading, and the presence of unforeseen stress redistributions within the structure. For example, the rotation capacity of a plastic hinge may be less than assumed. The variations in dimensional accuracy achieved in construction can increase the load on a member due, for example, to excessive thickness of a slab caused by the deflection of formwork; dimensional errors in spans and heights; deviation of columns from the vertical; and errors in setting out arch profiles. The ability of the structure to resist loading can also be reduced by, for example, inadequate or incorrect positioning of reinforcement and formwork, resulting in reduced lever arms and sizes of members or reduced steel cover.

Unduly flexible formwork can affect the relative stiffnesses of members which, in turn, can affect the magnitudes of the forces and bending moments acting throughout a structure. Notice that modifications due to design errors are partly interrelated with constructional errors: hence the reason for grouping them together.

Possible modification of design values

The design values for either materials or loading may require adjustment to take account of the following:

(i) *Nature of structure and its behaviour* (γ_{c1}). Structures or parts of structures in which partial or complete collapse can occur without warning, where reduction of internal forces is not possible, or where failure of a single element can lead to overall collapse.

(ii) *Seriousness of reaching limit state* (γ_{c2}). The seriousness of attaining a particular limit state with regard to economic consequences, danger to the community, etc. For example, the probability of collapse of a multi-storey block of flats should be much less than that for a small footbridge over a stream in a park.

Russian Code

The Russian Code groups γ_{l3} with γ_{c1} and γ_{c2} under a partial safety factor for *service*. This third partial safety factor can also allow for other factors which are not included in the calculations but which can either increase or decrease the carrying capacity. However, the justification for considering only two partial safety factors, as in the Unified Code, is given later.

Design load

The design load for a given type of loading and limit state is obtained from the appropriate characteristic load, W_k, and the partial safety factor for loads, γ_l, as follows:

$$\text{Design load, } {}_*W, = W_k\gamma_l \qquad (1.6.2)$$

In general a single coefficient, γ_l, can be used in the calculation where γ_l is a function of γ_{l1}, γ_{l2} and γ_{l3}.

The different types of loads and load effects which can occur are as follows:

(i) Direct loads: permanent and variable imposed loads.
(ii) Imposed deformations: thermal and shrinkage effects, settlement of supports, prestressing etc.
(iii) Impact, inertia or other loads due to exceptional events.

Recommended values for partial safety factors

The values for the partial safety factors for loads and strengths recommended by the Unified Code are given in table B1 in appendix B. These are usually in approximate agreement with the CEB-FIP recommendations [1.7] which are essentially more general.

The CEB-FIP recommendations for the ultimate limit states are as follows:

(i) Direct loads (maximum):

$${}_*W = 1{\cdot}5(W_d + W_v)$$

where W_v is the set of forces due to the most unfavourable set of characteristic variable loads. These include variable working loads, climatic loads (snow and wind), dynamic or centrifugal forces or vibrating construction loads, exceptional overloads (hurricanes, earthquakes, fires, shocks, etc.).

(ii) Direct loads (minimum):

$${}_*W = 0{\cdot}9W_d + 1{\cdot}5W_v$$

(iii) Combination of variable loads of different origin:

$${}_*W = 1{\cdot}5[W_d + 0{\cdot}9W_{v1} + 0{\cdot}8W_{v2} + 0{\cdot}7(W_{v3} + W_{v4} + \ldots)]$$

where $W_{v1} \geqslant W_{v2} \geqslant W_{v3} \geqslant \ldots.$

γ_m for the above conditions is 1·15 for steel and either 1·4 for carefully mixed and controlled precast concrete in a factory or 1·5 for concrete generally.

(iv) Exceptional loads of short duration

$$_{*}W = W_d + W_{v(permanent)} + W_{ex}$$

where $W_{v(permanent)}$ is the fraction of characteristic variable loads of a permanent nature, and W_{ex} is the exceptional load envisaged which falls outside the range of characteristic loads.

For the serviceability limit states $\gamma_m = 1$ in all cases and $\gamma_l = 1$ generally, except that for a combination of variable loads of different origin, the reduction coefficients γ_{l2} given above for W_v at ultimate should be used (i.e. values of 0·9, 0·8 and 0·7). The partial safety factors for concrete are applied to the standard cylinder (not the standard cube), hence $\gamma_m = 1$ for all serviceability limit states; this difference between the Unified and International Codes is reconciled for collapse limit states by differences in assumptions for the concrete stress blocks.

Application of design strengths

It is important to realize that the partial safety factor γ_m allows for material deficiencies only and is used to obtain the design strength, $_{*}\sigma$. The latter is not necessarily the *stress* which is intended, or is in fact permissible, for the appropriate limit state. For example, for the limit states of crack width the maximum desirable stress in the steel is $0{\cdot}75f_y$ (the crack width formula for beams becomes conservative at higher stresses). This value, $0{\cdot}75f_y$, is considerably less than the design tensile steel *strength* for crack widths (or local damage) in beams, which is f_y. Similarly, the maximum fibre stress in the concrete under working loads should not usually exceed $0{\cdot}5u_k$. The design cube strength for local damage from table B1 is $0{\cdot}77u_k$ which is appreciably higher. In this instance, the coefficient 0·5 automatically takes the partial safety factor of 1·3 into account. There are other cases, of course, when the design strength does equal the design stress considered for a particular limit state. This is so, for example, for both the steel and concrete stresses for the limit state of collapse, and for the steel stress for the limit states of crack width in a lightly reinforced wall. This observation will become clearer when each of the main limit states is considered in detail later.

1.7 General considerations of design philosophy

The twin concepts of characteristic strength and partial safety factor for material strength require to be kept carefully balanced, if some of the factors assessed by one of them are not to be obscured by an undue emphasis on the other. This and other general considerations which affected the recommendations of the CEB-FIP and the Unified Code are briefly discussed in this section.

Safety factors for concrete and steel

Concrete tests on cylinders or cubes cannot give the real strength of the concrete in the structure exactly. The results of tests on steel bar specimens, on the other hand, can correspond approximately to the strengths of the actual bars. In addition, since concrete is basically a brittle material, weakness in the concrete could lead to sudden failure of the structure without warning. For both these reasons the value of γ_m for concrete must be considerably greater than for steel.

Single value of safety factor

The value of γ_m for concrete at collapse, 1·50, is for the control conditions specified in the CEB-FIP Code for *in situ* concrete. For rigorous control in a factory the value could be reduced, whereas for poor control *in situ* it would need to be increased. In principle, the Unified Code assumes that variations in strength due to constructional faults or difficulties tend to increase with variations in inherent strength. This assumption enables a single value to be adopted for the partial safety factor for concrete strength at collapse.

Characteristic strength v *safety factor*

A structure should be designed in such a way that the probability of any limit state being attained is substantially constant for all component members and for the structure as a whole. Although the rational determination of structural safety should be based on the theory of probability it is at present necessary to adopt a semi-probability approach. Statistical theory can be used for determining the characteristic strength, whilst the partial safety factor for material strength allows for those aspects that are much more arbitrary in nature. Hence the guarantee that the actual strength does not fall below the value corresponding to the limit state is achieved in two stages. It is desirable not only to use both concepts, but also to maintain a certain balance between the relative importance of each for the following reasons:

(i) The importance of the concept of a characteristic strength, depending statistically on the variability of the strength, encourages manufacturers of materials to establish methods of control for reducing variability. This desirable effect is diminished if the value of z in equation 1.5.1 is reduced.

(ii) The normal statistical distribution of results is invalid for extreme values of strengths, so that the value of z should not be so great that there is an excessive discrepancy between the calculated and actual probabilities of the strength being less than σ_k. For example, when considering the design of concrete mixes it was shown that since the strength of fully compacted concrete varies primarily with variations in the c/w ratio, it is preferable to consider variations in this ratio rather than variations in the resulting cube strengths. A normal distribution in c/w ratio values will only result in a normal distribution in corresponding strength values if there is a linear relation between c/w ratio and strength. Since there are two distinct slopes to the strength-c/w-ratio curve, with a transition in between, a skew distribution will generally be obtained for the strength results (see reference 1.4). Such limitations in statistical theory indicate that the value of z should not be made too large.

A reasonable compromise between the foregoing effects is achieved if z is between about 1·0 and 2·0 (e.g. 1·67).

Nature of structure

The safety factor γ_l must be suitably increased for structures (or parts of structures) in which redistribution of moments is not possible, and in which partial failure might therefore result in complete collapse, or partial or complete failure occur without warning. The Unified Code approach means that the factor γ_{c2} must be effectively incorporated or "built in" where necessary: e.g. as in table C3 for critical shear stresses.

Seriousness of failure

The economic consequences of failure include not only the cost of repairing or replacing the structure, but also the cost of providing, while the structure is out of use, alternative means of fulfilling the function for which it was built. Personal considerations include danger to the lives of people in the vicinity of the structure at the time of failure and also the psychological effects of the failure. These considerations are at present covered by the statement that the partial safety factors must correspond to an acceptable probability of the limit state being reached, *having regard to the seriousness of failure should it occur*. If the seriousness of failure is taken as an indication of the care that must be taken in the calculations and in the execution of the work, it is generally possible to use a common value of 1·40 for γ_l. In that case, average accuracy of calculation and average workmanship are adequate where failure is not serious, but very good accuracy of calculations combined with very good workmanship should be used where failure would be very serious.

1.8 General assumptions for structural analysis

It is now convenient to consider some of the more general assumptions of the Unified Code for the different limit states. In particular, it should be noted that elastic analysis may be used as the basis for determining the force distribution *throughout the structure* for the limit states of collapse, deflection and local damage.

Analysis of structure

When analysing the structure, or part of the structure, to determine force distributions within the structure, the properties of materials are to be assumed to be those associated with their *characteristic* strengths *irrespective of which limit state is being considered.*

Relative stiffnesses

The relative stiffnesses may be based on any one of the following sections:

(i) The concrete section—the entire concrete cross-section, ignoring the reinforcement.
(ii) The equivalent gross section—the entire concrete cross-section, including the reinforcement on the basis of modular ratio.
(iii) The equivalent (or transformed) section—the compression area of the concrete cross-section combined with the reinforcement on the basis of modular ratio.

Analysis of sections

When analysing any section within the structure, the properties of the materials are to be assumed to be those associated with their *design* strengths *appropriate to the limit state being considered.*

Sustained loading

Where sustained loading is being considered the short-term stress-strain curves may need modification as follows:

Reinforcement: no modification
Prestressing tendons: due allowance for relaxation
Concrete: due allowance for shrinkage and creep.

REFERENCES

1.1 ROWE, R. E., CRANSTON, W. C. and BEST, B. C. New concepts in the design of structural concrete. *Struct. Eng.* vol. 43 London, Dec. 1965

pp 399–403. Also subsequent discussion, vol. 44 Apr. 1966 pp 127–33; vol 44 Nov. 1966 pp 411–21.
1.2 GOODMAN, R. *Teach yourself statistics* (The English Universities Press, London).
1.3 NEVILLE, A. M. and KENNEDY, J. B. *Basic statistical methods for engineers and scientists* (Int. Textbook Co., Pennsylvania 1964) 325 pp.
1.4 HUGHES, B. P. The rational design of high-quality concrete mixes. *Concrete* (Journal of the Concrete Society) London, May 1968. Also *Concrete* reprint no. 1.
1.5 BRITISH STANDARDS INSTITUTION (London) CP3 *Code of basic data for the design of buildings*, chapter V part 1:1967 *Dead and imposed loads* p 19.
1.6 BRITISH STANDARDS INSTITUTION (London) BS153 *Steel girder bridges* (undergoing revision).
1.7 CEB-FIP (Comité Européen du Beton-Fédération Internationale de la Précontrainte) *recommendations for the design and construction of concrete structures.* FIP sixth congress, Prague, June, 1970.

Exercise

1. The cube strength results (MN/m^2) for the first 40 separate batches of grade 60 concrete produced by the batching plant in a precast concrete works were as follows:

66·4	67·8	78·0	74·8	65·2	56·4	60·2	66·6
69·0	61·2	66·6	67·4	66·2	72·8	70·2	69·0
63·2	72·4	67·8	70·0	72·2	69·0	66·4	70·6
64·0	70·8	73·0	72·2	68·6	69·2	69·0	79·2
62·6	59·8	70·2	66·4	71·0	74·0	74·4	67·8

Calculate the mean strength and the standard deviation for the above data. Hence determine the characteristic strength for the concrete. Is this concrete satisfactory? You may use a class interval of 2·5 MN/m^2 for the frequency distribution table. Check that the distribution of the results is approximately normal by plotting the fractional cumulative frequency against strength on normal probability paper which is provided.

(UB3/H 1969, 45 min)

Chapter 2

Introduction to reinforced concrete

Reinforced concrete is by reason of its strength, durability and adaptability an eminently suitable and economical material for many types of permanent structures. Concrete is relatively strong in compression and can be made sufficiently massive to provide lateral stability. Steel, on the other hand, is strong in tension but often lacks adequate lateral stability in compression due to its slender proportions. A reinforced concrete member in which the concrete resists compression while the steel resists tension is therefore an ideal structural partnership.

Portland cement concrete is sufficiently impervious to moisture, and retains sufficient alkalinity, to inhibit corrosion of the reinforcement even when small cracks are present. Although the idea of reinforced concrete was suggested in 1830 [2.1], the earliest successful use of tensile reinforcement in Portland cement concrete is believed to be due to the French contractor J. L. Lambot. He used a mesh of iron bars to build a reinforced concrete rowing boat for the Paris International Exhibition of 1855. An Englishman, William B. Wilkinson of Newcastle upon Tyne, took out the first patent for reinforced concrete in 1854, which proposed embedding either flat iron rods or second-hand wire ropes in the tension zones of concrete beams. This patent revealed that Wilkinson also understood the need to develop good bond between the concrete and the reinforcement, and the fact that concrete floors possessed good fire resistance. Wilkinson, together with Joseph Monier, a French engineer who later took out several patents, are generally regarded as the inventors of reinforced concrete.

2.1 Basic considerations for reinforced concrete

Traditional elastic design is a deterministic design method using a single (stress) factor of safety which is based on elastic theory. That is, the basic parameters are treated as non-random. The stresses occurring under the maximum working loads are compared with fractions of the strengths of the materials, and stresses are assumed to be proportional to strain. Traditional elastic design has been widely used for many years and still has application, in the author's view, to very simple designs, even though it is omitted entirely from the draft Unified Code. Elastic theory, as distinct from elastic design, remains an extremely useful concept and provides an excellent introduction to the subject of reinforced concrete as well as an indication of the behaviour of structures *under working loads.* It therefore serves also to introduce the limit states of serviceability.

Tensile and compressive strengths of concrete

The use of plain concrete in most structures is limited because its ultimate tensile strength is small compared with its ultimate compressive strength. Consider, for example, a rectangular plain concrete

Fig 2.1.1 Failure of (*a*) plain beam, and (*b*) reinforced beam. In (*a*) the bottom fibre is critical, in (*b*) the top fibre is critical.

beam as shown in figure 2.1.1. According to elastic theory for a homogeneous beam the extreme fibre stress, f, is given by

$$M = fZ \tag{2.1.1}$$

where M = bending moment at the section
Z = section modulus.

If the strength of the concrete in tension is 3 MN/m^2 it is clear that the maximum moment which can be resisted by the plain beam is only $3Z$ MN m, even though the compressive strength of the concrete is of the order of 30 MN/m^2. If reinforcing bars are introduced into the bottom of the beam to take the tensile stresses then the moment which can be resisted becomes more like $30Z$ MN m, a very substantial increase.

Modulus of rupture

While the foregoing considerations are basically correct, it is essential to realize that elastic theory cannot be expected to predict the stress distribution in a beam *at failure.* If a plain concrete beam is actually tested in flexure then the moment at failure would be found to be nearer $4Z$ than the $3Z$ given above, for a concrete having a uniaxial tensile strength of 3 MN/m^2. This theoretical value is known as the *modulus of rupture* and must not be confused with a true value of stress. The increased value is due to plastic, rather than elastic, behaviour of concrete at high stresses. The behaviour of reinforced concrete at collapse is considered later. For the present it should be realized that elastic theory can only be expected to apply, in general, up to working load conditions.

Concrete quality

It is essential that all materials used in reinforced concrete conform to a certain minimum standard. As described previously, concrete can be produced in different *mixes* (proportions of cement, sand and aggregate), and different *grades* (qualities). The quality is usually measured in terms of the compressive strength of a 100-mm or 150-mm cube, u, or of a 300-mm long × 150-mm diameter cylinder, in which case the compressive strength is denoted by f_c'.

Reinforcement

There are ten standard sizes of steel reinforcement. The most common are from 6 to 40 mm in diameter but the standard sizes include a 50-mm diameter for use when a larger size is necessary. The characteristic strength, f_y, of the reinforcement is that value of the yield or proof stress below which not more than 5% of the test values fall. The values of f_y to be used in design are the specified characteristic strengths in the appropriate British Standard. These are included in table C1, appendix C.

Types of reinforcement

The main types of steel reinforcement are as follows:

(i) *Rolled mild steel bars.* Rolled mild steel bars to BS4449 [2.2] are usually circular although square section bars are also available. These square bars, chamfered or unchamfered, have a cross-sectional area which is the same as that of a standard round bar; for example, a 7·1-mm square bar is actually referred to as an 8-mm RABar (round area bar). The minimum yield stress is specified as 250 MN/m^2. The minimum elongation (on a length of 5·65 × bar diameter) is 22%. There is also a bend test. Mild steel may be made by the open hearth (acid or basic), acid Bessemer, top or bottom blown oxygen ($\ngtr$ 0·008% nitrogen) or electric processes. Mild steel has $\ngtr$ 0·25% carbon, $\ngtr$ 0·06% sulphur and $\ngtr$ 0·06% phosphorus.

(ii) *Hot rolled deformed bars.* A deformed bar has a surface which has been so shaped that its bond with concrete is increased by at least the amount required by the Unified Code. The deformation and consequent stretching raises the yield stress but reduces the elongation and the performance in the bend test. The composition is as for mild steel except for sulphur ($\ngtr$ 0·05%) and phosphorus ($\ngtr$ 0·05%). BS4449 specifies a minimum yield stress of 410 MN/m^2 (taken as 0·2% proof stress or as stress at a total strain of 5 millistrain). The minimum elongation for "high yield 410" bars, to which these figures apply, is 14%.

(iii) *Cold worked deformed bars.* Cold working can also raise the yield point (and reduce the elongation). The composition is as for mild steel. BS4461 [2.2] specifies a minimum yield stress (taken as 0·2% proof stress or as stress at a total strain of 4 millistrain) of 425 and 460 MN/m^2 for bars $>$ 16 mm and $\not>$ 16 mm diameter respectively. The minimum elongation is 14% and 12% for bars $>$ 16 mm and $\not>$ 16 mm respectively.

(iv) *High tensile bars.* High tensile steel bars with a 0·2% proof stress of about 900 MN/m^2 or higher are usually used in prestressed concrete only. However, some American tests using reinforcement having yield stresses up to 650 MN/m^2 have indicated that it can be satisfactory for use in reinforced concrete. Bars with a guaranteed minimum yield stress of 550 MN/m^2, and up to 20 mm in diameter, have been accepted for use in internal beams by some local authorities in Britain since 1965. A useful review of the use of steel reinforcement has been given by Lancaster [2.3].

(v) *Steel fabric.* Steel fabric has either an oblong or a square mesh in which the cross-wires are electrically welded to the main wires at the intersections. Expanded metal (steel) fabric has a diamond-shaped pattern and is formed from a metal sheet. Both types are covered by BS4483 [2.4]. The minimum 0·2% proof stress is specified in BS4482 [2.5] as 485 MN/m^2, or in BS4461 as 460 MN/m^2. Fabric usually has a wire diameter not exceeding 12 mm and an adequate bond strength even though the wires are not deformed. This is mainly due to the anchorage provided by the welded intersections although the relatively small bar diameters also help to reduce the average bond stress.

Steel bending

Reinforcing bars must usually be bent without heating before being placed in the structure prior to concreting. BS4449 and BS4461 therefore specify that sample lengths of bar shall withstand cold bending through 180° round a former of diameter $2D$ for mild steel, $3D$ for high-yield steel with yield strengths up to 460 MN/m^2, and $4D$ or $6D$ for stronger steels, where D is the nominal bar size or effective diameter. BS4466 [2.6] specifies that the minimum diameter of bend allowed in service is double that used in the foregoing cold bend test.

BS4449 and BS4461 also include an optional re-bend (i.e. bent and partially bent back) test which reveals any loss in strength due to the bending operation. Some deformed bars appear to perform unsatisfactorily in this test due to notch effects at intersections of ribs [2.3]. The test also indicates strain embrittlement which can occur in steels with an excessive nitrogen content.

The bearing stress increases with decreasing radius of bar, so the bearing stress for the concrete may be exceeded (see appendix C). It is clear that one disadvantage of high-yield reinforcement is that larger bend radii are necessary and this makes detail design more difficult.

Nominal size of deformed bars

The nominal size of deformed bars can be checked by determining the effective cross-sectional area. BS4449 and BS4461 state that if the pattern of deformation is such that the cross-sectional area appears, by visual inspection, to be substantially uniform along the length of the bar, the effective area shall be the gross area, determined by weighing and measuring to a precision of $\pm 0{\cdot}5\%$ a length $\not< 0{\cdot}5$ m. If the cross-sectional area varies along its length, it is necessary to weigh before (m) and after (m') removing the transverse ribs. If $1{\cdot}03\ m' \not> m$ the mass of the bar should be taken as $1{\cdot}03\ m'$; otherwise the area is obtained assuming mass m.

EXAMPLE 2.1.1

The cross-sectional area of a nominal 12-mm diameter hot rolled deformed bar appears to vary along its length. A sample length of 750 mm is weighed before and after removal of the projections and found to have a mass of 0·666 and 0·650 kg respectively. Determine the effective cross-sectional area of the bar.

Solution BS4449 assumes a density for steel of 7850 kg/m³.

$$1{\cdot}03\ \mathrm{m}' = 1{\cdot}03 \times 0{\cdot}650 = 0{\cdot}669$$

which is $>$ 0·666. Hence the effective cross-sectional area is

$$0{\cdot}666/(0{\cdot}75 \times 7850) = 113{\cdot}1 \times 10^{-6}$$

$$= 113{\cdot}1\ \mathrm{mm}^2 \qquad \text{(Answer)}$$

Note that rolling tolerances permit variations from the nominal size which should be taken into account. For example, 12-mm and larger hot rolled bars have a tolerance on the nominal mass per metre run of $\pm\ 2\tfrac{1}{2}\%$ for a batch, or $-\ 4\%$ for an individual bar. The effective area for an individual 12-mm bar can therefore vary by an unspecified positive value and by up to $-\ 4\%$ from the nominal area of 113 mm².

Specified characteristic strength for steel

BS4449 states the following:

"16.1 The specified characteristic strength of the steel is the value of the yield stress below which shall fall not more than 5% of the test results of the material complying with this standard.

16.2 The steel shall be deemed to comply with the specified characteristic strengths . . . provided that for each size and type of bar:

(1) not more than two test results of yield stress out of the last 40 consecutive results are less than the specified characteristic strength, and

(2) no test results are less than 0·93 of the specified characteristic strength."

EXAMPLE 2.1.2

Determine the significance of the BS4449 and BS4461 clauses: "no test results are less than 0·93 of the specified characteristic strength".

Solution The basic requirement for characteristic strength is given by

$$\sigma_k = \bar{\sigma} - 1{\cdot}64s \quad (5\% \text{ failures})$$

If σ is the strength of an individual specimen the additional requirement is given by

$$\sigma \nless 0{\cdot}93\sigma_k$$

that is

$$\sigma \nless 0{\cdot}93(\bar{\sigma} - 1{\cdot}64s) = \bar{\sigma} - (0{\cdot}07\bar{\sigma} + 1{\cdot}53s)$$

Now $V = s/\bar{\sigma}$, hence this condition becomes

$$\sigma \nless \bar{\sigma} - (0{\cdot}07/V + 1{\cdot}53)s$$

For example, if $V = 10\%$,

$$\sigma \nless \bar{\sigma} - 2{\cdot}23s \quad \text{i.e. } 1{\cdot}3\% \text{ failures}$$

or, if $V = 5\%$,

$$\sigma \nless \bar{\sigma} - 2{\cdot}93s \quad \text{i.e. } 0{\cdot}17\% \text{ failures.}$$

It can be expected, therefore, that if the coefficient of variation is 10%, over 1% of all batches of steel will be deemed not to comply with the Standard. The second clause therefore provides an additional incentive to improve the quality control. For example, if V is halved, to 5%, less than 0·2% of all batches can be expected to provide difficulty.

Further tension tests for steel

If a batch of steel does not comply with the Standard, or if a purchaser wishes to verify that a batch of steel is satisfactory, it is specified that $\nless 10$ bars shall be chosen from the batch and a tensile test specimen cut from each bar. If any one of the bars tested has a yield stress $< 0{\cdot}93\sigma_k$ then that bar shall be rejected. If another of the bars tested also has a yield stress $< 0{\cdot}93\sigma_k$, then the whole batch of steel shall be deemed not to comply with the requirements of the Standard. If $\nless 9$ results are $> 0{\cdot}93\sigma_k$ the steel shall be deemed to comply with the Standard.

Elastic modulus

Up to the limit of proportionality of an elastic material, stress is proportional to strain; the constant of proportionality being the elastic modulus, E, of the material. That is,

$$\text{Stress} = E \times \text{Strain}$$

where strain is the extension (change in length) of the material per unit length. E_s and E_c are used to denote the elastic moduli of steel and concrete respectively. Since concrete is not in fact an elastic material, E_c is taken to be the secant modulus at the working stress for a standard rate of loading. Creep effects are included if the rate of loading is extremely slow. CP114:1957 [2.7] assumed a constant secant modulus of 2×10^6 lbf/in^2 (13·8 GN/m^2) whereas the Unified Code assumes that the short-term E_c is given by

$$E_c = 4500\sqrt{u} \text{ (MN/m}^2\text{ units)} \qquad (2.1.2)$$

where u is the characteristic cube strength (MN/m^2 units) of normal density concrete. The long-term E_c is assumed to equal one-half of the above value.

The steel is considered to be perfectly elastic in both tension and compression. The elastic modulus is assumed to be 200 GN/m^2 for most steels except alloy steels and prestressing steels, where the values are slightly lower.

Modular ratio

Consider two similar bars, one concrete and one steel, which are compressed uniformly so that they deform together by equal amounts. If the overall compressions are the same, then the strains are also equal. Hence by equating these strains

$$f_s/f_c = E_s/E_c = m \qquad (2.1.3)$$

where f_s and f_c are the stresses in the steel and concrete respectively. If $E_s = 200$ GN/m^2 and E_c (long-term) $= 10$ GN/m^2 then the modular ratio, m, equals 20. CP114:1957 [2.7] assumed that $m = 15$ for all grades of concrete. The only merits of assuming a single value were simplicity and the fact that the secant modulus of concrete is variable with time as well as with grades of concrete.

Permissible stresses

The permissible stresses are the maximum stresses which are desirable *under the action of normal working loads.* For example, in beams it is desirable that the tensile stress in the steel should not be too high otherwise the cracking in the surrounding concrete becomes excessive. Similarly the maximum compressive stress in the concrete should be appreciably less than the design strength (which is given by u_w/γ_m) to ensure that no spalling or microcracking of the concrete can occur under working loads. The permissible stresses for beams in bending for elastic theory as given in CP114:1957 were $\frac{1}{2}f_y$ for steel and $\frac{1}{3}u_w$ for concrete.

Concrete tension cracks

It follows from the above that if no slipping occurs between a steel bar in the bottom of a loaded concrete beam and the surrounding concrete, the stress in the bar must be m times that in the concrete. If a mild steel bar is stressed to $\frac{1}{2} \times 250 = 125$ MN/m^2 under the working load, the stress in the adjacent concrete for $m = 15$ is 125/15 or 8·3 MN/m^2. Since concrete fractures at a tensile stress far less than this, it must have cracked. Such cracks are always present on the tension side of a reinforced beam or slab, even though they are not visible to the naked eye. The tensile strength of concrete is neglected in flexure since the concrete cracks long before the steel is fully stressed under the working load.

2.2 Basic assumptions for elastic theory

It is essential for a complete understanding of any theory to appreciate fully the assumptions which are made. The basic assumptions of the elastic theory are therefore summarized so that their full significance may become apparent subsequently as individual aspects of reinforced concrete are considered.

(i) Steel and concrete act in conjunction throughout, there being no relative movement of the bars in the concrete. This is attained by ensuring that adequate bond or frictional stress is developed between the steel and concrete.

(ii) Plane cross-sections of members before stressing remain plane under load.

(iii) Both steel and concrete are perfectly elastic.
(iv) Concrete resists no flexural tension stresses.
(v) Stresses due to thermal or shrinkage movement of the concrete may be neglected.
(vi) The cross-sectional area of the steel is small compared with that of the concrete.

These assumptions will now be discussed individually.

Bond

Adequate bond between concrete and steel is essential in reinforced concrete construction. Only a fraction of the potential strength of an RC member is realized if the concrete and steel do not act in conjunction. The bond is due to (*a*) adhesion and (*b*) grip, or frictional resistance. The concrete shrinks upon setting and therefore exerts a pressure upon the steel. This pressure is reduced slightly when the steel and surrounding concrete are stressed in tension, since Poisson's ratio for steel (about 0·29) is greater than that for concrete (0·10 to 0·20). Irregularities on the surface of the reinforcement mechanically lock the steel to the concrete, improving both adhesion and grip; it is this that justifies the indentation of cold-worked steel.

Bernoulli's theorem

The assumption that plane sections remain plane in bending implies that strains are proportional to the distance from the neutral axis.

Elasticity

While steel can be considered an elastic material, concrete is hardly elastic even below the working stress. This assumption is perhaps the least justified since creep in the concrete can cause very considerable stress redistribution with time in continuous beams and columns etc. Nevertheless, this assumption was the main basis used for designs for many years. They generally proved to be satisfactory in practice because due allowance was made in the light of experience for long-term effects such as creep, and because the overall safety margin was usually very high.

Tensile stresses

While the tensile strength of concrete is ignored in direct tension and in bending, it is still expected to resist diagonal tension stresses of the order of 1 MN/m^2 due to shear effects.

Secondary stresses

Stresses due to shrinkage and temperature variations are often ignored. Since these effects can lead to tensile stresses in excess of the tensile strength of the concrete, it is necessary to include nominal steel reinforcement in positions and directions where they would otherwise appear unnecessary. Relatively small amounts of reinforcement usually enable shrinkage cracks to be distributed as a series of very fine hair cracks at frequent intervals, instead of large cracks at infrequent intervals. Suitable amounts of nominal steel are included in appendix C.

Steel area

The cross-sectional areas of the steel bars are sufficiently small for the stress to be assumed constant across the steel section.

2.3 Practical design considerations

Design is based on either permissible stresses (for elastic theory) or design stresses (for limit state theory) in the concrete and steel reinforcement, as well as on certain rules governing the placing of the steel in the concrete. Elastic design is essentially simple and straightforward and factors such as shrinkage are not generally considered. Where practice indicated the desirability of taking other factors into account, various "design rules" became necessary and were incorporated in codes of practice such as CP114. The practical rules for limit state design and reinforced concrete design (generally as given in the Unified Code) are summarized, for convenience, in appendixes B and C respectively. These rules are now much less arbitrary than they used to be for designs based solely on elastic theory. The more elementary practical considerations are given below.

Accuracy of calculation

The assumptions given in section 2.2 emphasize the necessity of avoiding meticulous accuracy in design calculations. The elasticity and compressive strength of concrete differ very considerably from the design assumptions. Many variables are introduced due to inequalities of workmanship, materials and conditions. The compressive strength of concrete varies even in adjacent parts of the same structure, and thicknesses of members as actually placed in practice may well differ slightly from the dimensions given on the drawings, due to undetected errors in fixing formwork. Unpredictable secondary stresses are often set up by shrinkage and expansion of the concrete. These are some of the more important variables which limit state theory attempts to assess but which conventional elastic design virtually ignores except for the use of a single, global, safety factor. Fictitious precision in design calculations is to be avoided;

slide-rule accuracy is more than adequate, especially for elastic design.

Reinforcement arrangement

Beams and slabs are provided with bars near the faces where tensile stresses occur. Bars used in a beam run the length of the beam, whereas in a slab either steel bars or steel mesh fabric are used. Slabs must be provided with bars in two directions, even when the slab is intended to be one-way spanning: main bars running the length of the span and smaller transverse bars at right angles to the span. The former take the tensile stress in bending while the latter prevent large cracks forming due to shrinkage or temperature movement of the concrete. The transverse bars are known as distribution reinforcement and are wired to the main bars to form a mesh.

Size of members

A degree of standardization of the size of members within a structure is desirable from economical considerations such as re-use of forms, fewer bar types etc.

Cover

A certain minimum thickness or cover of concrete is necessary around all reinforcement in order to:

(i) Prevent moisture penetration to the steel causing corrosion and flaking off of the concrete around the bar.

(ii) Enable the bar to act in conjunction with the concrete without slip.

Bar sizes and applications

The general sizes and uses of round mild steel bars are as follows:

Stirrups and ties: 6 to 12 mm.
Common sizes for main reinforcement: 10 to 32 mm.
Very heavy construction only: 40 and 50 mm.

Use as few different diameters as possible when designing, since this assists identification on site and avoids unnecessary sorting: there should normally be no need to use more than the ten standard sizes recommended by the Unified Code.

Bar lengths

The normal maximum length of bars is about 9 m, or 6 m for sizes less than 10 mm. Longer bars up to 12 m are proportionately cheaper than shorter bars if the quantity is sufficient to justify the increased transport and handling costs.

Bar bending

Bars must be bent in order to conform to the geometry of the structure. Some bars have to be hooked at either end to increase their grip or anchorage in the concrete. Bends and hooks or other anchorages are of such forms, dimensions and arrangements as to ensure that the concrete is not over-stressed.

Wire ties

Bars are fixed rigidly in position relative to each other, in the formwork and prior to concreting, by securing some of the intersections of the bars with wire ties.

Effective span

The effective span of a simple beam or slab is taken as the lesser of

(i) The distance between centres of supports, and
(ii) The clear distance between supports plus the effective depth of the beam or slab.

Weight of reinforced concrete

The weight of dense reinforced concrete depends upon the type and proportions of its constituent materials. It also increases with the steel ratio. It usually varies, however, between 22 and 26 kN/m^3 for normal aggregates and is generally assumed for design purposes to be 24 kN/m^3.

REFERENCES

2.1 FAREBROTHER, J. E. C. Concrete—past, present and future. *Struct. Eng.* London, Oct. 1962 pp 337–41.
2.2 BRITISH STANDARDS INSTITUTION (London) BS 4449: 1969 *Hot rolled steel bars for the reinforcement of concrete (metric units)*. Also BS 4461: 1969 *Cold worked steel bars for the reinforcement of concrete (metric units)*.
2.3 LANCASTER, R. I. Steel reinforcement. *Structural concrete* vol. 3 London, July–Aug. 1966 pp 159–77, and discussion, pp 178–98.
2.4 BRITISH STANDARDS INSTITUTION (London) BS 4483:1969 *Steel fabric for the reinforcement of concrete (metric units)*.
2.5 BRITISH STANDARDS INSTITUTION (London) BS 4482:1969 *Hard drawn mild steel wire for the reinforcement of concrete.*
2.6 BRITISH STANDARDS INSTITUTION (London) BS 4466:1969 *Bending dimensions and scheduling of bars for the reinforcement of concrete.*
2.7 BRITISH STANDARDS INSTITUTION (London) CP114:1957 *Structural use of reinforced concrete in buildings* (Reset and reprinted in 1965).

Chapter 3

Elastic analysis of beams

The present chapter considers the elastic *analysis* of members such as beams, slabs and staircases which are essentially subjected to pure flexure. Analysis concerns the determination of certain unknowns *when the section is fully defined* (that is, A_{st}, b and d_1 are known). Two types of problem therefore arise in analysis:

(i) Determination of the stresses for a given loading

(ii) Determination of the maximum (permissible) loading for given permissible stresses.

3.1 Stress and strain distributions

The stress and strain distributions across a section of a member subjected to pure bending can now be determined from the six assumptions given in section 2.2. The more general considerations outlined in the previous chapter, however, must also be borne in mind.

Neutral axis

Consider a loaded rectangular concrete beam which is reinforced in one layer only as shown in figure 3.1.1(*a*). Notice that the neutral axis, which represents points at zero stress, is not necessarily at half the depth of the beam as it would be for a homogeneous beam.

Effective depth

Since all the concrete below the neutral axis is in tension, and is therefore assumed to take no stress, the depth of the beam that really affects the strength calculations is the depth from the compression face to the centroid of the tensile steel. This is known as the *effective depth*, d_1.

Strain profile

If no slip occurs between the concrete and the steel, the steel strain equals the strain in the concrete immediately surrounding the steel. The strain at any point is assumed proportional to its distance from the neutral axis, hence the strain profile is linear as shown in figure 3.1.1(*b*). The maximum compressive strain in the concrete, ε_{cb}, occurs at the top edge of the beam. The cross-sectional area of the steel is small, so the tensile strain in the steel is given by a mean value, ε_{st}. It follows from similar triangles that

$$\varepsilon_{st}/\varepsilon_{cb} = (d_1 - d_n)/d_n \tag{3.1.1}$$

Stress profile

The stress profile for the concrete is also linear as shown in figure 3.1.1(*c*), since stress varies as strain. The maximum compressive stress is f_{cb}. The steel stress, f_{st}, is given by:

$$f_{st} = E_s\varepsilon_{st} = m \times \text{(adjacent concrete stress)}$$

Conversely, the stress in the concrete around the steel equals f_{st}/m. This value is theoretical only since in general the concrete will have cracked, as was shown in chapter 2.

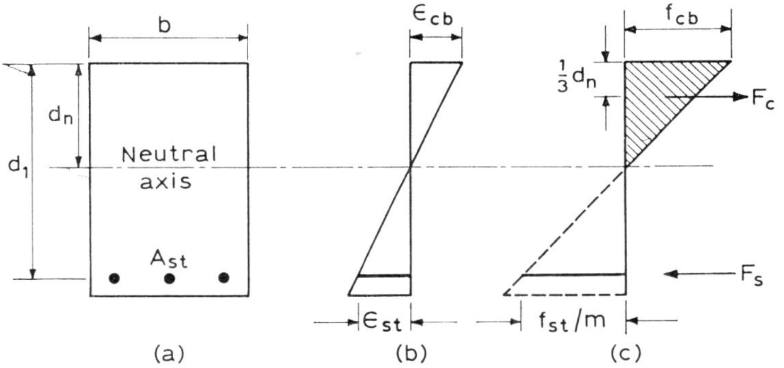

Fig 3.1.1 Beam cross-section showing strain distribution (concrete and steel) and theoretical concrete stress distribution.

Maximum stresses

By considering the strain profile and using Hooke's law,

$$\begin{aligned} f_{st}/f_{cb} &= E_s\varepsilon_{st}/E_c\varepsilon_{cb} = m(d_1 - d_n)/d_n \\ &= m(1 - n_1)/n_1 \end{aligned} \tag{3.1.2}$$

3.2 Moment of resistance

The moment of resistance of a section is the maximum moment which can be applied to the section without exceeding certain limiting stresses in *either* the concrete *or* the steel. For the elastic theory the limiting stresses are the permissible values. The moment of resistance is determined by considering the equilibrium of the internal forces and *both* permissible stresses.

Resultant compressive force

The compressive stress in the concrete varies uniformly from zero to f_{cb}, hence the average value is $f_{cb}/2$. This stress acts over a width b and depth d_n. The compressive force exerted by the concrete is therefore given by

$$F_c = \tfrac{1}{2} f_{cb} b d_n$$

This force acts at the centre of pressure, (i.e. the centroid of the shaded triangle in figure 3.1.1(*c*)) which is $d_n/3$ below the compressive face for a constant beam breadth.

Resultant tensile force

The total tensile force in the steel is given by

$$F_s = f_{st} A_{st} = f_{st} r b d_1$$

where r is the *steel ratio* and equals A_{st}/bd_1. The force acts at a depth d_1 from the compressive face.

Equilibrium requirements

If the beam is subjected to pure bending then the resultant compressive and tensile forces must be equal. Hence

$$F_s = F_c \quad \text{and therefore} \quad f_{st} r b d_1 = \tfrac{1}{2} f_{cb} b d_n$$

Re-arranging,

$$f_{st}/f_{cb} = d_n/2rd_1 = n_1/2r \qquad (3.2.1)$$

Neutral axis depth

The depth of the neutral axis, d_n, can be determined by eliminating the unknown stress ratio f_{st}/f_{cb} from equations 3.1.2 and 3.2.1 as follows:

$$m(1 - n_1)/n_1 = n_1/2r \qquad (3.2.2)$$

Re-arranging,

$$n_1^2 + 2mrn_1 - 2mr = 0 \qquad (3.2.3)$$

This equation is a quadratic in n_1, hence the neutral axis depth is given by

$$n_1 = -mr \pm \sqrt{[(mr)^2 + 2mr]}$$

Only the positive value of n_1 is appropriate, hence

$$n_1 = [\sqrt{(1 + 2/mr)} - 1]mr \qquad (3.2.4)$$

The position of the neutral axis for any given values of r and m can therefore be determined.

Lever arm

The parallel tension and compression forces form a couple which balances the applied bending moment. The bending moment is therefore given by either of the two equal forces multiplied by their distance apart. The arm of the couple is called the *lever arm*, a, of the beam and is given by

$$a = d_1 - d_n/3 = d_1(1 - n_1/3) \tag{3.2.5}$$

The stresses in the concrete and steel can therefore be determined from the following equations in terms of the applied bending moment, M, since $d_n(= n_1d_1)$ is known from equation 3.2.4.

$$\text{Steel} \quad M = f_{st}rbd_1(d_1 - d_n/3) \tag{3.2.6}$$

$$\text{Concrete} \quad M = \tfrac{1}{2}f_{cb}bd_n(d_1 - d_n/3) \tag{3.2.7}$$

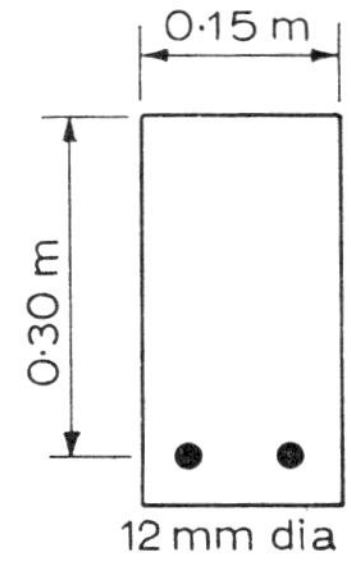

Fig 3.2.1 Example 3.2.1.

EXAMPLE 3.2.1

A rectangular reinforced concrete beam has a breadth of 150 mm and an effective depth of 300 mm. It is reinforced with two 12-mm diameter bars. Calculate the maximum stresses in the concrete and the steel if the beam is subjected to a bending moment of 7·5 kNm (*a*) for a short time (*b*) for a prolonged period.

Assume that

$u = 22{\cdot}5$ MN/m² $\quad f_{yt} = 250$ MN/m² $\quad E_s = 200$ GN/m²
Short-term $E_c = 4500\sqrt{u}$ (in MN/m² units)
Long-term $E_c = \frac{1}{2}$ short-term E_c.

Solution

(*a*) For short-term loading

$E_c = 4500\sqrt{22{\cdot}5}$ MN/m² $= 21{\cdot}4 \times 10^9$

$m = E_s/E_c = 200/21{\cdot}4 = 9{\cdot}35$

$r = A_{st}/bd_1 = 2 \times (\pi/4) \times 12^2/150 \times 300 = 0{\cdot}503\%$

$100mr = 4{\cdot}70$

Hence, from equation 3.2.4,*

$$n_1 = [\sqrt{(1 + 200/4{\cdot}70)} - 1] \times (4{\cdot}70/100) = 0{\cdot}263$$

Therefore

$$a_1 = 1 - 0{\cdot}263/3 = 0{\cdot}912$$

Hence from equations 3.2.6 and 3.2.7

$$7{\cdot}5 \times 10^3 = f_{st} \times 0{\cdot}00503 \times 0{\cdot}15 \times 0{\cdot}30^2 \times 0{\cdot}912$$

$$f_{st} = 121 \times 10^6 = 121 \text{ MN/m}^2 \qquad \text{(Answer)}$$

and

$$7{\cdot}5 \times 10^3 = \tfrac{1}{2}f_{cb} \times 0{\cdot}15 \times 0{\cdot}30^2 \times 0{\cdot}263 \times 0{\cdot}912$$

$$f_{cb} = 4{\cdot}63 \times 10^6 = 4{\cdot}63 \text{ MN/m}^2 \qquad \text{(Answer)}$$

(*b*) For long-term loading

$$E_c = \tfrac{1}{2} \times 21{\cdot}4 \times 10^9 = 10{\cdot}7 \times 10^9$$

$$m = 18{\cdot}7 \qquad 100\text{mr} = 9{\cdot}41$$

Therefore

$$n_1 = [\sqrt{(1 + 200/9{\cdot}41)} - 1](9{\cdot}41/100) = 0{\cdot}351$$

$$a_1 = 1 - 0{\cdot}351/3 = 0{\cdot}883$$

Therefore

$$7{\cdot}5 \times 10^3 = f_{st} \times 0{\cdot}00503 \times 0{\cdot}15 \times 0{\cdot}30^2 \times 0{\cdot}883$$

$$f_{st} = 125 \times 10^6 = 125 \text{ MN/m}^2 \qquad \text{(Answer)}$$

Also

$$7{\cdot}5 \times 10^3 = \tfrac{1}{2}f_{cb} \times 0{\cdot}15 \times 0{\cdot}30^2 \times 0{\cdot}351 \times 0{\cdot}883$$

$$f_{cb} = 3{\cdot}59 \times 10^6 = 3{\cdot}59 \text{ MN/m}^2 \qquad \text{(Answer)}$$

Notice that the reduced value for E_c has an appreciable effect upon n_1 and therefore upon f_{cb}, but that it has a relatively small effect upon a_1. The steel stress is therefore only slightly affected by the large change in E_c.

Moment of resistance—steel critical

The moment of resistance of the steel in tension, M_{rt}, is given by the maximum moment which can be resisted by the steel. Substituting M_{rt} for M and p_{st} for f_{st} in equation 3.2.6, therefore, gives

$$M_{rt} = p_{st}rbd_1(d_1 - d_n/3)$$

* The student should test his understanding of the text by calculating n_1 from first principles, as would be necessary in an examination. This applies to all worked examples which, for convenience, quote equations derived in the text.

Re-arranging,

$$M_{rt}/bd_1^2 = p_{st}r(1 - n_1/3) \qquad (3.2.8)$$

n_1 is known from equation 3.2.4, for given values of m and r, hence M_{rt}/bd_1^2 can be determined.

Moment of resistance—concrete critical

Similarly, the maximum moment which can be resisted by the concrete is given by

$$M_{rc}/bd_1^2 = \tfrac{1}{2}p_{cb}n_1(1 - n_1/3) \qquad (3.2.9)$$

EXAMPLE 3.2.2

A concrete beam of breadth 100 mm is reinforced with two 10-mm diameter bars such that the effective depth is 200 mm. Calculate the maximum moment that the section is capable of carrying for a short time, and the maximum steel and concrete stresses induced by that moment.

Assume that

$u = 22{\cdot}5\ \mathrm{MN/m^2}$ $\quad f_{yt} = 250\ \mathrm{MN/m^2}$

$p_{cb} = \tfrac{1}{3}u \quad p_{st} = \tfrac{1}{2}f_{yt}$

$E_c = 4500\sqrt{u}$ (in $\mathrm{MN/m^2}$ units) $\quad E_s = 200\ \mathrm{GN/m^2}$

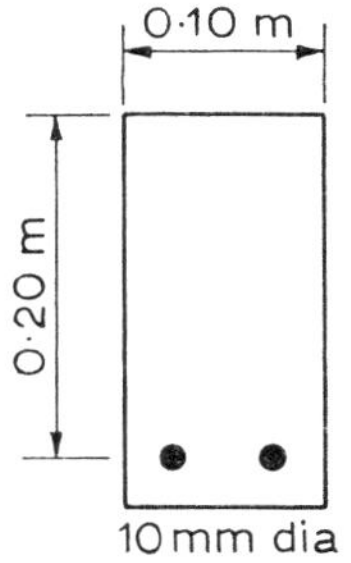

Fig 3.2.2 Example 3.2.2.

Solution For short-term loading, $E_c = 21{\cdot}4 \times 10^9$ and $m = 9{\cdot}35$, as for example 3.2.1.

$r = 10^{-6} \times 2 \times (\pi/4) \times 10^2/0{\cdot}1 \times 0{\cdot}2 = 0{\cdot}785\%$

$100mr = 9{\cdot}35 \times 0{\cdot}785 = 7{\cdot}34$

From equation 3.2.4:

$n_1 = [\sqrt{(1 + 200/7{\cdot}34)} - 1] \times 7{\cdot}34/100 = 0{\cdot}317$

$a_1 = 1 - 0{\cdot}317/3 = 0{\cdot}894$

$bd_1^2 = 0{\cdot}1 \times 0{\cdot}2^2 = 4 \times 10^{-3}$

$p_{st} = 125 \times 10^6 \quad p_{cb} = 7{\cdot}5 \times 10^6$

Hence from equations 3.2.8 and 3.2.9,

$$M_{rt} = 125 \times 0{\cdot}00785 \times 0{\cdot}894 \times 4 \times 10^{-3+6} = 3{\cdot}51 \times 10^3$$

$$M_{rc} = \tfrac{1}{2} \times 7{\cdot}5 \times 0{\cdot}317 \times 0{\cdot}894 \times 4 \times 10^{-3+6} = 4{\cdot}25 \times 10^3$$

Hence the steel is critical and the maximum moment is 3·51 kNm (Answer)

The maximum steel stress is

$$f_{st} = p_{st} = 125 \text{ MN/m}^2 \qquad \text{(Answer)}$$

The maximum concrete stress, f_{cb}, is given by equation 3.2.7. It is also given by

$$(M/M_{rc})p_{cb} = (3{\cdot}51/4{\cdot}25) \times 7{\cdot}5 \times 10^6$$
$$= 6{\cdot}19 \text{ MN/m}^2 \qquad \text{(Answer)}$$

EXAMPLE 3.2.3

A concrete beam is 100 mm wide and has an effective depth of 215 mm. It is reinforced by three 12-mm diameter bars, and carries a sustained applied moment of 7·543 kNm. Calculate the maximum stresses in the steel and concrete. What is special about this section? (Assume that the long-term value for E_c is half the value derived from the relationship below.)

Assume that

$$u = 22{\cdot}5 \text{ MN/m}^2 \quad f_{yt} = 250 \text{ MN/m}^2$$

$$p_{cb} = \tfrac{1}{3}u \quad p_{st} = \tfrac{1}{2}f_{yt}$$

$$E_c = 4500\sqrt{u} \text{ (in MN/m}^2\text{ units)} \quad E_s = 200 \text{ GN/m}^2.$$

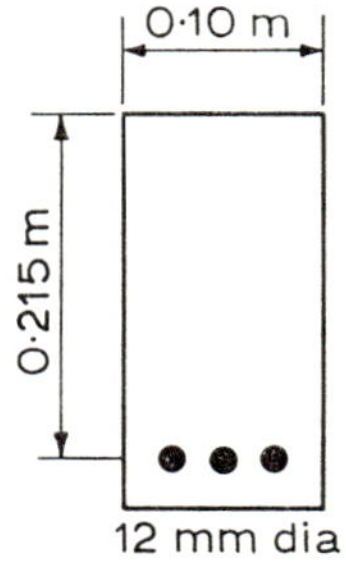

Fig 3.2.3 Example 3.2.3.

Solution For long-term loading $E_c = 10{\cdot}7 \times 10^9$ and $m = 18{\cdot}7$, as in example 3.2.1(*b*).

$$r = 10^{-6} \times 3 \times (\pi/4) \times 12^2/(0{\cdot}1 \times 0{\cdot}215) = 1{\cdot}58\%$$

$$100mr = 18{\cdot}7 \times 1{\cdot}58 = 29{\cdot}5$$

Therefore, from equation 3.2.4,

$$n_1 = [\sqrt{(1 + 200/29{\cdot}5)} - 1] \times 29{\cdot}5/100 = 0{\cdot}528$$

$$a_1 = 1 - 0{\cdot}528/3 = 0{\cdot}824$$

$$bd_1^2 = 0{\cdot}1 \times 0{\cdot}215^2 = 4{\cdot}62 \times 10^{-3}$$

Therefore

$$7{\cdot}543 \times 10^3 = f_{st} \times 0{\cdot}0158 \times 4{\cdot}62 \times 0{\cdot}824 \times 10^{-3}$$

$$f_{st} = 125 \text{ MN/m}^2 \qquad \text{(Answer)}$$

$$7{\cdot}543 \times 10^3 = \tfrac{1}{2} f_{cb} \times 4{\cdot}62 \times 0{\cdot}528 \times 0{\cdot}824 \times 10^{-3}$$

$$f_{cb} = 7{\cdot}5 \text{ MN/m}^2 \qquad \text{(Answer)}$$

This section is very special because the maximum steel and concrete stresses equal the permissible values *simultaneously*. In general, one of the permissible stresses is critical whilst the other is not.

3.3 The equivalent concrete section

The analysis given in section 3.1 and 3.2 derives expressions for the position of the neutral axis by considering first principles—that is, the longitudinal equilibrium of forces, the strain profile geometry and Hooke's law. An alternative and quicker approach is to use the concept of the *equivalent concrete section.*

Equivalent concrete area

The equivalent concrete area is the actual cross-sectional area represented entirely in terms of concrete having the same flexural rigidity, (i.e. the same values of the product EI). The concrete area below the neutral axis is ignored, since the concrete there is assumed to have cracked, and the steel area is multiplied by m to convert it into an equivalent concrete area, since it carries m times the concrete stress for the same strain.

Position of neutral axis

The neutral axis passes through the centre of area of the equivalent concrete section. Its position can therefore be obtained by considering the moments of the equivalent concrete areas about the neutral axis. Hence from figure 3.3.1, b being constant,

$$b \int_0^{d_n} y\,dy = mA_s(d_1 - d_n)$$

Alternatively,

$$\tfrac{1}{2}d_n^2 = mrd_1(d_1 - d_n) \tag{3.3.1}$$

Putting $d_n = n_1 d_1$ and rearranging gives equation 3.2.3, and solving for n_1 gives equation 3.2.4.

Equivalent second moment of area

If f_c and f_s are the concrete and steel stresses respectively at distance y from the neutral axis, and using appendix Y, sign convention (1), then

$$f_c/f_{cb} = -\,y/d_n \quad \text{and} \quad f_s/f_{cb} = my/d_n$$

(*Note:* y is positive downwards, hence f_c is positive when y is negative. d_n is always considered positive.)

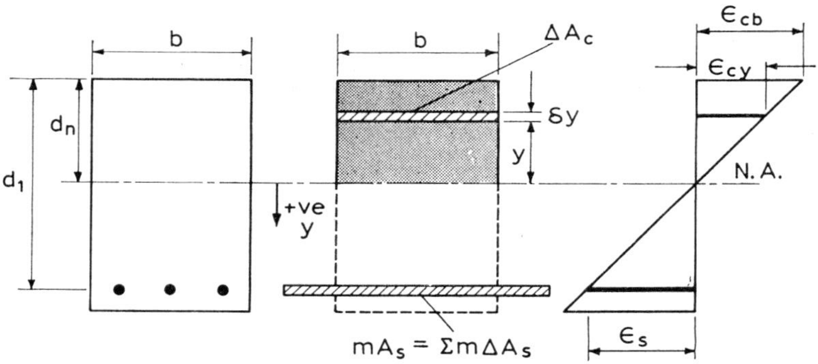

Fig 3.3.1 Cross-section for singly reinforced beam showing eqnivalent concrete section and strain profile.

Considering elemental strips as shown in figure 3.3.1:

$$\Delta F_c = f_c \Delta A_c = -\,y\Delta A_c f_{cb}/d_n$$

$$\Delta F_s = f_s \Delta A_s = my\Delta A_s f_{cb}/d_n$$

Taking moments about the neutral axis gives

$$M = [\textstyle\sum y^2 \Delta A_c + \sum my^2 \Delta A_s] \times (f_{cb}/d_n)$$

The equivalent second moment of area of the section, I_e, is given by

$$I_e = \textstyle\sum y^2 \Delta A_c + \sum my^2 \Delta A_s \tag{3.3.2}$$

Hence

$$M/I_e = f_{cb}/d_n \tag{3.3.3}$$

Similarly

$$M/I_e = f_{st}/m(d_1 - d_n) \tag{3.3.4}$$

Equations 3.3.4 and 3.3.3 correspond to equations 3.2.6 and 3.2.7 respectively. For the rectangular beam shown in figure 3.3.1:

$$I_e = bd_n^3/3 + mA_{st}(d_1 - d_n)^2$$

$$I_e/bd_1^3 = n_1^3/3 + mr(1 - n_1)^2 \qquad (3.3.5)$$

EXAMPLE 3.3.1

Repeat example 3.2.1(*a*) using the concept of the equivalent concrete section.

Solution $E_c = 21{\cdot}4 \times 10^9$, $m = 9{\cdot}35$, $r = 0{\cdot}503\%$ and $mr = 0{\cdot}0470$, as before.

From equation 3.3.1

$$\tfrac{1}{2}d_n^2 = 0{\cdot}0470\,d_1(d_1 - d_n)$$

$$\tfrac{1}{2}n_1^2 + 0{\cdot}0470\,n_1 - 0{\cdot}0470 = 0$$

from which $n_1 = 0{\cdot}263$ as before.

$$bd_1^3 = 0{\cdot}15 \times 0{\cdot}30^3 = 4{\cdot}05 \times 10^{-3} \quad \text{and} \quad 1 - n_1 = 0{\cdot}737$$

From equation 3.3.5

$$I_e = [0{\cdot}263^3/3 + 0{\cdot}0470 \times 0{\cdot}737^2] \times 4{\cdot}05 \times 10^{-3}$$

$$= 128 \times 10^{-6}$$

From equations 3.3.3 and 3.3.4

$$f_{cb} = 0{\cdot}263 \times 0{\cdot}30 \times 7{\cdot}5 \times 10^{3+6}/128 = 4{\cdot}63 \times 10^6$$

$$= 4{\cdot}63\ \text{MN/m}^2 \qquad \text{(Answer)}$$

$$f_{st} = 9{\cdot}35 \times 0{\cdot}737 \times 0{\cdot}30 \times 7{\cdot}5 \times 10^{3+6}/128 = 121 \times 10^6$$

$$= 121\ \text{MN/m}^2 \qquad \text{(Answer)}$$

Moments of resistance

The moments of resistance of the concrete and of the steel are obtained by substituting the permissible stresses into equations 3.3.3 and 3.3.4 and transposing as follows:

$$M_{rc} = p_{cb}I_e/d_n \qquad (3.3.6)$$

$$M_{rt} = p_{st}I_e/m(d_1 - d_n) \qquad (3.3.7)$$

EXAMPLE 3.3.2

If the area of steel in the beam in example 3.2.3 is doubled, calculate the percentage increase in sustained moment which the section could carry. Use the equivalent area method.

Solution $E_c = 10{\cdot}7 \times 10^9$ and $m = 18{\cdot}7$ as before, but $r = 2 \times 1{\cdot}58\%$

$$100mr = 18{\cdot}7 \times 3{\cdot}16 = 59$$

Therefore from equation 3.2.4

$$n_1 = [\surd(1 + 200/59) - 1] \times 0{\cdot}59 = 0{\cdot}646$$

$$bd_1^3 = 0{\cdot}1 \times 0{\cdot}215^3 = 0{\cdot}994 \times 10^{-3}$$

From equation 3.3.5

$$I_e = [0{\cdot}646^3/3 + 0{\cdot}59(1 - 0{\cdot}646)^2] \times 0{\cdot}994 \times 10^{-3}$$

$$= 163 \times 10^{-6}$$

From equations 3.3.6 and 3.3.7

$$M_{rc} = 7{\cdot}5 \times 163 \times 10^{6-6}/0{\cdot}646 \times 0{\cdot}215 = 8800$$

$$M_{rt} = 125 \times 163 \times 10^{6-6}/18{\cdot}7 \times 0{\cdot}354 \times 0{\cdot}215 = 14320$$

Hence M_{rc} is critical and the actual moment of resistance is $8{\cdot}8 \times 10^3$. This represents a percentage increase of

$$100 \times (8{\cdot}8 - 7{\cdot}543)/7{\cdot}543 = 17\% \qquad \text{(Answer)}$$

Notice that the increase in M_{rc} due to the extra steel is relatively small since the tension steel can only have the secondary effect of increasing the neutral axis depth d_n. In these circumstances it is preferable to add compression steel as explained in the next section.

3.4 Doubly reinforced beams and slabs in flexure

Doubly reinforced beams (and slabs) are reinforced in compression as well as in tension. The use of compression steel should generally be restricted as much as possible in practice since it can cause congestion and tends to increase the cost of the section for a given resistance moment. Its use is often unavoidable, however, and in such cases doubly reinforced beams are desirable or even essential. There is a limit to the amount of tension steel which can profitably be used in a singly reinforced beam.

Applications

Compression steel is justified when the dimensions of a beam are restricted due to requirements of headroom, space, appearance etc. If a high bending moment exists over a relatively short length of the beam only (e.g. over supports), then the cost of additional steel is small compared with the overall saving along the full length of a beam of constant cross-section.

Neutral axis depth

The equivalent concrete area for the doubly reinforced beam shown in figure 3.4.1 is given by

$$A_e = bd_n + (m - 1)A_{sc} + mA_{st}$$

Taking first moments of area about the neutral axis gives

$$\tfrac{1}{2}bd_n^2 + (m - 1)A_{sc}(d_n - d_2) = mA_{st}(d_1 - d_n) \qquad (3.4.1)$$

This quadratic can be solved for d_n and corresponds to equation 3.3.1 for a singly reinforced beam.

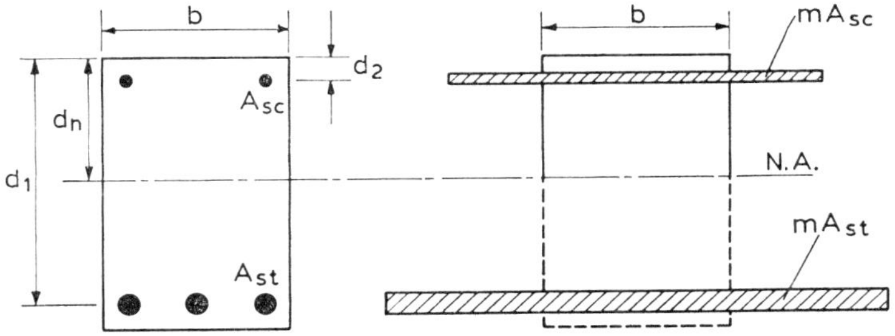

Fig 3.4.1 Cross-section for doubly reinforced beam showing equivalent concrete section.

Concrete and steel stresses

The equivalent second moment of area about the neutral axis is given by

$$I_e = \tfrac{1}{3}bd_n^3 + (m - 1)A_{sc}(d_n - d_2)^2 + mA_{st}(d_1 - d_n)^2 \qquad (3.4.2)$$

The stresses for the doubly reinforced beam are given by

$$f_{cb} = Md_n/I_e \qquad (3.4.3)$$

$$f_{sc}(= -f_s) = mM(d_n - d_2)/I_e \qquad (3.4.4)$$

$$f_{st}(= +f_s) = mM(d_1 - d_n)/I_e \qquad (3.4.5)$$

Notice the sign convention: the symbol f_{sc} specifically implies that the steel stress is compressive and positive values only are considered. Similarly, f_{st} specifically implies that the steel stress is tensile and positive values only are again considered. However, if a reversal of stresses could occur then it would be necessary to consider the general case, for which steel stresses are assumed to be tensile. That is, to consider f_s, which can be either positive (tension) or negative (compression). The quantities m, d_1, d_2, d_n and I_e all have positive values. M is assumed to be a sagging (i.e. positive) moment for convenience. If M is actually a hogging moment (i.e. negative) the section must be inverted, i.e. tension steel at the top of the section and compression steel at the bottom as shown in figure 3.4.2.

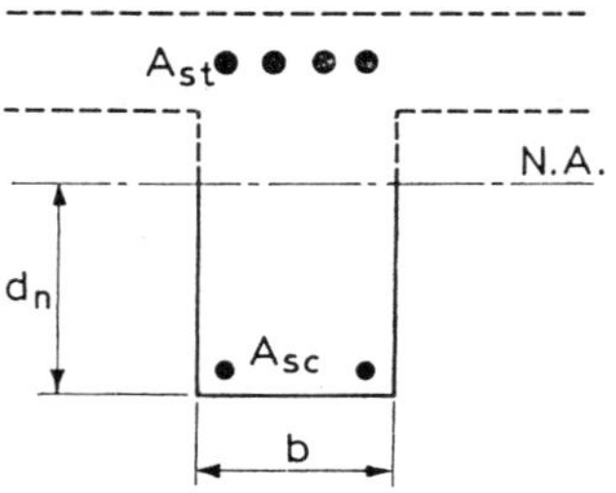

Fig 3.4.2 Doubly reinforced section at support.
Note: presence of any slab at support section is immaterial since concrete shown in broken outline is in tension.

Moments of resistance

The moments of resistance can be obtained by transposing equations 3.4.3 and 3.4.5 and substituting into them the permissible values as follows:

$$M_{rc} = p_{cb}I_e/d_n \tag{3.4.6}$$

$$M_{rt} = p_{st}I_e/m(d_1 - d_n) \tag{3.4.7}$$

It is not necessary to consider equation 3.4.4 since the compression steel is unlikely to reach its permissible value before the concrete fibre stress reaches p_{cb}. Notice also that equations 3.4.6 and 3.4.7 are exactly the same as equations 3.3.6 and 3.3.7 since the value for I_e allows for the compression steel when present. The equivalent area method is certainly the most convenient to use when dealing with other than singly reinforced sections.

EXAMPLE 3.4.1

Calculate the maximum stresses in an RC beam, 200 mm wide and of 500 mm effective depth, having a tensile steel area of 1 960 mm^2 and a compressive steel area of 628 mm^2 at a depth of 75 mm below the compressive face of the beam. The applied bending moment is 100 kNm. Assume that E_c = 10·7 GN/m^2 and E_s = 200 GN/m^2.

Solution

$$m = E_s/E_c = 200/10{\cdot}7 = 18{\cdot}7$$

From equation 3.4.1

$$\tfrac{1}{2} \times 0{\cdot}20d_n^2 + 17{\cdot}7 \times 628(d_n - 0{\cdot}075)10^{-6} - 18{\cdot}7 \times 1960(0{\cdot}50 - d_n) \times 10^{-6} = 0$$

$$d_n^2 + 0{\cdot}478d_n - 0{\cdot}192 = 0$$

$$d_n = \tfrac{1}{2}[- 0{\cdot}478 \pm \sqrt{(0{\cdot}478^2 + 4 \times 0{\cdot}192)}] = 0{\cdot}26$$

(since d_n must be $+ve$). Therefore from equation 3.4.2

$$I = \tfrac{1}{3} \times 0{\cdot}20 \times 0{\cdot}26^3 + 17{\cdot}7 \times 628(0{\cdot}26 - 0{\cdot}075)^2 \times 10^{-6} + 18{\cdot}7 \times 1960(0{\cdot}50 - 0{\cdot}26)^2 \times 10^{-6}$$

$$= 3{\cdot}67 \times 10^{-3}$$

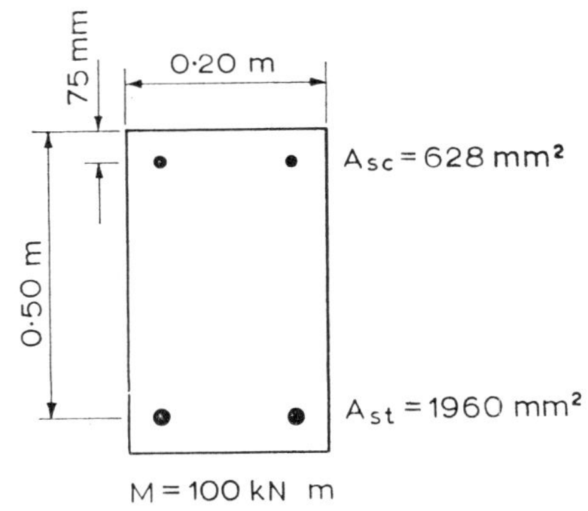

Fig 3.4.3 Example 3.4.1.

Therefore

$$f_{cb} = \frac{0{\cdot}26 \times 100 \times 10^{3+3}}{3{\cdot}67} = 7{\cdot}1 \text{ MN/m}^2 \qquad \text{(Answer)}$$

$$f_{st} = \frac{18{\cdot}7 \times 0{\cdot}24 \times 100 \times 10^{3+3}}{3{\cdot}67} = 122 \text{ MN/m}^2 \qquad \text{(Answer)}$$

$$f_{sc} = \frac{18{\cdot}7(0{\cdot}26 - 0{\cdot}075) \times 100 \times 10^{3+3}}{3{\cdot}67} = 94{\cdot}5 \text{ MN/m}^2$$

(Answer)

EXAMPLE 3.4.2

Show that the neutral axis depth d_n of a doubly reinforced rectangular concrete beam for which $d_2 = d_n/3$ is given by

$$n_1/mr = \sqrt{\left[\left(1 + \frac{2r_c}{3r}\right)^2 + \frac{2}{mr}\right]} - \left(1 + \frac{2r_c}{3r}\right)$$

where $n_1 = d_n/d_1$ and it is assumed that $(m - 1)$ approximately equals m.

Solution Taking first moments of area about the neutral axis,

$$\tfrac{1}{2}b(n_1d_1)^2 + (m-1)r_cbd_1^2(1-\tfrac{1}{3})n_1 = mrbd_1^2(1-n_1)$$

or, since $m - 1 \simeq m$,

$$\tfrac{1}{2}n_1^2 + m(r+\tfrac{2}{3}r_c)n_1 - mr = 0$$

$$n_1 = -m(r+\tfrac{2}{3}r_c) \pm \sqrt{[m^2(r+\tfrac{2}{3}r_c)^2 + 2mr]}$$

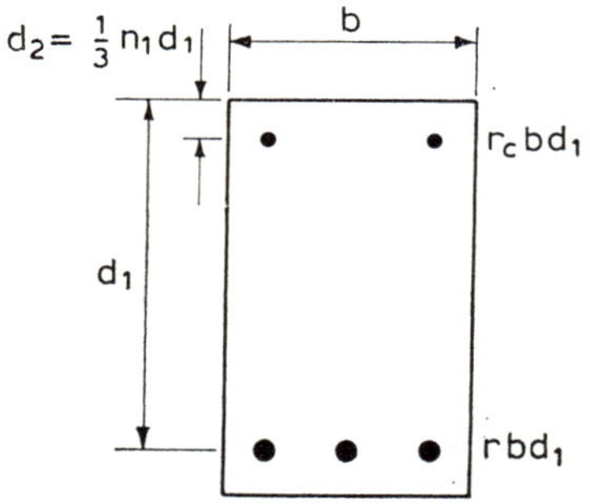

Fig 3.4.4 Example 3.4.2.

Neglecting the negative root

$$n_1/mr = \sqrt{\left[\left(1+\frac{2r_c}{3r}\right)^2 + \frac{2}{mr}\right]} - \left(1+\frac{2r_c}{3r}\right) \qquad \text{(QED)}$$

EXAMPLE 3.4.3

Use the result of example 3.4.2 to calculate the moment of resistance of a reinforced concrete beam with a total steel area, $A_{st} + A_{sc}$, of 4000 mm² for the following values of r_c/r: 0, 0·3 and 0·6. Assume that

$$p_{st} = 125 \text{ MN/m}^2 \quad p_{cb} = 7{\cdot}5 \text{ MN/m}^2 \quad p_{sc} = 110 \text{ MN/m}^2$$

$$m = 18{\cdot}7 \quad b = 0{\cdot}30 \text{ m} \quad d_1 = 0{\cdot}45 \text{ m}$$

Solution

$$r + r_c = 4000 \times 10^{-6}/0{\cdot}30 \times 0{\cdot}45 = 29{\cdot}6 \times 10^{-3}$$

If $r_c = kr$ then

$$r = 29{\cdot}6 \times 10^{-3}/(1+k)$$

The values are most conveniently tabulated as follows:

k	$r(\%)$	$r_c(\%)$	$1+\frac{2}{3}k$	$(1+\frac{2}{3}k)^2$	mr	$2/mr$	$\sqrt{[(1+\frac{2}{3}k)^2+2/mr]}$
0	2·96	0	1·0	1·00	0·553	3·61	2·15
0·3	2·275	0·685	1·2	1·44	0·425	4·70	2·48
0·6	1·85	1·11	1·4	1·96	0·346	5·78	2·78

n_1 is obtained using the result of example 3.4.2.

$$I_e/bd_1^3 = \tfrac{1}{3}n_1^3 + (m-1)r_c(\tfrac{2}{3}n_1)^2 + mr(1-n_1)^2$$

$$bd_1^2 = 0{\cdot}30 \times 0{\cdot}45^2 = 60{\cdot}8 \times 10^{-3}$$

$$M_{rc} = (p_{cb}/n_1)(I_e/bd_1^3)bd_1^2 = (456/n_1)(I_e/bd_1^3) \times 10^3$$

$$M_{rt} = [p_{st}/m(1-n_1)](I_e/bd_1^3)bd_1^2$$

$$= [407/(1-n_1)](I_e/bd_1^3) \times 10^3$$

Further tabulation can be made as follows, using units of 10^{-3} except where shown otherwise:

n_1	$\frac{1}{3}n_1^3$	$(m-1)r_c(\frac{2}{3}n_1)^2$	$mr(1-n_1)^2$	I_e/bd_1^3	M_{rc}	M_{rt}
636	85·7	0	73·3	159·0	114×10^3	178×10^3
544	53·6	15·9	88·4	157·9	132×10^3	141×10^3
478	36·4	20·0	94·3	150·7	144×10^3	118×10^3

Therefore M_r = 114, 132 and 118 kNm for r_c/r = 0, 0·3 and 0·6 respectively. (Answer)

Notice that there is a certain optimum amount of compression reinforcement. This is considered further in chapter 4.

3.5 T-beams in flexure

Reinforced concrete beams often carry a floor slab such that the whole construction is monolithic. Each beam can therefore include part of the slab in its structural strength and so produce a T-shaped section. Such beams are clearly very common in practice and even if a reinforced slab is not involved monolithically it is sometimes economical to use a T-beam, despite the additional shuttering costs.

Advantages

The relatively wide horizontal flange of concrete provides the resistance to compression where it is most effective, while the vertical rib gives adequate depth for the lever arm. A relatively narrow rib can contain sufficient steel to provide the tensile resistance. The wide flange can also be useful in providing increased lateral stability. An RC beam may require additional reinforcement in order to reduce the concrete stresses if the breadth is very narrow, i.e. if the slenderness of the beam is above the limit given in appendix C.

Compression flange

The breadth b of the flange which effectively acts with the rib and contributes to its strength in bending depends upon the T-beam span, slab thickness, rib breadth, and distance between adjacent T-beams. A certain minimum quantity of transverse reinforcement is also necessary in the flange. A combination of theory and practice has enabled the design rules given in appendix C to be formulated for both the flange breadth and the minimum amount of transverse reinforcement.

EXAMPLE 3.5.1

A continuous slab, 150 mm deep, spans between beams at 750 mm centres as shown in figure 3.5.1. Calculate the effective width of the

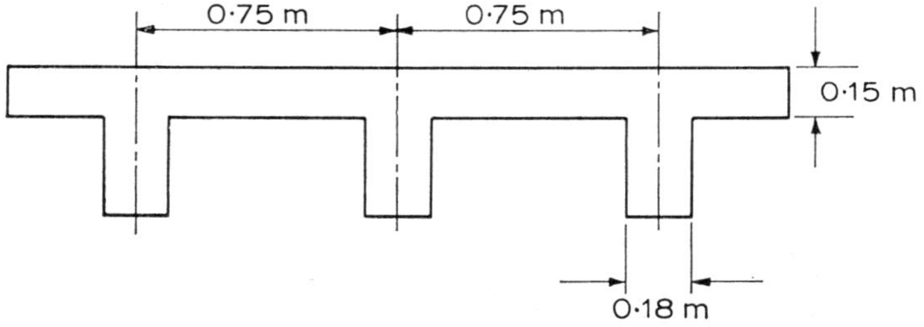

Fig 3.5.1 Example 3.5.1.

flange if the beams are simply supported on a span of 10 m and have a rib breadth of 180 mm. What is the minimum amount of transverse reinforcement required by the Unified Code in the top of the flange? Is the slenderness of the beam satisfactory if the effective depth is 400 mm?

Solution From appendix C

$$b \not> 0{\cdot}750 \quad \text{(i.e. actual flange width)}$$

and

$$b \not> 0{\cdot}18 + 0{\cdot}2 \times 10 \quad \text{(i.e. web width} + XY\text{)}$$

Hence

$b = 0{\cdot}750$ (Answer)

Transverse reinforcement $\nless 0{\cdot}3 \times 0{\cdot}15 \times 1{\cdot}0/100$

$\nless 450 \times 10^{-6}$

The minimum amount of transverse reinforcement is 450 mm² per metre run of beam. (Answer)

$Ld_1/b^2 = 10 \times 0{\cdot}4/0{\cdot}75^2 = 7{\cdot}1$

$\ll$ maximum limit, i.e. 250

$L/b = 10/0{\cdot}75 = 13{\cdot}3$

$\ngtr 30$

so no additional reinforcement is required.

Deep flanges

Notice that when the neutral axis lies within the depth of the slab, the section resembles a rectangular beam of breadth b, since the concrete below the neutral axis is irrelevant to the analysis. Only the case where the neutral axis lies below the soffit of the slab need therefore be considered.

Neutral axis depth

The equivalent concrete area for the T-beam shown in figure 3.5.2 is given by

$$A_e = bd_n - (b - b_r)(d_n - d_s) + mA_{st} \tag{3.5.1}$$

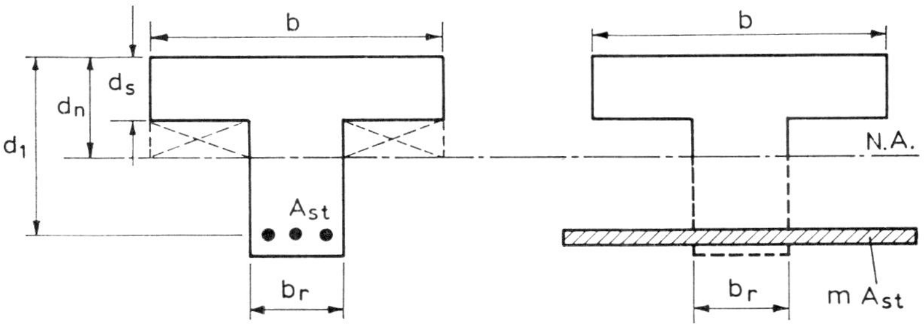

Fig 3.5.2 T-beam. (*a*) Actual section. (*b*) Equivalent concrete section.

Taking first moments of the equivalent areas about the neutral axis gives

$$\tfrac{1}{2}bd_n^2 - \tfrac{1}{2}(b - b_r)(d_n - d_s)^2 = mA_{st}(d_1 - d_n)$$

or

$$b_r d_n^2 + 2[d_s(b - b_r) + mA_{st}]d_n - (b - b_r)d_s^2 - 2mA_{st}d_1 = 0 \quad (3.5.2)$$

This quadratic in d_n can be solved in the normal way.

Concrete and steel stresses

The equivalent second moment of area about the neutral axis for the T-beam is given by

$$I_e = \tfrac{1}{3}bd_n^3 - \tfrac{1}{3}(b - b_r)(d_n - d_s)^3 + mA_{st}(d_1 - d_n)^2 \quad (3.5.3)$$

Hence the concrete and steel stresses are again given by

$$f_{cb} = Md_n/I_e \qquad \text{and} \qquad f_{st} = mM(d_1 - d_n)/I_e$$

EXAMPLE 3.5.2

A T-beam has a flange width of 750 mm, a rib width of 150 mm, a flange depth of 150 mm, an effective depth of 600 mm and a steel area of 5000 mm². If the beam is simply supported on a span of 10 m, and the total uniformly distributed load is 25 kN/m, calculate the maximum stresses in the beam assuming that $E_c = 10{\cdot}7$ GN/m² and $E_s = 200$ GN/m².

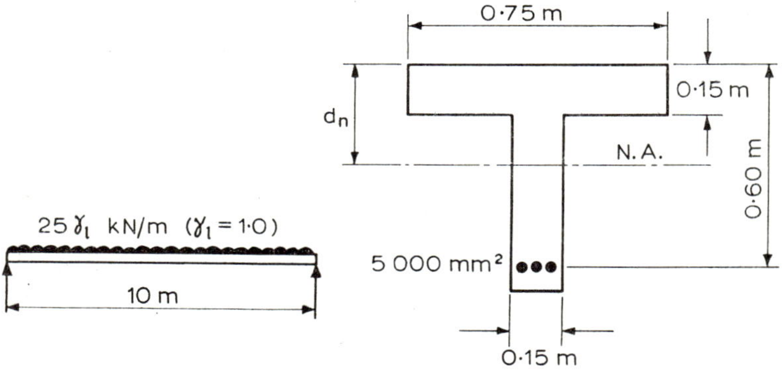

Fig 3.5.3 Example 3.5.2.

Solution

$$m = 200/10{\cdot}7 = 18{\cdot}7$$

The midspan moment is

$$wL^2/8 = (25 \times 10^2/8) \times 10^3 = 313 \times 10^3$$

Assume that $d_n > 0{\cdot}15$ and take first moments of area about the neutral axis.

$$\tfrac{1}{2} \times 0{\cdot}75 d_n^2 - \tfrac{1}{2}(0{\cdot}75 - 0{\cdot}15)(d_n - 0{\cdot}15)^2 = 18{\cdot}7 \times \times 5(0{\cdot}60 - d_n) \times 10^{-3}$$

$$0{\cdot}15 d_n^2 + 0{\cdot}367 d_n - 0{\cdot}126 = 0$$

Hence $d_n = 0{\cdot}307$ and $d_n > 0{\cdot}15$ as assumed.

$$I_e = \tfrac{1}{3} \times 0{\cdot}75 d_n^3 - \tfrac{1}{3}0{\cdot}60(d_n - 0{\cdot}15)^3 + 18{\cdot}7 \times 5(0{\cdot}60 - d_n)^2 \times 10^{-3}$$

$$= 14{\cdot}5 \times 10^{-3}$$

$$f_{cb} = 313 \times 0{\cdot}307 \times 10^{3+3}/14{\cdot}5 = 6{\cdot}62\ \text{MN/m}^2 \qquad \text{(Answer)}$$

$$f_{st} = 18{\cdot}7 \times 313 \times 0{\cdot}293 \times 10^{3+3}/14{\cdot}5 = 118\ \text{MN/m}^2 \qquad \text{(Answer)}$$

Compression steel

Although it is theoretically possible to have an undersize compression flange, it is preferable to increase the effective depth in such cases since compression steel should generally be avoided for the following reasons.

(i) The stress in the compression steel is low due to a fairly shallow neutral axis depth.

(ii) The compression steel must be tied by stirrups. This blocks the top of the rib more than usual and accentuates the difficulties in placing the concrete.

(iii) Even more compression steel than usual is required at the supports.

L-beams

An L-beam is similar to a T-beam except that the slab, and therefore the flange, is available on one side of the rib only: for example, an end bay. The analysis is similar so that the same formulae apply. The allowable breadth of the compression flange is again covered by arbitrary design rules as given in appendix C.

Torsion reinforcement

Since neither an L-beam nor its loading is symmetrical, a bending moment applied in a vertical plane produces torsion. This torsion is resisted by the floor slab and any beams which frame into the L-beam. The torsional resistance is greatly increased by providing top steel along the length of the L-beam, in conjunction with a fairly strong system of stirrups. L-beams are usually located at the outside of a building, so that the top steel is also very effective in resisting the stresses due to the high temperature gradient between the inside and outside of the building. Beams in torsion are considered in detail for the limit state of collapse.

EXAMPLE 3.5.3

Calculate the central deflection for the T-beam in example 3.5.2.

Solution One method of determining the elastic deflection is to consider the bending moment for the conjugate beam (i.e. the beam loaded with $(1/EI) \times$ bending moment diagram).

$$E_c I_e = 10{\cdot}7 \times 14{\cdot}5 \times 10^{9-3} = 155 \times 10^6$$

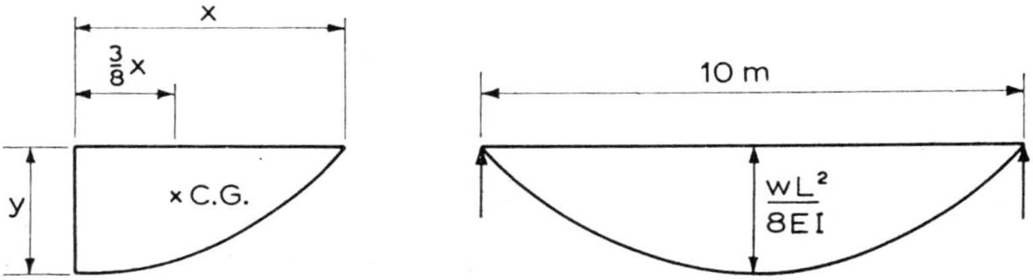

Fig 3.5.4 Example 3.5.3.

The BM diagram is parabolic, hence the deflection at midspan is

$$\tfrac{2}{3} \cdot \tfrac{1}{2}L(wL^2/8EI)(L/2 - 3L/16) = 5wL^4/384EI = 21 \times 10^{-3}$$

i.e. the deflection is 21 mm (Answer)

Although the limit state of deflection is considered in detail later, notice how the concept of the equivalent concrete section can be very conveniently applied.

3.6 Steel beam theory

The *steel beam theory* is a very approximate method of calculation which ignores the strength of the concrete and which can be useful because of its simplicity. It is used mainly as a rough check of the strength of an existing structure, where the reinforcement is known, or to give a rough estimate of the steel required for a job before the detailed design is available.

In the case of doubly reinforced beams having a very high proportion of compression steel it is possible for the steel beam theory to indicate a smaller quantity of steel than the elastic theory for a particular moment and size of section. However, its use other than as a quick guide is not to be recommended since collapse theory, which is considered later, is always preferable for accurate analysis or design.

Theory

The basis of the steel beam theory is that a beam having equal steel top and bottom, as shown in figure 3.6.1, may be treated as a steel beam, and that its strength may be calculated on the assumption that the concrete merely holds the steel apart. On this basis, the stress in the compression steel equals that in the tension steel. The strength of the concrete is entirely neglected and the lever arm is simply the distance the bars are apart irrespective of the steel ratio or grade of concrete.

Hence referring to figure 3.6.1,

$$M = A_{st}f_{st}(d_1 - d_2) = A_{sc}f_{sc}(d_1 - d_2) \quad (3.6.1)$$

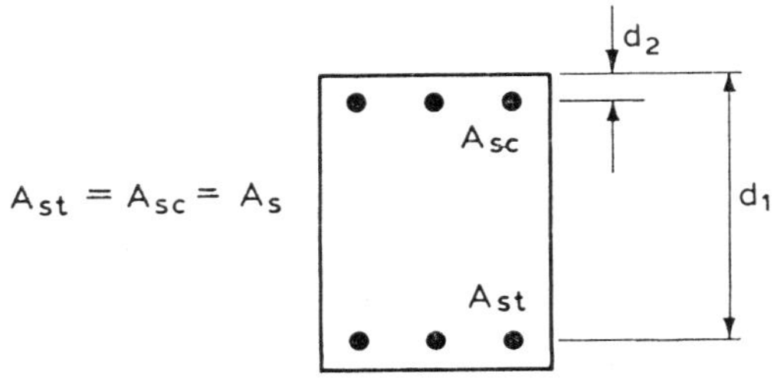

Fig 3.6.1 "Steel beam".

EXAMPLE 3.6.1

Check the solutions to (*a*) example 3.2.2 and (*b*) example 3.3.2 assuming that the additional (imaginary) compression steel is 20 mm from the top surface. What are the percentage errors?

Solution

(*a*) $A_{st} = 2 \times (\pi/4) \times 10^2 \times 10^{-6} = 157 \times 10^{-6}$

$a = 180 \times 10^{-3} \qquad p_{st} = 125 \times 10^6$

Hence for steel-beam

$$M_r = p_{st}A_{st}a = 3{\cdot}53 \times 10^3$$

The percentage error is

$$(3{\cdot}53 - 3{\cdot}51) \times 100/3{\cdot}51 = +1\% \quad \text{(Answer)}$$

(*b*) $A_{st} = 6 \times (\pi/4) \times 12^2 \times 10^{-6} = 679 \times 10^{-6}$

Hence for steel-beam

$$M_r = 125 \times 679 \times 195 \times 10^{6-6-3}$$
$$= 16{\cdot}6 \times 10^3$$

The percentage error is

$$(16{\cdot}6 - 8{\cdot}8) \times 100/8{\cdot}8 = +89\% \qquad \text{(Answer)}$$

It is clear that the steel-beam theory can give a good check for well-proportioned beams but can be hopelessly wrong where the concrete is critical, because the beam is over-reinforced (in tension).

Exercises

1. A singly reinforced rectangular concrete beam has a breadth of 300 mm and an effective depth of 600 mm. The area of steel is 1 800 mm². Calculate the maximum working load moment the beam could sustain according to elastic theory if $u = 30$ MN/m², $f_y = 250$ MN/m², $p_{cb} = u/3$, $p_{st} = \frac{1}{2}f_y$ and $m = 16$.

2. A concrete beam is 250 mm wide and has an effective depth of 500 mm. The tensile reinforcement consists of four 25-mm diameter mild steel bars. Calculate the maximum moment that the beam could carry under working load conditions (*a*) by elastic theory, (*b*) by collapse theory using the rectangular stress block. Assume that

$$u_w = 37{\cdot}5 \text{ MN/m}^2 \quad f_y = 250 \text{ MN/m}^2$$

For elastic analysis $p_{cb} = \frac{1}{3}u_w \quad p_{st} = \frac{1}{2}f_y \quad m = 14{\cdot}5$
For collapse analysis $\gamma_l = 1{\cdot}6$
Maximum steel stress $= 0{\cdot}87 f_y$
Mean concrete stress in the rectangular stress block $= 0{\cdot}40 u_w$
(UB2 1969, 40 min)

(*Note:* attempt part (*b*) only after chapter 8 has been studied.)

Chapter 4

Elastic design of beams

Design is concerned with the selection of suitable values for b, d_1 and A_{st} so that the beam can resist a given bending moment. This is very different to a problem in *analysis*, where the section is completely defined and the required unknowns are concerned with the stresses in the concrete and steel. An infinite number of different sections can be chosen to resist a given moment, hence good design aims to select only those which are the most economical. This often involves making a further assumption *in addition* to those already made in the analysis.

The reader must distinguish clearly between analysis and design. A very common mistake among students who are presented with a problem in elastic analysis is to assume that it has been designed in a special way when this need not be the case. A section is, in general, only at ESR (see below) *if it has been designed accordingly.*

4.1 Equal-strength steel ratio

If the steel ratio is so adjusted that the steel and concrete stresses reach their permissible values simultaneously it is termed the *equal-strength steel ratio*, or ESR. A design is clearly efficient, theoretically, if it ensures that when the maximum bending moment is applied at a section both the steel and the concrete are stressed to their maximum permissible values. For such a design

$$f_{cb} = p_{cb} \quad \text{when} \quad f_{st} = p_{st}$$

Neutral axis depth at ESR

If the steel and concrete stresses are both made equal to their respective permissible values, substitution in equation 3.1.2 gives

$$f_{st}/f_{cb} = p_{st}/p_{cb} = m(1 - n_1)/n_1$$

Re-arranging,

$$n_1 = \frac{m}{m + (p_{st}/p_{cb})} \tag{4.1.1}$$

Beam dimensions

The beam dimensions can be so adjusted that the maximum concrete stress just equals the permissible stress. If the beam is too large the concrete is understressed and this may be uneconomic. Conversely, if too small, the concrete is overstressed and this is not permissible. Similarly, if the tensile steel ratio exceeds the ESR value the steel is understressed, whilst if it is less than the ESR value the steel is overstressed.

ESR values

Equating the tensile and compressive forces, and assuming that the stresses in both the concrete and steel equal their respective permissible values,

$$p_{st}A_{st} = \tfrac{1}{2}p_{cb}bd_n$$

Re-arranging gives the equal-strength (tensile) steel ratio, r_e, as follows:

$$r_e(= A_{st}/bd_1) = \tfrac{1}{2}n_1 p_{cb}/p_{st} \qquad (4.1.2)$$

Moment of resistance at ESR

The moment of resistance of the concrete for a beam at ESR is given by

$$M_{rc}/bd_1^2 = \tfrac{1}{2}p_{cb}bd_n a/bd_1^2 = \tfrac{1}{2}p_{cb}n_1 a_1 \qquad (4.1.3)$$

The moment of resistance of the steel for a beam at ESR is given by

$$M_{rt}/bd_1^2 = p_{st}r_e a_1 \qquad (4.1.4)$$

Notice that $M_{rc} = M_{rt}$ for an ESR beam (by definition). Notice also that equation 3.2.5 still applies, that is,

$$a_1(= a/d_1) = 1 - \tfrac{1}{3}n_1$$

Permissible stresses for elastic design

The elastic theory is concerned with stresses at working loads. It has overall safety factors which are applied to the material strengths in order to obtain the permissible stresses for the steel and the concrete. In terms of limit state theory this means that the partial safety factor for the loads is always unity. The overall (stress) safety factors previously used in conventional elastic design were 2 for the steel yield strength and 3 for the concrete cube strength (for the extreme compressive fibre stress in flexure). The permissible values were therefore given by

$$p_{st} = \tfrac{1}{2}f_{yt} \quad \text{and} \quad p_{cb} = u/3$$

Typical coefficients for ESR beams

Consider the typical values of $f_{yt} = 250\ \text{MN/m}^2$ and $u = 22{\cdot}5\ \text{MN/m}^2$. If it is assumed that

$$p_{st} = \tfrac{1}{2}f_{yt} \quad p_{cb} = u/3$$

$$E_s = 200\ \text{GN/m}^2 \quad \text{and} \quad E_c = 4500\sqrt{u}\ (\text{in MN/m}^2\ \text{units})$$

then:

$$p_{st} = 125\ \text{MN/m}^2 \quad p_{cb} = 7{\cdot}5\ \text{MN/m}^2 \quad p_{st}/p_{cb} = 16{\cdot}7$$

$$E_c = 4500 \times 22{\cdot}5^{\frac{1}{2}}\ \text{MN/m}^2 = 21{\cdot}4\ \text{GN/m}^2$$

and

$$m = 200/21{\cdot}4 = 9{\cdot}35$$

These values can then be substituted into equations 4.1.1, 3.2.5, 4.1.2 and 4.1.3 (or 4.1.4) to give n_1, a_1, r_e and M_r/bd_1^2 as shown in table 4.1.1(*a*). Values have also been given for two other grades of concrete, $u = 30$ and $37{\cdot}5\ \text{MN/m}^2$, as well as for the combination of cold worked deformed bars ($f_{yt} = 425\ \text{MN/m}^2$) with concrete grade 30.

Table 4.1.1 Typical elastic design coefficients for various modular ratios

f_{yt} (MN/m²)	u (MN/m²)	E_c (GN/m²)	m	p_{st} (MN/m²)	p_{cb} (MN/m²)	n_1	a_1	r_e (%)	M_r/bd_1^2 (MN/m²)
250	22·5	21·4	9·35	125	7·5	0·359	0·880	1·08	1·19
250	30·0	24·7	8·10	125	10·0	0·393	0·869	1·57	1·70
250	37·5	27·6	7·25	125	12·5	0·420	0·860	2·10	2·25
425	30·0	24·7	8·10	212	10·0	0·276	0·908	0·65	1·25
(*a*) Short-term E_c									
250	22·5	10·7	18·7	125	7·5	0·528	0·824	1·59	1·63
250	30·0	12·4	16·2	125	10·0	0·565	0·812	2·26	2·29
250	37·5	13·8	14·5	125	12·5	0·592	0·803	2·96	2·97
425	30·0	12·4	16·2	212	10·0	0·433	0·856	1·02	1·85
(*b*) Long-term E_c(= 0·5 × short-term E_c)									
250	22·5	13·3	15·0	125	7·5	0·475	0·844	1·43	1·50
250	30·0	13·3	15·0	125	10·0	0·545	0·818	2·18	2·23
250	37·5	13·3	15·0	125	12·5	0·600	0·800	3·00	3·00
425	30·0	13·3	15·0	212	10·0	0·414	0·829	0·98	1·72
(*c*) $m = 15$ (as in CP 114:1957)									

Coefficients for long-term loading

Concrete continues to strain with time if the loading is sustained. This means that for a given beam the concrete creeps under the working loads such that its effective "elastic" modulus decreases and the modular ratio m increases. As m increases, the area of concrete in compression increases, hence n_1 increases and $a_1(= 1 - n_1/3)$ decreases. It follows that the concrete stress, f_{cb}, decreases with time and that the steel stress, f_{st}, increases with time. This is illustrated in figure 4.1.1 for the case $f_{yt} = 250$ MN/m^2

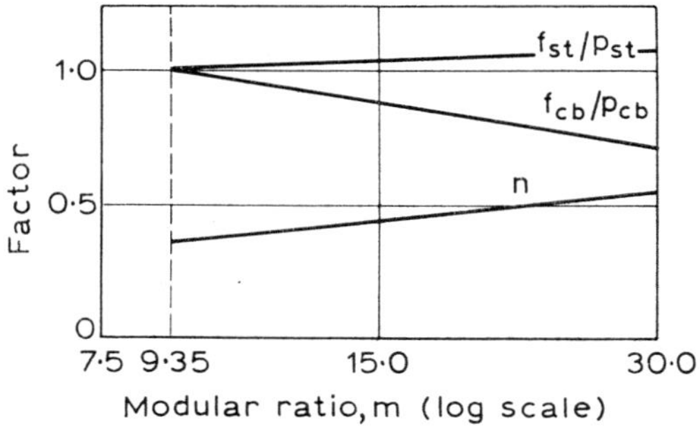

Fig 4.1.1 Variations in factors for n and the stresses as the modular ratio increases.

and $u = 22{\cdot}5$ MN/m^2 and where $p_{st} = \frac{1}{2}f_{yt}$ and $p_{cb} = \frac{1}{3}u$ are stresses which are assumed to exist initially before any creep has taken place in the concrete. If the creep strain for long-term loading of the concrete is approximately equal in magnitude to the initial elastic strain, then the long-term modulus for concrete equals half the short-term modulus. Hence the modular ratio m for long-term loading is doubled. Table 4.1.1(*b*) shows typical values of the various coefficients assuming that m is twice the short-term value given in table 4.1.1(*a*).

Modular ratio given in CP114:1957

The old code of practice assumed a constant value of 15 for the modular ratio, irrespective of the grade of concrete or duration of loading. The values obtained for a constant value of 15 are shown in table 4.1.1(*c*). Comparison with tables 4.1.1(*a*) and (*b*) shows that the old code corresponded approximately with the long-term conditions in many cases. This was intended to ensure that neither the (initial) concrete stress, nor the (final) steel stress, was too high. This over-simplification was possible because the overall safety factors were relatively large.

Timber and concrete

It is interesting to compare the moment of resistance of a timber beam with that for reinforced concrete. For a homogeneous rectangular beam

$$M_r = fZ = fbd^2/6$$

The permissible stress in bending for high-quality timber is often about 8 MN/m^2, hence

$$M_r = 1{\cdot}33bd^2$$

A good-quality timber beam is therefore comparable with a singly reinforced concrete beam at ESR when $p_{cb} = 7{\cdot}5$ MN/m^2.

EXAMPLE 4.1.1

Design a singly reinforced concrete beam, 200 mm wide, to withstand a bending moment under working loads of 15 kNm. Assume that:

$$p_{st} = 125 \text{ MN/m}^2 \quad p_{cb} = 7{\cdot}5 \text{ MN/m}^2 \text{ and } m = 15.$$

Solution From table 4.1.1, $M_r = 1{\cdot}50bd_1{}^2$ MNm. Therefore, for an ESR beam of breadth 0·2, the required d_1 is given by

$$15 \times 10^3 = 1{\cdot}50 \times 0{\cdot}2d_1{}^2 \times 10^6$$

$$d_1 = \sqrt{0{\cdot}05} = 0{\cdot}224$$

Assume a bar diameter $\not>$ 12 mm and a nominal cover of 20 mm. Hence let the overall depth be 260 mm and the effective depth be

$$0{\cdot}260 - 0{\cdot}020 - 0{\cdot}012/2 = 0{\cdot}234$$

The required A_{st} is given by

$$A_{st}(= M/p_{st}a) = 15 \times 10^{3-6}/125 \times 0{\cdot}844 \times 0{\cdot}234$$
$$= 608 \times 10^{-6}$$

since $a_1 = 0{\cdot}844$ for ESR beam, from table 4.1.1. Hence use eight 10-mm diameter bars (area = 628 mm^2).

Use a 200-mm-wide × 260-mm-deep beam with eight 10-mm diameter bars at an effective depth of 234 mm (Answer)

It is essential to appreciate both the problem and the method of solution illustrated by this example. The basic problem is to determine suitable values of d_1 and A_{st} to resist a given bending moment, and the elastic design solution is obtained by using the ESR beam.

4.2 Effect of varying the steel ratio

The previous section describes the special case when both the steel and the concrete are stressed to their full permissible values. Although it would appear that the most efficient elastic design is given by the ESR beam, this purely theoretical concept of an efficient section must often be modified in practice. It is not always practicable to design a section at approximately ESR. Simple shuttering necessitates a constant cross-section so that at positions along the beam where the applied moment is low the sections can have less steel. Conversely, the beam dimensions are usually limited by space considerations so that at some sections additional steel is necessary where the moment exceeds M_r for the ESR beam.

General values of the coefficients n_1, a_1 and M_r/bd_1^2

It is clear that since the ESR section need not always be employed a knowledge of the effect of varying the steel ratio upon n_1, a_1 and M_r/bd_1^2 for a beam of constant cross-section is essential for intelligent design. Table 4.2.1 gives values for the typical case when

$$p_{st} = 125 \text{ MN/m}^2 \quad p_{cb} = 7{\cdot}5 \text{ MN/m}^2$$

and

$$E_c = 4500\sqrt{u} \text{ (in MN/m}^2 \text{ units).}$$

It is apparent that as r increases, n_1 also increases although a_1 decreases only slightly.

Table 4.2.1 Values of coefficients n_1, a_1 and M_r/bd_1^2 for different steel ratios when $p_{cb} = 7{\cdot}5$ MN/m², $p_{st} = 125$ MN/m² and $m = 9{\cdot}35$ (short-term E_c)

r	n_1	a_1	M_{rc}/bd_1^2	M_{rt}/bd_1^2	Critical stress
0·05	0·088	0·971	0·32	0·06	Steel
0·1	0·128	0·957	0·46	0·12	
0·2	0·176	0·941	0·62	0·23	
0·5	0·250	0·917	0·88	0·57	
1·075	0·359	0·880	1·19	1·19	Steel and concrete
1·5	0·408	0·864	1·32	1·62	Concrete
2·0	0·452	0·849	1·44	2·12	
5·0	0·588	0·804	1·77	5·03	

Variation in moment of resistance

Figure 4.2.1 shows M_r/bd_1^2 plotted against r, and clearly indicates that the moment of resistance increases rapidly with increasing steel ratios up to the ESR but increases much more slowly above this value. The ESR therefore represents a critical value. High steel

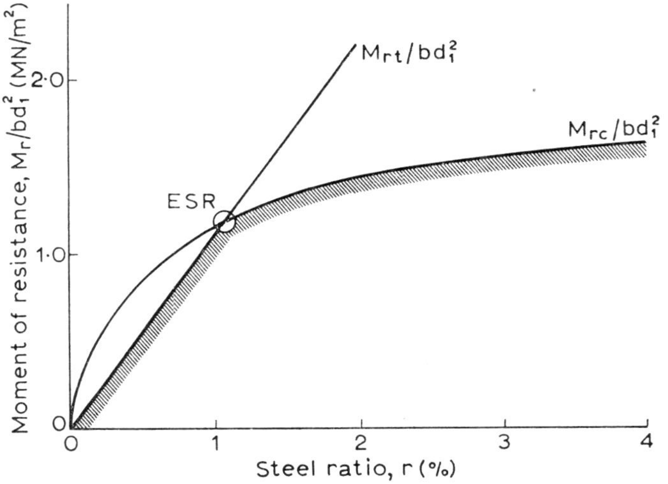

Fig 4.2.1 Typical variations in moments of resistance with steel ratio (short-term E_c; $p_{cb} = 7{\cdot}5$ MN/m^2, $p_{st} = 125$ MN/m^2, $m = 9{\cdot}35$).

ratios tend to be extravagant in steel, low ratios extravagant in concrete. Economical designs would therefore appear to be given by values which do not vary greatly from ESR.

Small variations from ESR

In practical beam design the calculated depths are rounded off to the nearest 10 mm. Similarly, the areas of reinforcement are increased slightly so as to provide a whole number of bars. If r varies slightly from the ESR value the effect on a_1 is very small and can be neglected. Even if r varies by 50% from the ESR value the change in a_1 is sufficiently small to be neglected for most purposes.

Large variations from ESR

It is sometimes impractical to design near the ESR. Some typical examples are as follows:

(i) Stiffness considerations may necessitate larger beam dimensions (e.g. to avoid excessive deflections).

(ii) At locations near the ends of beams, bars may have either been curtailed or bent up to provide for shear and reversed bending moments.

(iii) Different beams may have to possess the same dimensions, even though they are subjected to different maximum bending moments, for the sake of appearance or standardization of formwork.

It is preferable in such cases to use r-values less than the ESR, in which case the moment of resistance for the steel is the ruling factor. A greater value is not normally desirable (chiefly because of collapse considerations) but in such cases the concrete becomes critical.

EXAMPLE 4.2.1

Reconsider the beam which was designed in example 4.1.1. The quantity of steel has been halved at a particular section nearer the supports since the applied moment is only 50% of the maximum (7·5 kN m). Calculate the actual moment of resistance of the section according to elastic theory and hence comment on the significance of the error for practical design purposes.

Solution 50% of theoretical area of steel $= \frac{1}{2} \times 612 = 306\ \text{mm}^2$. The actual n_1 is given by equation 3.2.4.

$$r = 306 \times 10^{-6}/0{\cdot}2 \times 0{\cdot}232 = 0{\cdot}66\% \qquad mr = 99 \times 10^{-3}$$

Therefore

$$n_1 = 99 \times 10^{-3}[\sqrt{(1 + 2000/99)} - 1] = 0{\cdot}357$$

Hence

$$a_1 = 1 - 0{\cdot}357/3 = 0{\cdot}881$$

$$M_{rt} = 125 \times 306 \times (0{\cdot}881 \times 0{\cdot}232) = 7810$$

Hence the error is

$$(7{\cdot}81 - 7{\cdot}5)/7{\cdot}5 = +4{\cdot}1\%$$

The error is only 4·1% and is on the safe side. Hence the coefficients for the ESR beam can be used for practical design purposes even for a section where the steel ratio is only 50% of the ESR value. (Answer)

4.3 Singly reinforced beam and slab design

Floor and wall slabs are usually designed as rectangular beams reinforced in tension only. The following recommended procedure for designing a singly reinforced beam or slab indicates the factors which must be considered:

(*a*) Assume b, d_1 and d. For a slab, consider a metre-wide strip, i.e. $b = 1$ m.
(*b*) Calculate the total loading, i.e. if the load is uniformly distributed, determine the total load per metre run.
(*c*) Determine the effective span (see appendix C).
(*d*) Calculate the maximum bending moments.
(*e*) Determine M_r/bd_1^2 for the selected grade of concrete (for an ESR beam).
(*f*) Determine minimum d_1 from $M_r = M$. Adjust assumed d_1 (and hence d and therefore M) if necessary.
(*g*) Derive a from d_1.
(*h*) Calculate A_{st} (from $M = p_{st}A_{st}a$).
(*i*) Determine a suitable size and number of bars, checking that the bars fit into the assumed breadth of beam.
(*j*) Decide actual total depth, d, from minimum d_1, bar size and assumed cover. The depth is rounded off to the nearest 10 or 20 mm greater than the calculated value.
(*k*) If actual d_1 differs appreciably from the assumed d_1 in (*a*), repeat steps (*b*) to (*j*).
(*l*) In slabs only, decide size and spacing of distribution bars from $r \nless 0{\cdot}12\%$ or $0{\cdot}15\%$ (see appendix C).
(*m*) Check for shear and bond and determine shear reinforcement.

EXAMPLE 4.3.1

Design a floor slab to carry a superimposed loading of 4·0 kN/m^2 on a simple span of 4 m. Use concrete of grade 30 MN/m^2, cold worked deformed steel reinforcement ($f_{yt} = 425$ MN/m^2) and assume long-term loading.

Solution The design procedure is implemented as follows:

(*a*) Span/depth ratio $\ngtr 60$ (appendix C: deflection)
Hence minimum depth $= 4/60 = 0{\cdot}067$
The superimposed loading is fairly high, hence try $d = 150$ mm
Assume bar diameter $\ngtr 12$ mm
Hence actual cover $\nless 15$ mm and nominal cover $= 15 + 5 = 20$ mm (appendix C)
$\therefore d_1 = 150 - 20 - 12/2 = 124$ mm
(*b*) Self-weight $= 24 \times 0{\cdot}150 \times 10^3 = 3600$ N/m^2
$\therefore$ total loading for metre-wide strip $= (4000 + 3600) = 7600$ N/m

(*c*) Effective span = 4·0 m

(*d*) Midspan moment = $1{\cdot}0 \times 7600 \times 4^2/8 = 15\,200$ Nm (partial safety factor for load = 1·0 for elastic theory)

(*e*) $M_r/bd_1^2 = 1{\cdot}85 \times 10^6$ for $f_{yt} = 425 \times 10^6$, $u = 30 \times 10^6$ and $E_c = \frac{1}{2} \times 4500\sqrt{u}$ (MN/m² units) (table 4.1.1)

(*f*) Hence $1{\cdot}85 \times 10^6 bd_1^2 = 15200$, or
$d_1 = 10^{-3}\sqrt{(15200/1{\cdot}85)} = 0{\cdot}091$
Assumed d_1 of 0·124 is therefore satisfactory

(*g*) $a_1 = 0{\cdot}856$ (from table 4.1.1)
$\therefore a = 0{\cdot}856 \times 0{\cdot}124 = 0{\cdot}106$

(*h*) $A_{st} = 15{\cdot}2 \times 10^{3-6}/212 \times 0{\cdot}106 = 677 \times 10^{-6}$

(*i*) Use 12 mm diameter at 150 mm spacing; area = 754×10^{-6} (appendix G)

(*j*) Assumed d = 150 mm is satisfactory

(*k*) Not necessary

(*l*) Distribution steel:
$r \nless 0{\cdot}12\%$, hence $A_{st} \nless 0{\cdot}12 \times b \times 0{\cdot}124/100 = 149 \times 10^{-6}$
Use 6 mm diameter high-yield bars at 175 mm spacing; area = 162×10^{-6} (appendix G)

(*m*) Shear and bond are considered later in chapter 5. However, shear reinforcement is usually neither necessary nor desirable in slabs.

General design considerations

Notice that in design the load due to self-weight must be estimated before the bending moment can be determined. Notice also that the depth of the slab was taken on the generous side (i.e. $d_1 = 0{\cdot}124$ is considerably greater than 0·091, the ESR value for a singly reinforced section). This ensured that the values assumed initially were found to be adequate regarding cover, with the bars finally found to be necessary, etc. This approach, which is necessary in design if numerous attempts are to be avoided, can lead to uneconomic solutions if carried to extremes. This tended to be the case with elastic theory since some limit states were not specifically considered. The maximum span/depth ratio given in CP114:1957 for this simply supported slab would have been $0{\cdot}9 \times 30 = 27$, from which the minimum overall depth is $4/27 = 0{\cdot}148$: virtually the value which has been chosen. Although greater economies are now possible with limit state concepts, the student must appreciate that meticulous accuracy *in only one aspect of design* is usually a waste of time, for all manual solutions, since several aspects must generally be considered. For example, deflections and shear strength as well as flexural strength must be satisfactory.

4.4 Doubly reinforced beam design

The moment of resistance of a singly reinforced beam at ESR can be increased slightly by increasing the tensile steel, as shown previously. It is much better, however, to add compressive steel as well. If the size of a beam is restricted, or a very high bending moment occurs along a relatively short length of the beam, it is possible to achieve a satisfactory design by using a doubly reinforced section.

Low compression-steel stress

Doubly reinforced beams are normally designed so that the tension steel and the concrete are stressed to their permissible values, as in the case of singly reinforced beams. The compression steel, however, is only stressed to $m \times$ (adjacent concrete stress). For example, if

$$d_2 = d_n/3 \quad m = 15 \quad \text{and} \quad p_{cb} = 10\ \mathrm{MN/m^2}$$

then f_{sc} is given by

$$f_{sc} = \tfrac{2}{3} m p_{cb} = 100\ \mathrm{MN/m^2}$$

This is less than the permissible stress in tension for mild steel and only about half that for deformed bars. It cannot be increased without overstressing the concrete. However, despite the generally lower stresses in the compression steel, the *total* quantity of steel required in a singly reinforced beam almost always exceeds that for a doubly reinforced beam of the same size and moment of resistance. The only exception is when d_2 becomes very large and nearly equals d_n.

Design procedure for doubly reinforced beams

The procedure for practical design of doubly reinforced beams is therefore as follows:

(i) Determine M_r for a singly reinforced beam at ESR having the same dimensions and permissible stresses.

(ii) Subtract M_r for the ESR beam from the applied moment.

(iii) Add sufficient compression steel and additional tension steel to take the excess moment.

Notice that since the assumed stresses in the compression and tension steel are not the same, greater areas of compression steel are required in order that the additional compressive force should equal the additional tensile force. Notice also, however, that the areas of steel are added in just the right proportions to keep the position of the neutral axis constant.

Maximum quantity of steel

Theoretically, although it is extravagant in steel, the moment of resistance of a beam of given dimensions can be increased almost indefinitely by increasing both the tension and compression steel ratios. In practice, however, the more steel that is crowded into the concrete member, the more difficult it is to ensure good concreting and the higher the shear and local bond stresses. Difficulties in compacting the concrete between the mass of steel result in air pockets which reduce the effective bond between steel and concrete, as well as reducing the concrete strength. The total steel in doubly reinforced beams should therefore be restricted to about 8%, i.e. 4% compression steel ratio and 4% tension steel ratio (appendix C).

Moments of resistance

Equations 3.4.6 and 3.4.7 are not suitable for design since I_e is not known until the section is known. It is therefore more convenient to obtain expressions for M_{rc} and M_{rt} in terms of the lever arm, a. It is also convenient to assume that $d_2 \simeq d_n/3$ so that the compression steel force coincides with the concrete force. That is

$$a \simeq d_1 - d_2 \simeq d_1 - d_n 3$$

The simplified expressions for the moments of resistance are then given by

$$M_{rc} = p_{cb}[\tfrac{1}{2}bd_n + 2A_{sc}(m - 1)/3]a \qquad (4.4.1)$$

$$M_{rt} = p_{st}A_{st}a \qquad (4.4.2)$$

EXAMPLE 4.4.1

Design a doubly reinforced concrete beam to resist a bending moment of 132 kNm if the breadth and effective depth are restricted to 300 mm and 450 mm respectively. Assume that

$$p_{st} = 125 \text{ MN/m}^2 \quad p_{cb} = 7{\cdot}5 \text{ MN/m}^2 \quad p_{sc} = 110 \text{ MN/m}^2$$

$$m = 18{\cdot}7 \quad \text{and} \quad d_2 = \tfrac{1}{3}d_n$$

Solution From table 4.1.1(*b*) (or by calculation), M_r for a singly reinforced ESR beam is

$$1{\cdot}63 \times 10^6 bd_1^2 = 1{\cdot}63 \times 0{\cdot}30 \times 0{\cdot}45^2 \times 10^6$$

$$= 99{\cdot}1 \times 10^3$$

Lever arm for ESR beam is

$$0{\cdot}824 \times 0{\cdot}45 = 0{\cdot}371$$

Hence A'_{st} for ESR beam is

$$99{\cdot}1 \times 10^{3-6}/125 \times 0{\cdot}371 = 2140 \times 10^{-6}$$

The additional moment is

$$(132 - 99{\cdot}1) \times 10^3 = 32{\cdot}9 \times 10^3$$

$$f_{sc} = (2/3) \times 18{\cdot}7 \times 7{\cdot}5 \times 10^6 = 93{\cdot}5 \times 10^6$$

$$< p_{sc} = 110 \times 10^6$$

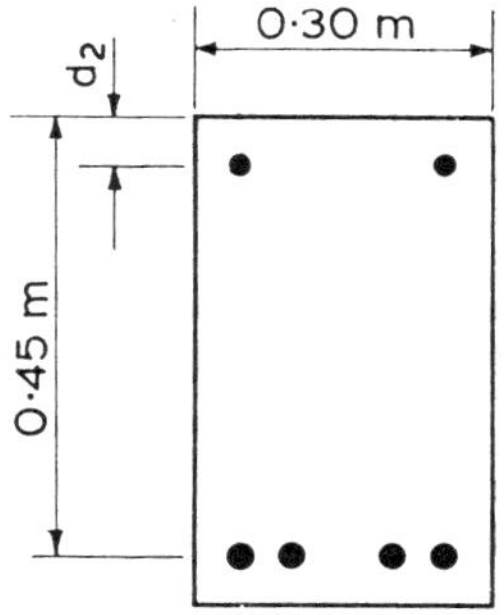

Fig 4.4.1 Example 4.4.1.

Therefore

$$A_{sc} = 32{\cdot}9 \times 10^{3-6}/93{\cdot}5 \times 0{\cdot}371 = 949 \times 10^{-6} = 949 \text{ mm}^2$$ (Answer)

since the compression steel force coincides with concrete force, i.e. $d_2 = \frac{1}{3}d_n$, and

$$A'_{st} = 32{\cdot}9 \times 10^{3-6}/125 \times 0{\cdot}371 = 710 \times 10^{-6}$$

Therefore

$$A_{st} \text{ (total)} = 2850 \text{ mm}^2$$ (Answer)

Note:

$$r_c/r = A_{sc}/A_{st} = 0{\cdot}333 \quad \text{and} \quad A_{st} + A_{sc} = 3849 \text{ mm}^2.$$

Comparing this result with the answers to example 3.4.3 shows that the design approach for doubly reinforced beams gives the optimum amount of compression reinforcement.

4.5 T-beam design

T-beams, unlike rectangular beams, are usually designed for steel ratios which are much lower than the ESR value. The large compressive flange area usually requires a relatively small effective depth in order to produce the required resistance moment in compression. A very substantial steel area, however, may be necessary to produce an equal tensile resistance moment for such a small effective depth. The steel may become overcrowded in the comparatively narrow rib, resulting in difficulties in concrete placing. In such cases it is preferable to increase the lever arm, by increasing the effective depth, so reducing the required area of steel for the given applied moment. This means that the concrete is no longer fully stressed and that in general the steel ratio does not approximate to the ESR value.

Assumptions

Since ESR conditions do not usually apply, it is necessary to reconsider the general analysis of T-beams and introduce simplifying assumptions as follows:

(i) The small compressive force due to the concrete area in the rib between the neutral axis and the soffit of the slab is negligible.

(ii) The additional lever arm due to the compressive force in the flange acting just above the mid-depth of the flange, instead of actually at the mid-depth, is also negligible.

Moment of resistance of the steel

The T-beam is usually considerably under-reinforced, hence the steel area is given simply by

$$M_{rt} = p_{st}A_{st}(d_1 - \tfrac{1}{2}d_s) \tag{4.5.1}$$

Moment of resistance of the concrete

The strength of the concrete is usually more than adequate in the case of T-beams. However, if the moment is high then this may be checked from equation 4.5.3 which is derived as follows.

The average compressive stress in the flange is given by

$$f_{cb}(d_n - \tfrac{1}{2}d_s)/d_n$$

Equating longitudinal forces, for equilibrium of the section, gives

$$f_{st}A_{st} = f_{cb}(d_n - \tfrac{1}{2}d_s)bd_s/d_n$$

Substituting $f_{st}/f_{cb} = m(d_1 - d_n)/d_n$, putting $A_{st} = rbd_1$ and rearranging, gives

$$d_n = (mrd_1^2 + \tfrac{1}{2}d_s^2)/(mrd_1 + d_s) \tag{4.5.2}$$

Once d_n is known, and since the lever arm is simply $(d_1 - \frac{1}{2}d_s)$, then

$$M_{rc} = p_{cb}\frac{d_n - \frac{1}{2}d_s}{d_n} bd_s(d_1 - \tfrac{1}{2}d_s) \quad (4.5.3)$$

where $M_{rc} \gg M_{rt}$ generally.

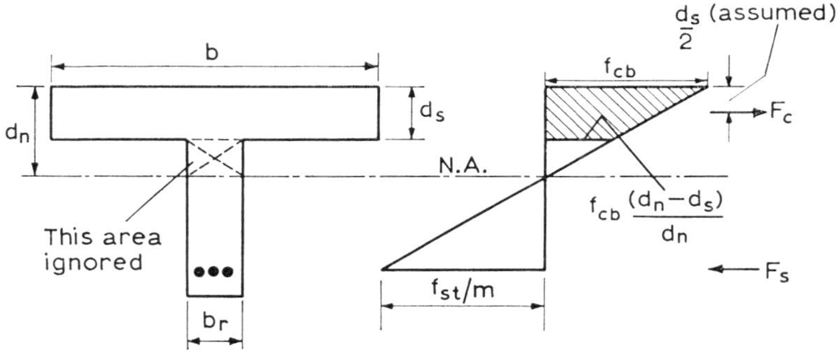

Fig 4.5.1 Cross-section and concrete stress profile for a typical T-beam.

EXAMPLE 4.5.1

Calculate the required reinforcement for a simply supported T-beam with a web (or rib) 275 mm wide and extending 450 mm below a 150 mm deep slab. The midspan moment = 200 kNm. The span is 5·4 m and the distance between centres of beams is 2·4 m. Assume an effective depth of 520 mm. Refer to appendix C for flange width and use the usual assumptions for elastic design. Let p_{st} = 212 MN/m², p_{cb} = 10·0 MN/m² and m = 16·2.

Solution From appendix C,

$$b \not> 0{\cdot}275 + 0{\cdot}2 \times 5{\cdot}4 = 1{\cdot}355$$

Assuming that the section is under-reinforced, that the resultant concrete force acts at mid-depth of the slab and ignoring the concrete in the web:

$$A_{st} = 200 \times 10^{3-6}/212(0{\cdot}520 - 0{\cdot}075) = 2{\cdot}12 \times 10^{-3}$$
$$= 2120 \text{ mm}^2 \quad \text{(Answer)}$$

To check the concrete strength at ESR:

$$d_n/(d_1 - d_n) = mp_{cb}/p_{st} \quad \text{or} \quad d_n = d_1(1 + p_{st}/mp_{cb})^{-1}$$

$$p_{st}/mp_{cb} = 212/162 = 1{\cdot}31,$$

therefore

$$d_n = 0{\cdot}520/2{\cdot}31 = 0{\cdot}225$$

Hence M_{rc} at ESR for the flange only is

$$1{\cdot}355 \times 0{\cdot}150 \times 10{\cdot}0[(0{\cdot}225 - 0{\cdot}075)/0{\cdot}225] \times (0{\cdot}520 - 0{\cdot}075) \times 10^6$$

i.e.

$$M_r = 603 \times 10^3$$

This is more than adequate.

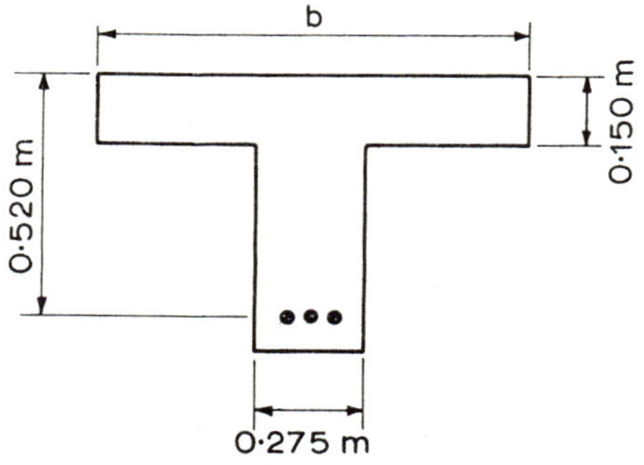

Fig 4.5.2 Example 4.5.1.

4.6 Continuous beam and slab design

Beams continuous over three or more supports frequently occur in floors where secondary beams frame into main beams or main beams frame into columns. Slabs spanning between T-beams are taken as continuous over the T-beam ribs. Similarly, bridge beams may span over several piers. When determining the distribution of moments it is often sufficiently accurate to assume that the beam or slab is free to rotate about the supports. That is, the stiffness of the column or any other member is ignored.

Theorem of three moments

A loaded continuous beam or slab takes a shape as shown in figure 4.6.1, the curvature changing in direction at points such as K, L, M and N. This change in curvature means that a negative (or hogging) bending moment occurs at the supports while a positive or sagging bending moment occurs near midspan.

The negative bending moments at the three supports A, B and C, can be obtained using the theorem of three moments. This theorem can be expressed as follows for any two continuous spans on fixed supports:

$$\frac{M_A L_1}{I_1} + 2M_B\left(\frac{L_1}{I_1} + \frac{L_2}{I_2}\right) + \frac{M_C L_2}{I_2}$$

$$= -6\left[\frac{A_1 \bar{x}_1}{I_1 L_1} + \frac{A_2(L_2 - \bar{x}_2)}{I_2 L_2}\right] \qquad (4.6.1)$$

I_1, I_2 are the moments of inertia of the beams in spans L_1, L_2; A_1, A_2 are the areas of the free bending moment diagrams for the loading on spans L_1, L_2 respectively; and $\bar{x}_1$, $\bar{x}_2$ are the distances of the centroids of A_1, A_2, respectively, from the left-hand supports.

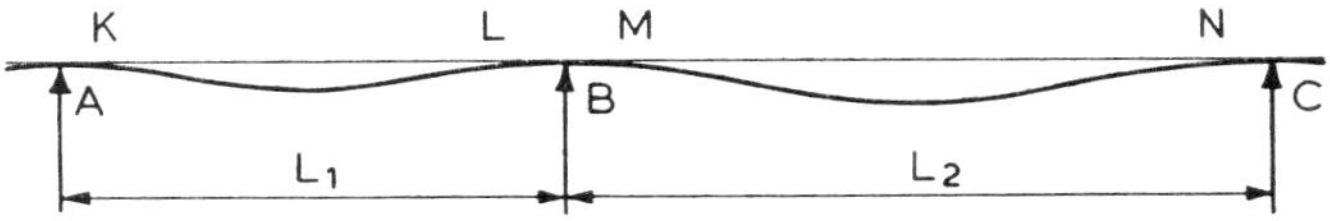

Fig 4.6.1 Deformed shape of loaded continuous beam.

Uniform loading and sections

If the loading is uniformly distributed throughout and the moment of inertia of the beam is constant, then equation 4.6.1 reduces to

$$M_A L_1 + 2M_B(L_1 + L_2) + M_C L_2 = \tfrac{1}{4}w(L_1^3 + L_2^3) \qquad (4.6.2)$$

If the beam consists of two spans only, and both ends are free, then $M_A = M_C = 0$ and M_B can be calculated directly. If the beam spans across several supports, the theorem can be applied to groups of three adjacent supports in turn along the beam, and simultaneous equations obtained for all the support moments. Where all spans are equal then, by symmetry, pairs of support moments are equal. If span L_1 is loaded and L_2 unloaded then equation 4.6.2 becomes

$$M_A L_1 + 2M_B(L_1 + L_2) + M_C L_2 = \tfrac{1}{4}wL_1^3 \qquad (4.6.3)$$

Bending moment diagram

Consider a uniformly loaded beam continuous over two supports as shown in figure 4.6.2. Let all three spans be equal. The "free" bending moment diagram is parabolic with a maximum value of $wL^2/8$. The resultant BMD is given by the difference between the negative BMD (given by the support moments) and the free BMD.

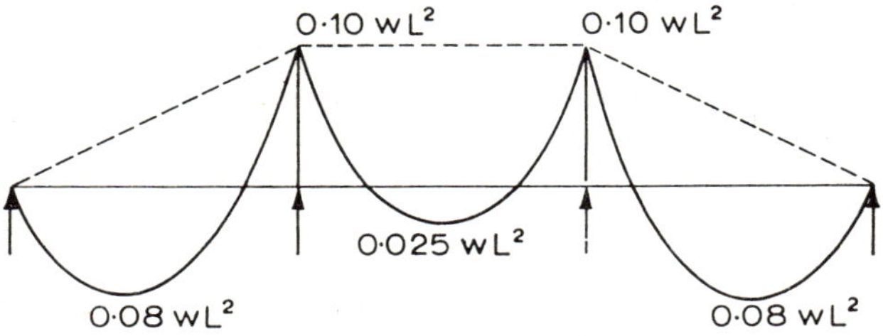

Fig 4.6.2 Bending moment diagram for a uniformly distributed load on beam with three equal spans.

Maximum moment adjustment

It has long been recognized that the following factors tend, in practice, to modify the elastic support moments in an RC beam as follows:

(i) Concrete does not behave as a purely elastic material. This moderates the moments throughout a beam in such a way that high moments are reduced and low moments increased.

(ii) Supports have a finite width. This always reduces the support moment.

(iii) The beam is partially fixed at the supports, and is not completely free to rotate about them as is assumed in deriving the theorem of three moments. This effect, like (ii), always reduces the support moment.

Inelastic behaviour of concrete

Collapse load considerations for reinforced concrete indicate that very large plastic deformations, and therefore redistribution of moments, can occur before failure. A distribution of moments

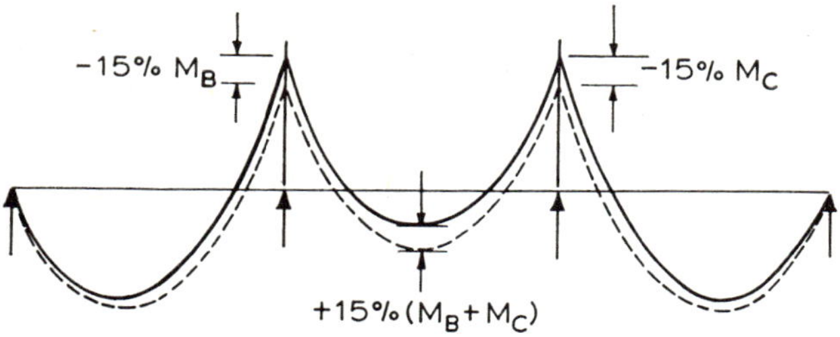

Fig 4.6.3 Bending moment diagram after 15% moment redistribution.

based solely on elastic theory can be misleading for conditions at failure. The old code, CP114, recognized this fact and permitted a maximum adjustment allowance of 15% for moments based on elastic behaviour. However, if the maximum theoretical support moment *is* reduced by 15%, then the positive moments in the two adjacent spans must be increased accordingly, as shown in figure 4.6.3. Similarly, the support moments can be increased by 15% if the span moment is critical for a particular loading condition.

Loading conditions

The self-weight (or *dead load*) of a structure is such that all spans are always loaded. The superimposed (or *live*) load is such that it can vary in magnitude between zero and its maximum value at any location, and must therefore be considered separately. However, extreme bending moments are produced by entire spans being either fully loaded or fully unloaded. The most severe moments are assumed to be given by one of the following cases:

(i) Alternate spans loaded, all other spans unloaded
(ii) Any two adjacent spans loaded, all other spans unloaded.

Case (i) gives the highest midspan moments, while case (ii) gives the highest support moments. It should be noticed, however, that even higher support moments can be achieved for four or more spans if, in addition to the two adjacent spans being loaded, alternate spans on both sides are also loaded (e.g. spans 1, 3, 4 and 6 for a 6-span beam produces the maximum possible moment at the support between spans 3 and 4). However, the moments for this case never exceed those for case (ii) by more than 4%, and can be ignored. CP114 in effect allowed an adjustment of 19% for this theoretically possible form of loading.

Width of supports

If the continuous member frames into very massive supports—for example, a very thin slab resting on a very wide main beam—then it is preferable to assume that the slab is fixed-ended and not free to rotate about the support. However, it is far more usual to ignore any fixity at an intermediate support, whether it is a slab supported by beams or a beam framing into columns.

Bending moment coefficients

The different spans of a continuous beam or slab are frequently equal to within 15% of the longest span. If all spans are assumed equal, then a bending moment analysis assuming elastic distribution and free rotation about supports can be carried out for a given number of spans. Each support or midspan moment can therefore be represented by some coefficient multiplied by the total load on the appropriate span. Standard coefficients for a given number of approximately equal spans can therefore be tabulated for both dead and superimposed loads as shown in appendix I, tables I1 and I2.

The standard coefficients can be modified by making adjustments to the support moments. Both 10% and 15% reductions have been included in appendix I. The use of standard coefficients is very satisfactory for most purposes, except that it is usually advisable to allow for some fixity at the end supports (25% of the full fixed end moment is often assumed).

Elastic shear forces

The maximum shears for a continuous beam or slab differ slightly from the shear determined for a number of single separate spans. Standard coefficients can therefore be used as for bending moments, as shown in table 6.3.2, which is sufficiently accurate for most purposes.

Distribution steel in slabs

If the ratio of the longer to the shorter side of a rectangular slab is greater than about $2\frac{1}{2}$, the longitudinal bending is relatively small and the slab effectively spans in one direction only. Distribution steel, given by appendix C as $r \nless 0{\cdot}12\%$, should always be provided for this nominal longitudinal bending. Also, since high concentrated loads can produce some bending in the longitudinal direction, it is wise to check that the distribution steel is also $\nless 10\%$ of the main steel.

End supports

Unlike the case for intermediate supports, fixity at end supports should be allowed for. The moments induced at the intermediate supports of a continuous beam of equal spans, when a unit moment is applied at the end support, are as given in table I3. The moment induced over each support is found by multiplying the end moment by the

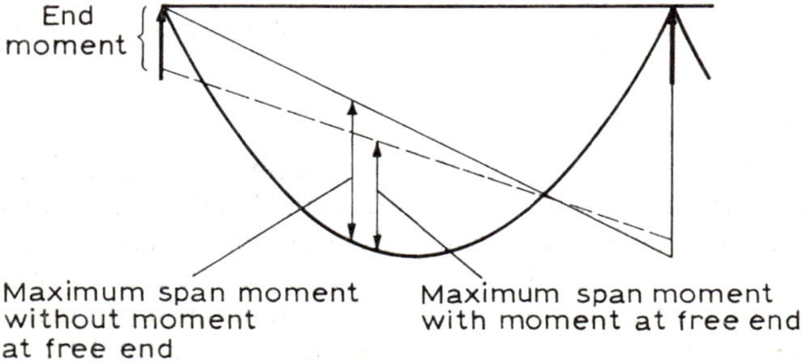

Fig 4.6.4 Effect of moment at free end.

appropriate coefficient. These values are then combined with negative moments obtained from tables I1 and I2 to give the actual support moments. The maximum positive moments in the spans are not materially affected by partial fixing of the nominally free ends unless the end moment is large, as shown in figure 4.6.4, although the position is displaced slightly.

The end moment is calculated from the relative stiffnesses of the various members at the junction. This is considered in more detail in chapter 9 for beams framing into columns. If full end fixity can be assumed then the appropriate end moment for one end fixed can be obtained from appendix L. Alternatively, an approximate probable value can be assumed (e.g. $wL^2/24$ is typical).

Continuous beam design

A beam spanning three or more columns or other supports is treated as a continuous beam. If it is a rectangular beam monolithic with a floor slab then the midspan and support sections are different. It is a T-beam section for positive bending moments near midspan and a rectangular section for negative bending moments at the supports. The top flange of the beam does not assist the resistance to bending if it is in tension (which it is over the supports of a continuous beam). Bond stresses should be checked for the main reinforcement and all beams must contain at least nominal shear reinforcement (see chapter 5).

Continuous slab design

A slab spanning three or more T-beam ribs or other supports is treated as a continuous slab and the bending moments and shears are calculated as for continuous beams. The negative moments over the supports can be provided for by either of the following methods:

(i) Tensile bars in the bottom of the slab, when no longer required for positive bending moment, are bent up and carried over the supports in the top of the slab as shown in figure 4.6.5(*a*). It is common practice to bend up half the number of tensile bars and carry the remaining half through to the support in the bottom half of the slab.

(ii) Two separate mats of reinforcement run continuously over all supports, one in the bottom and one in the top of the slab as shown in figure 4.6.5(*b*). The slab is, in fact, doubly reinforced. This method increases the steel cost but reduces bar bending. It is the most convenient method if the live load exceeds about twice the dead load, since negative moments can exist at midspan in this case.

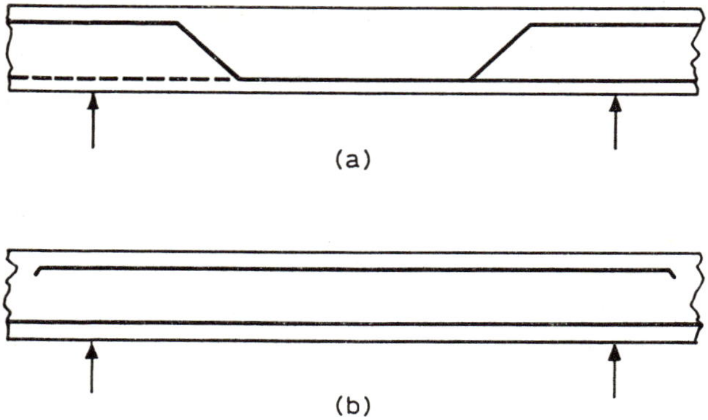

Fig 4.6.5 Alternative methods of reinforcing continuous slabs.

Shear and bond stresses in continuous slabs

Shear reinforcement is usually neither necessary nor desirable in slabs, whether continuous or not. If a slab is not strong enough in shear (see chapter 5) the depth should be increased to obtain the required strength. Bond stresses (see chapter 5) should be checked as for a single span.

EXAMPLE 4.6.1

A floor slab 150 mm deep overall carries a superimposed load of 4·8 kN/m² over continuous spans of 4·9 m.

If $p_{cb} = 7{\cdot}5$ MN/m² and $p_{st} = 125$ MN/m², what areas of steel are required at midspan and supports for the central spans? Assume a maximum bar diameter of 12 mm and a nominal cover of 20 mm. Refer to appendix I for the bending moment coefficients and assume 15% moment redistribution.

Solution

$$\text{Self-weight of slab} = 24 \times 0{\cdot}150 = 3{\cdot}6 \text{ kN/m}^2$$

Reference to appendix I indicates coefficients of 0·058 and 0·067 for dead load, and 0·080 and 0·091 for superimposed load, at midspan and supports respectively.

Midspan moment (per metre width):

$$\begin{aligned} 0{\cdot}058 \times 3{\cdot}6 \times 4{\cdot}9^2 \times 10^3 &= 5020 \\ 0{\cdot}080 \times 4{\cdot}8 \times 4{\cdot}9^2 \times 10^3 &= 9220 \\ \hline & 14240 \end{aligned}$$

Support moment:

$$0{\cdot}067 \times 3{\cdot}6 \times 4{\cdot}9^2 \times 10^3 = 5790$$
$$0{\cdot}091 \times 4{\cdot}8 \times 4{\cdot}9^2 \times 10^3 = 10500$$
$$\overline{16290}$$

Effective depth = 150 − 20 − 6 = 124 mm

For an ESR slab (see table 4.1.1.(*b*)), $a_1 = 0{\cdot}824$, hence $a = 102$ mm

At midspan:

$$A_{st} = 14{\cdot}24 \times 10^{3-6+3}/125 \times 102 = 1117 \times 10^{-6}$$
$$= 1117 \text{ mm}^2 \qquad \text{(Answer)}$$

At support:

$$A_{st} = 16{\cdot}29 \times 10^{3-6+3}/125 \times 102 = 1280 \times 10^{-6}$$
$$= 1280 \text{ mm}^2 \qquad \text{(Answer)}$$

Exercises

1. What is meant by the equal strength steel ratio (ESR) for a reinforced concrete beam? Calculate the moment of resistance for a rectangular beam 300 mm wide and of 450 mm effective depth, at ESR, for $p_{st} = 205$ MN/m², $p_{cb} = 12{\cdot}5$ MN/m² and $m = 14{\cdot}5$. What is the ESR value for this beam?

2. One support section of the beam described in question 1 is subjected to a bending moment of 182 kNm. Determine the areas of main reinforcement required at this support section. Assume $d_2 = 20$ mm and $p_{sc} = 170$ MN/m².

Chapter 5

Shear and bond

Shear forces must necessarily accompany any changes in bending moment which occur along the length of a beam. The effects of these shear forces include both diagonal tension and the formation of bond stresses between the reinforcement and the concrete. The behaviour of beams subjected to combined bending and shear is therefore much more complex than the behaviour of beams subjected to pure bending only. Elastic theory provides a useful introduction to the subject even though all the expressions for shear and bond given in the Unified Code are based on the limit state of collapse.

5.1 Elastic shear stress distribution

The vertical and horizontal shear stresses which exist in a reinforced concrete beam can be determined by using elastic theory assumptions. It is therefore convenient to reiterate some of the conventional results for homogeneous materials before applying elastic theory specifically to reinforced concrete beams.

Complementary shear stress

Consider a small cube of concrete in the web of a beam subjected to a vertical shear force. Let the cube sides be of length δx as shown in figure 5.1.1 and let the shear force, Q, produce a shear stress, q,

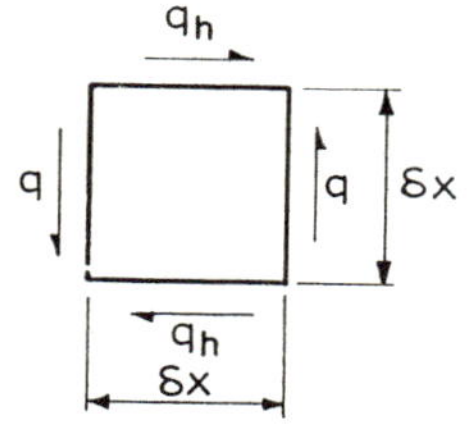

Fig 5.1.1 Complementary shear stresses.

in the vertical plane. If there is no change in Q over the length δx, the shear forces on opposite faces of the cube must be equal in magnitude and opposite in sign, and therefore form a couple. The shear force on each vertical face is $q(\delta x)^2$. To maintain equilibrium there must also be a pair of shear forces on the horizontal faces of the cube. If the shear stress on each horizontal face is q_h, the shear forces are given by $q_h(\delta x)^2$. Taking moments about one corner of the cube

$$q(\delta x)^2\delta x = q_h(\delta x)^2\delta x \quad \text{i.e.} \quad q = q_h \tag{5.1.1}$$

That is, a vertical shear stress at any point produces an equal and opposite horizontal shear stress at that point.

Shear force and bending moment

Consider a length δx of a beam which is symmetrical about a vertical axis. Let the loading be w per unit length, and let the changes in shear and bending moment be δQ and δM respectively, as shown in figure 5.1.2. Taking moments about the steel at the right hand

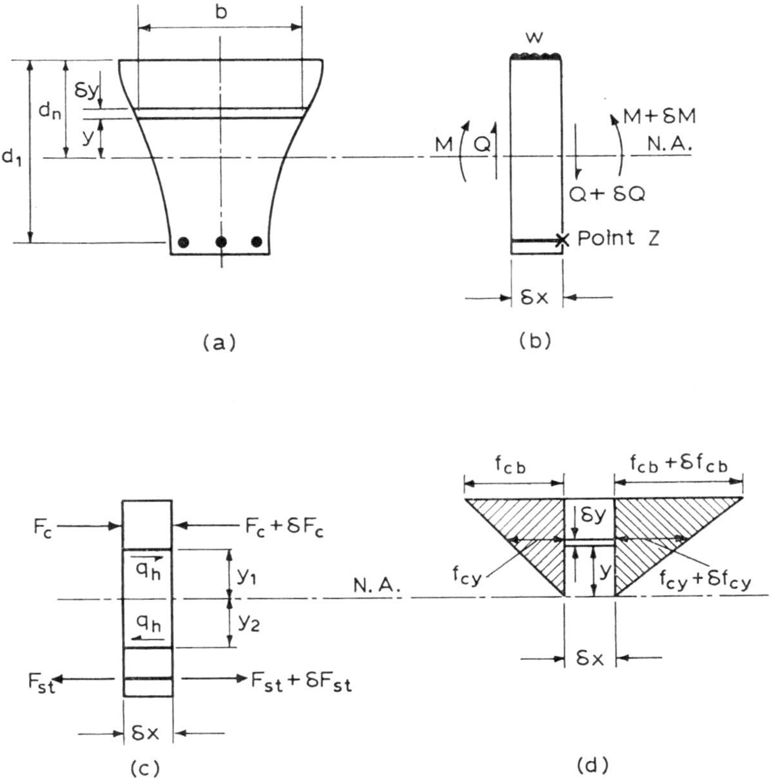

Fig 5.1.2 Shear stresses and normal stresses in singly reinforced concrete beam. (*a*) Cross-section. (*b*) External loads, bending moments and shear forces. (*c*) Internal forces and shear stresses. (*d*) Normal stresses in concrete.

section (point Z), for the external forces acting on the short length of beam, gives

$$(M + \delta M) + \tfrac{1}{2}w(\delta x)^2 = M + Q\delta x$$

Neglecting second order terms in the limit gives

$$Q = dM/dx \tag{5.1.2}$$

That is, shear force equals rate of change of bending moment: a well-known result.

General shear stress equation

The net horizontal force acting on the elemental strip δy (see figure 5.1.2(*d*)) due to the change in flexural strain along length δx of beam is given by

$$\delta F_{cy} = \delta f_{cy} b \delta y \tag{5.1.3}$$

Now the concrete stress due to bending at the level y in figure 5.1.2 is given by

$$f_{cy} = (M/I_e)y$$

where I_e is the second moment of area of the equivalent concrete section. Similarly, the change in stress at level y with change in moment is given by

$$\delta f_{cy} = (\delta M/I_e)y \tag{5.1.4}$$

Substituting for δf_{cy} from equation 5.1.4 and putting $\delta M = Q\delta x$ in equation 5.1.3 enables δF_{cy} to be expressed in terms of Q as follows:

$$\delta F_{cy} = (\delta M/I_e)by\delta y = (Q\delta x/I_e)by\delta y \tag{5.1.5}$$

The equilibrium of the portion of the beam above level y_1 in figure 5.1.2(*c*) is given by summing the resultant horizontal forces on the strips above this level, and equating this force to the horizontal shear force at level y_1:

$$\sum_{y_1}^{d_n} \delta F_{cy} = q_h b_1 \delta x$$

where b_1 is the breadth at level y_1. Substituting for δF_{cy} from equation 5.1.5 gives

$$(Q\delta x/I_e) \sum_{y_1}^{d_n} by\delta y = q_h b_1 \delta x$$

Putting $q_h = q$ and $\sum by\delta y = A\bar{y}$, where $A\bar{y}$ is the first moment of area of the portion above level y_1, gives

$$q = \frac{QA\bar{y}}{b_1 I_e} \tag{5.1.6}$$

Rectangular section

Plotting equation 5.1.6 for a rectangular *homogeneous* section gives a parabolic shear stress distribution as shown in figure 5.1.3(*a*).

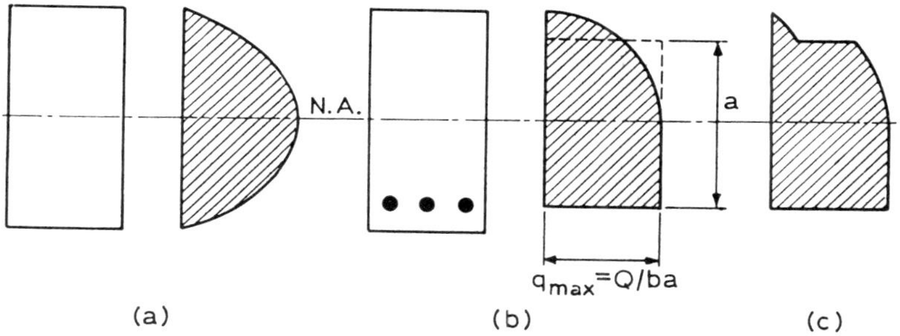

Fig 5.1.3 Elastic shear stress distributions in beams. (*a*) Homogeneous beam. (*b*) RC beam. (*c*) Doubly reinforced beam.

In the case of reinforced concrete, the concrete is assumed to have cracked below the neutral axis, due to tensile bending stresses. Reference to figure 5.1.2(*c*) indicates that the resultant horizontal force in the steel is simply δF_{st}, so by considering the equilibrium of the portion below some level y_2:

$$\delta F_{st} = q_h b_2 \delta x \tag{5.1.7}$$

Now $M = F_{st}a$, hence $\delta F_{st} = \delta M/a$. But $\delta M = Q\delta x$ and $q = q_h$, hence

$$q = Q/b_2 a \tag{5.1.8}$$

This equation is independent of y_2, so the shear stress below the neutral axis is constant for a constant beam breadth.

Maximum shear stress

The shear stress distribution above the neutral axis has been shown to be parabolic and given by equation 5.1.6. The maximum value is therefore at the neutral axis. For a rectangular beam

$$A\bar{y} = \tfrac{1}{2}bd_n^2$$

at the neutral axis, hence

$$q_{max} = Qd_n^2/2I_e \tag{5.1.9}$$

Now from equations 3.2.5, 3.2.7 and 3.3.3,

$$M = \tfrac{1}{2} f_{cb} b d_n a = f_{cb} I_e / d_n$$

Hence

$$ba = 2I_e / d_n^2$$

Substituting for d_n in equation 5.1.9 gives

$$q_{max} = Q/ba \qquad (5.1.10)$$

This value agrees with the value given by equation 5.1.8 for all positions at or below the neutral axis. The shear stress distribution for a singly reinforced beam is therefore as shown in figure 5.1.3(*b*). For a T- or L-beam, then, the equation becomes

$$q_{max} = Q/b_r a \qquad (5.1.11)$$

Difficulties of elastic theory

Equations 5.1.6 and 5.1.8 were derived assuming that shear stresses act on the vertical faces considered in figure 5.1.2. However, it has already been shown that extensive vertical cracking of the concrete below the neutral axis must occur even at working loads. The shear must therefore be transferred to the reinforcement via the "struts" of concrete between the cracks. The shear stresses, at least below the neutral axis, must be much less uniform and continuous than those indicated by elastic theory.

Nominal shear stress

In view of the above difficulties equation 5.1.10 gives a misleading precision which is not justified. Furthermore, at or near to collapse, the distribution of stresses is affected by extensive cracking throughout the beam and other factors become significant; hence this value *by itself* can only give a very approximate indication of the strength. The Unified Code has therefore discarded the apparently precise elastic value in favour of a "nominal shear stress" given by

$$q = Q/bd_1 \qquad (5.1.12)$$

EXAMPLE 5.1.1

A cantilever 150 mm wide has an effective depth of 250 mm and is reinforced with four 16-mm diameter bars. It carries a design shear force of 75 kN at the support. Calculate the nominal shear stress.

Solution The nominal shear stress is

$$Q/bd_1 = 75 \times 10^3/0{\cdot}15 \times 0{\cdot}25 = 2 \text{ MN/m}^2 \qquad \text{(Answer)}$$

Notice the simplicity of this solution. It would have been necessary to calculate the lever arm, *a*, in order to obtain the elastic shear stress value, *Q*/*ba*.

5.2 Diagonal tension

One effect of shear forces is to produce tension across a diagonal plane. This *diagonal tension*, as it is called, is now considered in more detail and shown to be more significant than the shear stresses themselves.

Shear strength

Concrete is much stronger in compression and pure shear (sometimes called punching shear) than it is in tension. It follows, therefore, that the principal tensile stresses *induced* by the vertical and horizontal shear stresses are much more critical than the shear stresses themselves. Consider, for example, the beam distorted by the load W in figure 5.2.1. The effect of the load W is to shorten the diagonals

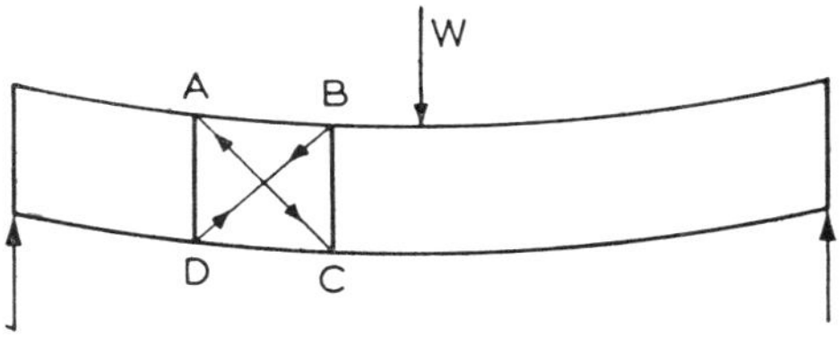

Fig 5.2.1 Diagonal tension and compression in deformed beam.

BD and to extend the diagonals AC: i.e. compressive and tensile stresses act normal to the 45° planes indicated by AC and BD respectively.

Principal stresses

The maximum shear force in a simply supported beam occurs near the supports. The ends of such a beam are under tensile stress on one diagonal plane and under compressive stress on the other, which is at right angles to the former. The principal stress contours for a homogeneous beam are shown in figure 5.2.2. Notice that the bending

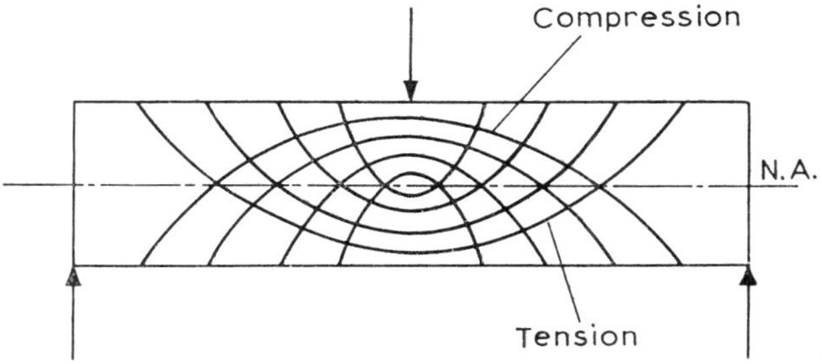

Fig 5.2.2 Principal stress contours for a homogeneous beam.

stresses are a maximum at the extreme fibres and decrease to zero at the neutral axis, whereas the shear stress is a maximum at the neutral axis and decreases to zero at the extreme fibres. The maximum principal tensile stresses are given at the points of maximum shear and minimum compressive bending stresses, i.e. at or below the neutral axis.

Diagonal tension and compression

Consider small triangular portions of concrete at the neutral axis as shown in figure 5.2.3. It is clear that the shear stresses on the

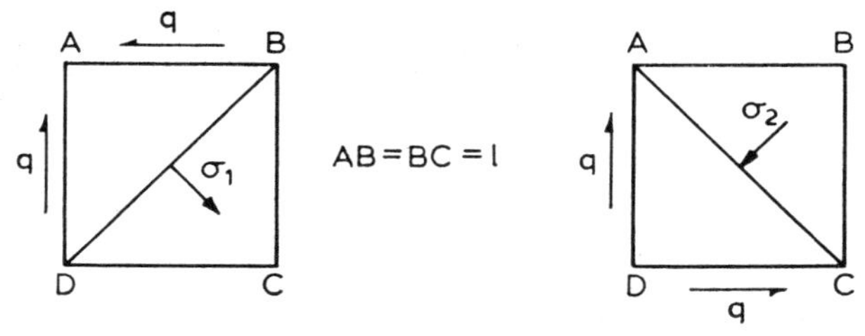

Fig 5.2.3 Principal stresses at neutral axis of beam.

faces of the cube produce a principal tensile stress σ_1 on the diagonal plane BD given by

$$ql\mathrm{AB}/\sqrt{2} + ql\mathrm{AD}/\sqrt{2} = \sigma_1 l\mathrm{BD}$$

Hence

$$\sigma_1 = q \qquad (5.2.1)$$

Similarly, the principal compressive stress σ_2 on the diagonal plane AC also equals q. However, since concrete is weak in tension it tends to fail across the plane BD. It is important to remember that the diagonal tension stress is equal in magnitude to the shear stress. Critical shear stresses really represent critical diagonal tension stresses of the same magnitude.

Notice also that the diagonal tension is the principal tensile stress *at the level of the neutral axis.* Above the neutral axis the principal tensile stress is reduced by the flexural compressive stress, and the plane on which it acts becomes flatter. Below the neutral axis the principal tensile stress is increased by the flexural tensile stress, and the plane on which it acts becomes more nearly vertical.

5.3 Bond between concrete and steel

One of the basic assumptions made in the analysis of reinforced concrete is that good bond exists between the concrete and the reinforcement. Reinforced concrete members can act satisfactorily as composite units only if the bond between the concrete and the steel is adequate. Shear forces are present wherever the bending moment changes, and such changes imply a change in the tensile force in the main reinforcement. This in turn means that bond stresses must be present to transmit forces from the reinforcement to the concrete.

Flexural bond stress

If the applied bending moment, and hence the stress in the main steel, changes rapidly along the beam it is possible for the flexural (or local) bond stress to become undesirably high. Considering the equilibrium of the portion of the beam below the neutral axis in figure 5.1.2(*c*) and putting $q = Q/ba$ gives

$$\delta F_{st} = qb\delta x = Q\delta x/a \tag{5.3.1}$$

This horizontal shear stress produces the change in tension of the steel. It must therefore be transmitted into the steel by the action of the bond stress between the concrete and the steel. No *bond* stress problem arises in the upper compressive zone of the beam, although the horizontal shear stress acts between successive layers of the beam in a similar manner.

If q_b is the flexural bond stress and $\sum o$ is the sum of the perimeters of the *tensile* reinforcement, then

$$\delta F_{st} = q_b \sum o \delta x = Q\delta x/a$$

Re-arranging, and putting $a = 0{\cdot}8d_1$ as in the Unified Code:

$$q_b = Q/(0{\cdot}8d_1 \sum o) \tag{5.3.2}$$

The maximum flexural bond stress values are shown in table C6 of appendix C and are generally at least 50% higher than the anchorage bond values for tension which are given in table C5.

Anchorage bond stress

The axial stress at the end of any bar must be zero. The stress at any intermediate section, distance L from the end, must be resisted by an anchorage (or average) bond stress f_b, between the concrete and steel, which acts over the length L. Equating the tensile force at the section to the bond force

$$F_{st} = f_b \pi DL$$

where D is the bar diameter. The anchorage bond stress is therefore given by

$$f_b = F_{st}/\pi DL = f_{st}D/4L \tag{5.3.3}$$

Development length

The minimum development length is given by re-arranging equation 5.3.3 as follows

$$L_{\min} = (f_{st}/4f_b)D \qquad (5.3.4)$$

The development length is therefore given by the bar diameter multiplied by an appropriate coefficient. If

$f_{st} = 250\ \mathrm{MN/m^2}$ and $f_b = 1{\cdot}3\ \mathrm{MN/m^2}$ then

$L = 48D$

The maximum bond stress, and therefore the bond length, varies with the grade of concrete as shown in table C5. The grip of the bars can be increased by about 80% by indenting the surface, or by about 40% by using a square twisted bar, thanks to the additional mechanical bond so provided. Where the full bond distance cannot be provided, the assumed stress (tensile or compressive) in the bar must be reduced.

EXAMPLE 5.3.1

Calculate the flexural bond stress for the main bars at the support section for the beam in example 5.1.1. What bond length would be required at the support if the anchorage bond stress is not to exceed 1·3 MN/m² and the design stress in the bars is 217 MN/m² (i.e. 250/1·15)?

Solution

$$\textstyle\sum o = 4 \times \pi \times 16 \times 10^{-3} = 0{\cdot}201$$

$$q_b = Q/(0{\cdot}8d_1\textstyle\sum o) = 75 \times 10^3/0{\cdot}8 \times 0{\cdot}25 \times 0{\cdot}201$$

$$= 1{\cdot}87 \times 10^6$$

$$= 1{\cdot}87\ \mathrm{MN/m^2} \qquad \text{(Answer)}$$

From equation 5.3.4

$$L = 217 \times 16 \times 10^{6-6-3}/4 \times 1{\cdot}3 = 0{\cdot}670 = 670\ \mathrm{mm}$$

(Answer)

End anchorage

Consider a simply supported beam subjected to a single point load as shown in figure 5.3.1. The shear force between A and C is constant. If the perimeter of the tensile reinforcement is constant then the flexural bond stress is also constant. From the bending moment diagram the theoretical tensile stress in the steel is a maximum at midspan and zero at the supports A and B. It is clear that if the maximum flexural bond stress can exceed the maximum anchorage bond stress, the development length requirement is such that the bar must extend beyond the point of theoretical zero stress. This additional length of bar is known as the *end length* or *end anchorage*. The maximum required end length is given when the flexural bond stress is constant at the maximum value.

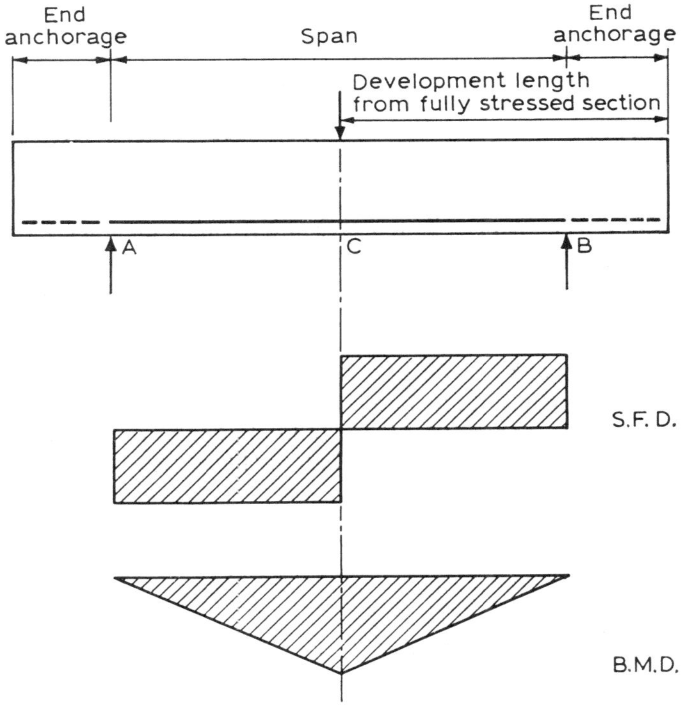

Fig 5.3.1 End anchorage for steel reinforcement.

The minimum end anchorages for various grades of concrete and steel which are adequate for the worst possible cases are then given by equation 5.3.5 (below). Typical values are shown in table 5.3.1. It is therefore convenient (and advisable) to provide these end lengths in all cases and then simply check that the flexural bond stress is not exceeded at the most critical section (which is usually adjacent to a support, as explained in the next section).

$$L_a \not< {}_*f_{yt}(1/f_b - 1/q_b) \times \tfrac{1}{4}D \qquad (5.3.5)$$

The Unified Code specifies also that the minimum end anchorage should never be less than $12D$, hence the bracketed values in table 5.3.1. It should be noted that the end anchorages for plain bars have been increased by up to 40% compared with the former CP114 values. The values for square twisted and high bond bars

Table 5.3.1 Minimum end lengths to ensure satisfactory development lengths in all cases

Characteristic cube strength:	15	22·5	30	37·5	≮45
Plain bars	26*D*	19*D*	17*D*	15*D*	13*D*
Square twisted bars	25*D*	18*D*	16*D*	15*D*	13*D*
High bond bars	19*D*	13*D*	(11*D*)	(11*D*)	(10*D*)

are generally slightly higher, and over 20% lower, respectively, than the CP114 values. If the minimum values in table 5.3.1 are not provided bars can only be curtailed if the shear force is $\not> \frac{1}{2} \times$ shear capacity of the section, or if the moment is very low (see appendix C).

Types of anchorage

The common forms of anchorage for both tension and compression bars are shown in figure 5.3.2 for mild steel. The requirements for other angles or for high yield steel are included in appendix C.

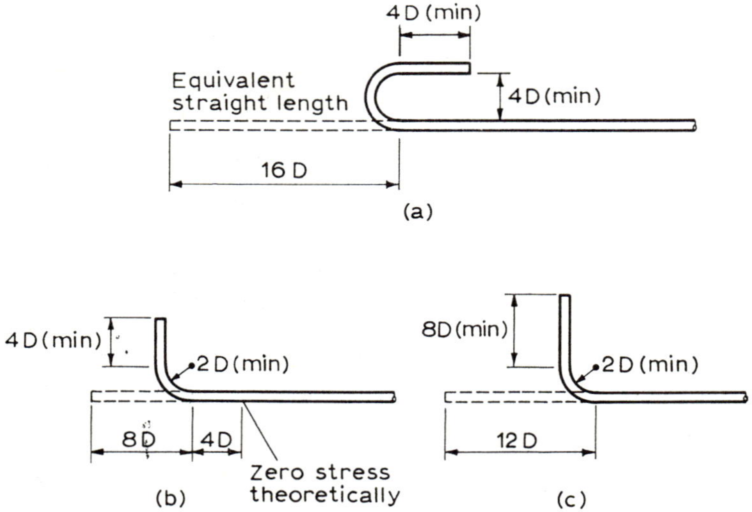

Fig 5.3.2 Typical anchorages for mild steel. (*a*) Standard U hook (tensile bars only). (*b*) Standard L hook. (*c*) L hook with equivalent straight length of 12*D*.

Laps in bars

If the normal maximum length of a bar is exceeded, or if the ease of placing the steel so requires, one or more overlaps must be introduced. The stress in the bars at the lap must be transmitted from one bar to the other through the bond stress between the bars and the surrounding concrete. The lap length must therefore be at least equal to the required bond length for the smaller diameter bar. In addition it should be noted that the axial forces in the reinforcing bars at a lap are displaced slightly, which produces a lateral bursting force which could cause the concrete cover to break away. This must be prevented by the provision of stirrup reinforcement, which is considered later.

5.4 Curtailment of reinforcement

The section of a beam and the quantity of main reinforcement is determined from the maximum bending moment only. It is clear, however, that where the bending moment is considerably less than the maximum some of the bars are superfluous. Such bars can be curtailed if suitable end anchorage is provided beyond the point at which they can be cut off theoretically according to the bending moment diagram.

Use of bending moment diagram

The points at which main tensile bars are no longer required in the bottom of a beam can be obtained very conveniently from the bending moment diagram. At the points of curtailment of successive sets of bars the area of the remaining bars should provide sufficient moment of resistance to take the bending moment at that point. If the lever arm is assumed to remain constant as bars are curtailed, then the bending moment diagram can be divided into equal increments of the bending moment at midspan as shown in figure 5.4.1. This assumption is on the safe side since the lever arm increases as the steel ratio decreases for either elastic theory or collapse theory.

Maximum flexural bond stress

The flexural bond stress is a maximum at points of maximum shear (i.e. maximum slope of the BMD) combined with minimum tensile reinforcement. The maximum flexural bond stress for a simply supported beam therefore occurs at the supports. It is clear that if too many bars are curtailed, the value of $\sum o$ becomes too small and q_b becomes excessive in equation 5.3.2. It is usually found that nearly 50% of the tensile bars need to be run through

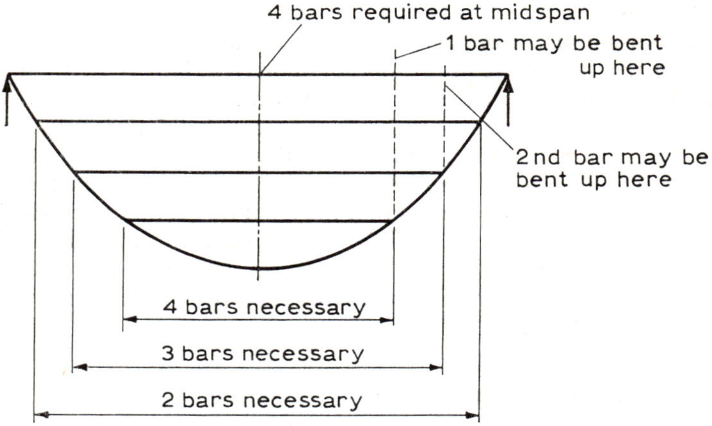

Fig 5.4.1 A simply supported beam with four bars for midspan moment.

to the support in a simple beam and about 33% for a continuous beam. Compare these values with the simplified Code requirements given later.

Standard parabolic BMD

The bending moment diagram for a uniformly loaded beam is parabolic. The points at which bars may be either curtailed or bent up for such a beam can therefore be obtained from a standard curve. Furthermore, if the negative moment at the support is assumed to be equal to the positive moment at midspan, this simple approach can be extended to the end and interior spans of a continuous beam, as shown in appendix J.

Bent-up bars

In a continuous beam the maximum amounts of main steel are required in the bottom of the beam at midspan and in the top of the beam at the supports. It is therefore often convenient to bend up some of the bars from the bottom of the beam into the top. The number and spacing of bars available for bending up are determined from the bending moment diagram and the maximum flexural bond stress. The actual positions at which the bars are bent up usually differ slightly from the positions at which the bars first become available, in order to provide a more uniform spacing to suit shear strength requirements (see later).

Bond length—tension bars

The bond length must be considered for those bars running through the support in the bottom of a beam or slab, as well as those bars which are stopped short. Similarly, the curtailment of tensile bars running through the top of a continuous beam at the supports is also affected by the required bond length.

Bond length—compression bars

The considerations for compression bars in beams are similar to those for tension bars, except that the maximum anchorage bond stress in compression is about 25% greater than the tensile value (see appendix C) since the lateral expansion, due to the Poisson's ratio effect, improves the bond for compression bars. Notice also that since the stresses in the reinforcement are lower in compression than in tension, the required anchorage lengths are appreciably lower for compression bars.

Stopped-off bars

Bars which are stopped off, rather than bent up, require an adequate end length beyond the points at which bars become available from the bending moment diagram.

Simplified Code requirements for curtailment

The curtailment of reinforcement is governed by the bending moment diagrams and bond considerations as discussed above. The Unified Code also permits, however, an alternative simplified approach which, for either beams or one-way slabs, merely requires compliance with one of the requirements given in appendix C.

Bond length—lightweight concrete

Bond strength varies according to the inclination of the reinforcement, for all concretes, due to internal water-gain effects [5.1]. At present, the Unified Code only differentiates between horizontal and vertical bars for lightweight concrete (see appendix C), even though equally large differences would appear to be possible for normal concrete also.

5.5 Beams without shear reinforcement

It will have been noticed that the maximum bond stresses given by the Unified Code are in terms of collapse values rather than permissible values. The same bond lengths could equally well have been obtained by considering permissible stresses in bond and tension, so the difference is largely immaterial in this instance. However, the shear strengths of beams according to elastic theory and collapse theory cannot be made directly comparable. This is largely because some factors which affect the collapse strength are ignored completely by elastic theory. All the expressions for shear strength are therefore in terms of the limit state of collapse.

Mutual effect of stresses

It is clear that a beam fails, in general, due to the combined effects of bending and shear. This should be allowed for if these effects are arbitrarily separated by considering shear stresses and bending stresses individually. There is no doubt that the presence of shear stresses can decrease the ultimate moment of resistance of a beam but the problem is very complex. However, it has been shown experimentally that within limits the effect of bending stresses on the shear strength of a section is sufficiently small to enable this effect to be virtually ignored in normal design. For example, Taylor has shown [5.2] that when the maximum applied moment varied between 50% and 100% of the nominal flexural strength of the member, the shear strength only decreased by about 10%.

Quantity of main steel

The nominal shear strength increases with the quantity of longitudinal reinforcement, irrespective of type. It follows, therefore, that beams with ordinary mild steel bars can sustain slightly higher shears than corresponding beams with an equivalent (that is, reduced) quantity of high-yield steel bars. This increase in strength is primarily due to dowel action of the bars which is considered later.

Concrete strength

The nominal shear strength increases with the cube strength, but at a reduced rate. The Unified Code gives values for the critical shear stress, q_c, which allow for different quantities of main steel and different cube strengths of concrete as shown in table C3 in appendix C. q_c is the maximum nominal shear stress (or diagonal tension stress) for unreinforced beams which can be taken by the concrete section without shear reinforcement when the member is subjected to the maximum design loads. That is

$$q_c = {}_{*}Q/bd_1$$

where ${}_{*}Q$ is the *design* shear force for collapse, given by $\gamma_l Q_k$.

Partial safety factors for shear

Shear-compression failures of individual members cannot be predicted with the same precision as a flexural failure, hence the safety margin against this type of failure should err on the high side. Again, shear failures are brittle and tend to occur suddenly, whereas an under-reinforced beam gives ample warning of imminent flexural failure. Reference has been made previously to the Russian Code which has a partial safety factor for "service" which can allow for these considerations. The American Code also has a factor which can allow specifically for deficiencies in knowledge or special considerations in service: that is, the particular location of the member or the particular stresses being considered. The Unified Code reduces the possibility of a shear failure by increasing the partial safety factor for material strength of concrete when subjected to shear. This increased value of γ_m (i.e. $\gamma_{l3}\gamma_{c1}$) is already included in table C3 for critical shear stress in order to avoid confusion with the values of γ_m used for other calculations such as flexure.

EXAMPLE 5.5.1

A beam 150 mm wide has an effective depth of 250 mm and a steel area of 924 mm². It carries a design shear force of 30 kN. Is shear reinforcement required if $u = 45$ MN/m²? Refer to table C3 in appendix C.

Solution The steel ratio is

$$r = 924/150 \times 250 = 2{\cdot}46\%$$

Hence from table C3, for $u = 45$ MN/m², $q_c = 1{\cdot}1 \times 10^6$,

$$q = 30 \times 10^3/0{\cdot}15 \times 0{\cdot}25 = 0{\cdot}8 \times 10^6$$

No shear reinforcement is required, theoretically, since $q_c > q$. However, nominal stirrups should be used, as discussed later.

5.6 Shear reinforcement for diagonal tension

It has been shown already that the diagonal tension which results from shear forces is much more critical for the concrete than the direct effect of the shear itself. If the critical shear stress values (which are really critical "diagonal tension" stress values) are exceeded shear reinforcement must be provided to take the excess tension.

Types of shear reinforcement

Ideally, the tensile reinforcement in a beam should follow the lines of principal tension shown in figure 5.2.2. However, since this is impractical for normal RC beams, the two main types of shear reinforcement are as follows:

(i) *Stirrups* Vertical stirrups are the most usual form of shear reinforcement. They are relatively light bars which are bent into a rectangular shape so as to pass round the tensile reinforcement in the bottom of the beam as well as the compression reinforcement in the top. Where no compression reinforcement is present, top bars of light section, termed *hanger bars*, should generally be provided. The diameter of such bars is usually governed by the distance they must "hang" while the steel is being fixed.

(ii) *Bent-up bars* It is often possible to bend up the main tensile bars, from the tensile flange into the compression flange, near the ends of the beam where the shear is large and the bending moment is reduced. The angle of bend is normally 45° to the line of the main bars. The bars must be bent up through a depth equal to, or greater than, the spacing. Bent-up bars can produce excessive compression and hence bursting of the concrete near the bends which could only be restrained adequately by providing ties at appropriate heights across the legs of the stirrups. This is usually impracticable. Hence bent-up bars should only be used in moderation and only in association with stirrups.

Vertical stirrups

Vertical stirrups are commonly used as shear reinforcement, either alone or in conjunction with bent-up bars. A vertical stirrup system of pitch (or spacing), s, is shown in figure 5.6.1. If it is assumed that

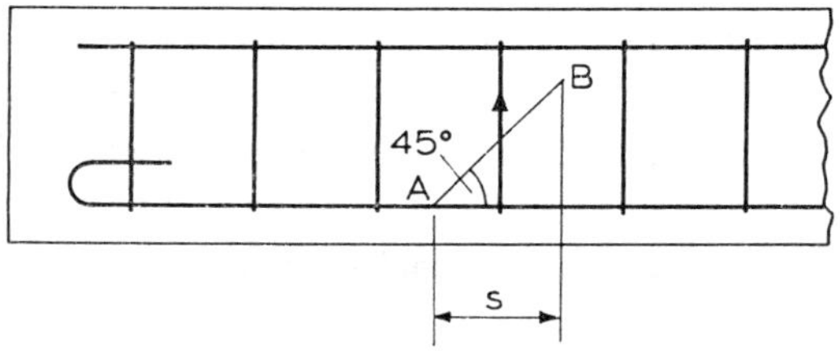

Fig 5.6.1 Vertical stirrups across an uncracked section.

the vertical stirrups act in conjunction with the main horizontal bars, then *the stirrups need only take the vertical component of the diagonal tension.* Hence the force in each stirrup, F_{sq}, is given by

$$F_{sq} = qb\text{AB}\cos 45° = qsb.$$

since q equals the diagonal tension stress, and s is the spacing of the stirrups. Substituting for $q = q_{\text{max}} = Q/ba$ from equation 5.1.10, putting $F_{sq} = f_{sq}A_{sq}$, and re-arranging, gives

$$Q = (a/s)f_{sq}A_{sq} \qquad (5.6.1)$$

This is the result to be expected from elastic theory assuming that $s \not> a$ and that the concrete is incapable of taking any shear at all.

However, since concrete can normally resist some diagonal tension unless it is cracked, it is preferable to consider the actual cracked section.

Formation of diagonal crack

Instead of using elastic theory it is possible to consider the mechanism of the diagonal crack as the basis for the analysis. When the diagonal tension stress reaches the ultimate tensile strength of the concrete the latter fails and a diagonal crack forms. The portions

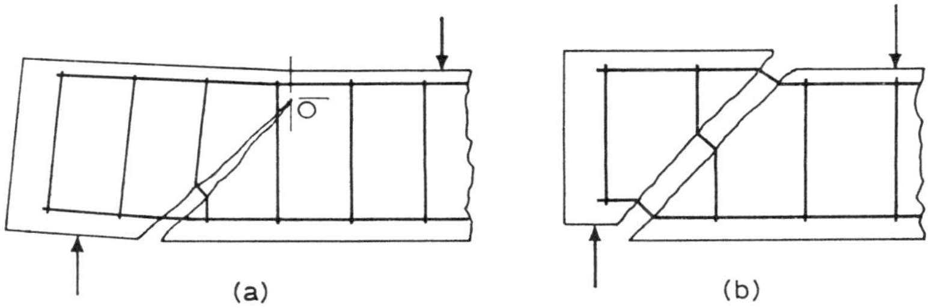

Fig 5.6.2 Vertical stirrups across cracked sections.

of the beam on either side of the inclined crack tend to rotate in opposite directions about some point O in the compression zone above, as shown in figure 5.6.2(*a*). Although shear-compression failures actually follow a complex curve, it is convenient to assume that the failure plane follows a straight line.

Vertical stirrups at collapse

The concrete in the compression zone can be assumed to shear or crush, finally, as shown in figure 5.6.2(*b*). Just before this occurs the shear force Q at the head of the section is resisted by the vertical forces in both the steel and the concrete. If it is assumed that the maximum average tensile stress which can be resisted by the compression zone of the concrete is given by an average value of q_c acting across the entire diagonal section (see figure 5.6.3) then the

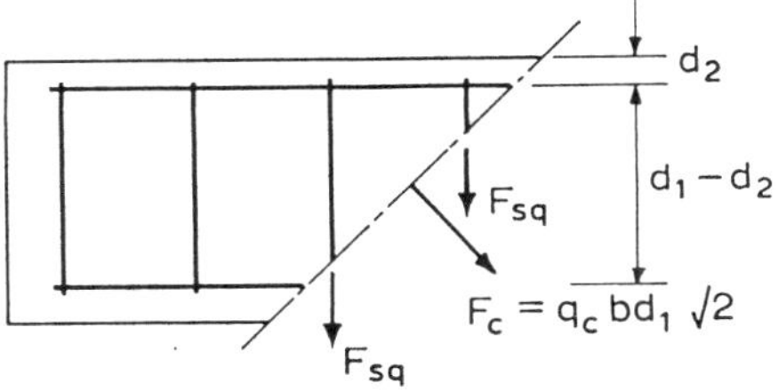

Fig 5.6.3 Combined resistance of stirrups and concrete across a diagonal section.

vertical components of the forces in both the steel and the concrete can be equated to the total shear force as follows:

$$Q = q_c b(d_1\sqrt{2})/\sqrt{2} + F_{sq}(d_1 - d_2)/s$$

Dividing throughout by bd_1 and transposing gives

$$q - q_c = 0{\cdot}87 f_{yq} A_{sq}(d_1 - d_2)/bd_1 s \qquad (5.6.2)$$

where A_{sq} = total cross sectional area of stirrups (i.e. both legs for 2-leg stirrups)
f_{yq} = characteristic strength of stirrup reinforcement ($\not> 425\ \mathrm{MN/m^2}$)
b = width of beam (or web)
s = spacing of shear steel along the longitudinal axis of member, which should not exceed $0{\cdot}75 d_1$.

Notice that this derivation is based upon the vertical equilibrium of forces at the crack. It does not depend upon elastic theory assumptions in any way. It does assume, however, that the main steel is more than adequate to take the horizontal component of the diagonal tension force.

Inclined stirrups at collapse

The approach is the same as for vertical stirrups. That is, the main steel takes the horizontal component of the diagonal tension, and the vertical components of the design strengths of the steel and concrete equal the vertical component (i.e. the design shear force).

Consider triangle ABC in figure 5.6.4.

$$\frac{z}{\sin\theta} = \frac{s'}{\sin[180 - (\theta + 45)]} = \frac{s'}{\sin(\theta + 45)}$$

or

$$z = s' \sin\theta/(\sin\theta\cos 45 + \cos\theta\sin 45).$$

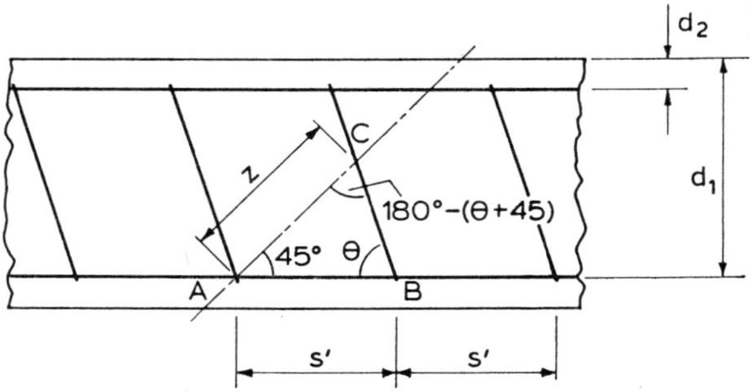

Fig 5.6.4 Inclined stirrups across a diagonal section.

The number of bars intersected by a plane at 45° is

$$(\sqrt{2})(d_1 - d_2)/z = (d_1 - d_2)(\sin\theta + \cos\theta)/s' \sin\theta$$

Hence the vertical component of the design yield strength of the stirrups which intersects a plane at 45° is given by

$$(0{\cdot}87 f'_{yq} A'_{sq} \sin\theta)(d_1 - d_2)(\sin\theta + \cos\theta)(s' \sin\theta)^{-1}$$

The shear stress to be resisted by the stirrups is the actual shear stress less the shear stress which can be resisted by the concrete. Since $q = Q/bd_1$,

$$q - q_c = (0{\cdot}87 f'_{yq} A'_{sq})(d_1 - d_2)(\sin\theta + \cos\theta)(s'bd_1)^{-1}$$

The draft Code assumes that $(d_1 - d_2)/d_1$ can be taken as unity if the shear which is assumed to be taken by the concrete is reduced to $0{\cdot}8q_c$ (that is, $0{\cdot}2q_c$ allows for neglecting d_2). Hence for safe design

$$q - 0{\cdot}8q_c \ngtr (0{\cdot}87 f'_{yq} A'_{sq})(\sin\theta + \cos\theta)(s'b)^{-1}$$

The expression in the proposed Unified Code substitutes $1/\sin\theta$ for $(\sin\theta + \cos\theta)$. Although this gives the same values when $\theta = \pi/2$ and $\pi/4$ it gives incorrect values at any other angle. At small angles

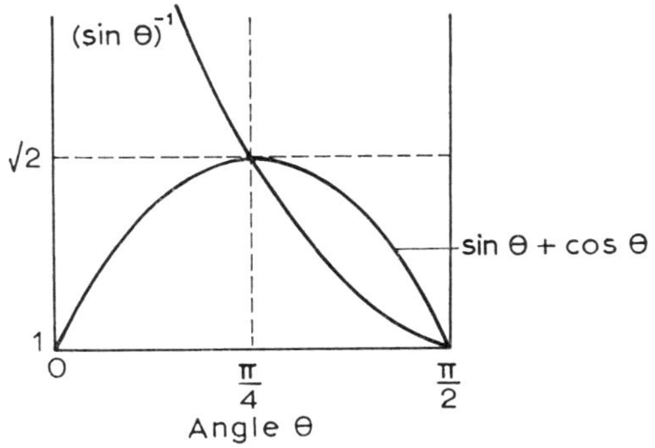

Fig 5.6.5 Comparison between functions $(\sin\theta)^{-1}$ and $(\sin\theta + \cos\theta)$.

it becomes infinitely large; hence the necessity to arbitrarily require that an angle of 45° gives a maximum value while an angle of 44°, which could easily occur due to tolerances in placing, must be rejected! Figure 5.6.5 shows how inaccurate and misleading this substitution is since it is incorrect at all values other than $\pi/4$ and $\pi/2$.

EXAMPLE 5.6.1

The steel ratio for the web reinforcement, r_q, for stirrups is given by the total cross-sectional area of all the arms of a stirrup, A_{sq}, divided by the area of concrete appropriate to the stirrup and normal to it. Determine the optimum angle of inclination of the stirrups such that r_q is a minimum for a given shear.

Solution Using the usual notation and, referring to figure 5.6.4 for convenience,

$$A'_{sq} = \frac{Q's'}{0{\cdot}87f'_{yq}(d_1 - d_2)(\sin\theta + \cos\theta)}$$

where Q' is the excess shear, and

$$r_q = \frac{A'_{sq}}{bs'\sin\theta} = \frac{Q'}{0{\cdot}87f'_{yq}b(d_1 - d_2)\sin\theta\,(\sin\theta + \cos\theta)}$$

The optimum inclination of the stirrups is given by considering f'_{yq}, b, $d_1 - d_2$ and Q as constants. Let

$$X = \sin\theta\,(\sin\theta + \cos\theta)$$

$$\frac{dr_q}{d\theta} = 0 \quad \text{when} \quad \frac{dX}{d\theta} = 0$$

i.e. when

$$\sin\theta\,(\cos\theta - \sin\theta) + \cos\theta\,(\sin\theta + \cos\theta) = 0$$

$$\tan 2\theta = -1 \quad \text{i.e.} \quad \theta = 67\tfrac{1}{2}^\circ \qquad \text{(Answer)}$$

Notice that this solution again assumes that there is ample main steel to take the horizontal component of the diagonal tension. In practice it is far more convenient to use vertical stirrups for ease of positioning, and inclined stirrups are seldom used.

Minimum stirrups

The concrete alone may be strong enough to take the shear, especially in the midspan portion. However, nominal (or minimum) stirrups in accordance with the Unified Code (see appendix C), should always be used in beams and similar members for the following reasons:

(i) Premature failure occurs if the concrete cover bursts away and the bond to the main bars is lost. This is prevented by nominal stirrups.

(ii) Shrinkage and thermal stresses, which are normally ignored, can produce appreciable tension. Shallow sections such as slabs and footings have adequate longitudinal steel suitably positioned whereas beams may have little steel in the top (or bottom). The shear strength of a vertically cracked plain concrete section is nil! Stirrups should always take a proportion of the total shear and help to transfer the longitudinal stresses to the face containing the main steel.

(iii) Stirrups act as ties to hold the beam reinforcement in position while the concrete is being placed.

(iv) Shear failures in concrete beams without secondary reinforcement are essentially brittle and can occur very suddenly. Stirrups help to provide increased warning of failure.

The minimum stirrups should be increased for slender beams (see appendix C). However, minimum shear reinforcement need not be provided in members where the breadth is large (i.e. $b > d_1$) and $q < q_c$ since the shear strength is greater than that for deeper members and so provides some reserve of strength. Similarly, in walls and members of minor importance and where the shear stress is $< \frac{1}{2}q_c$, minimum stirrups can also be omitted.

EXAMPLE 5.6.2

Calculate the maximum nominal shear stresses for collapse for the beam in example 8.2.2 (chapter 8, below) and hence check the strength in shear at the following positions using the Unified Code recommendations:

(i) At midspan ($r = 3{\cdot}40\%$)

(ii) At a section 0·6 m from a support, without any stopping-off of bars

(iii) At a section 0·6 m from a support, where the longitudinal reinforcement has been stopped off such that $r = 1{\cdot}50\%$

Hence determine the required spacing for 6-mm diameter mild steel stirrups at sections (i), (ii) and (iii) if $f_{yq} = 250$ MN/m².

Solution

$$R_A = 1{\cdot}6 \times 36(3{\cdot}6 - x) \times 10^3/3{\cdot}6 = 16(3{\cdot}6 - x) \times 10^3$$

Hence $x = 1{\cdot}8$ and $0{\cdot}6$ for maximum shear at midspan and 0·6 m from support, respectively.

(i) At midspan:

$${}_*Q = 16 \times 1{\cdot}8 \times 10^3 = 28{\cdot}8 \times 10^3$$

Therefore

$$q = 28{\cdot}8 \times 10^3/0{\cdot}15 \times 0{\cdot}25 = 0{\cdot}768 \times 10^6$$

and, for $r = 3{\cdot}40\%$ and $u = 37{\cdot}5 \times 10^6$, from appendix C:

$$q_c = 1{\cdot}14 \times 10^6$$

Therefore only nominal stirrups are necessary. Hence

$$s \not> 0{\cdot}87 \times 250 \times 2 \times \tfrac{1}{4}\pi \times 6^2 \times 10^{+6-6}/0{\cdot}150 \times 0{\cdot}4$$
$$= 0{\cdot}205$$

Also

$$s \not> \tfrac{3}{4}d_1 = 0{\cdot}75 \times 0{\cdot}250 = 0{\cdot}188$$

Hence let $s = 185$ mm (to nearest 5 mm) (Answer)

(ii) 0·6 m from support:

$$_*Q = 16 \times 3{\cdot}0 \times 10^3 = 48 \times 10^3$$

Therefore

$$q = 48 \times 10^6/37{\cdot}5 = 1{\cdot}28 \times 10^6$$

$q_c = 1{\cdot}14 \times 10^6$ as before; hence stirrups are necessary.

$$q - 0{\cdot}8q_c = (1{\cdot}28 - 0{\cdot}8 \times 1{\cdot}14)10^6 = 0{\cdot}368$$

This is less than 0·4, hence use nominal stirrups as in (i). (Answer)

(iii) $q = 1{\cdot}28 \times 10^6$ as in (ii).

$q_c = 0{\cdot}90 \times 10^6$ for $r = 1{\cdot}5\%$.

Therefore

$$q - 0{\cdot}8q_c = 0{\cdot}56 \times 10^6$$

hence

$$s \not> 0{\cdot}87 \times 250 \times 2 \times \tfrac{1}{4}\pi \times 6^2 \times 10^{+6-6}/0{\cdot}150 \times 0{\cdot}56$$
$$= 0{\cdot}146$$

$s = 145$ mm (to nearest 5 mm) (Answer)

Notice that all the Unified Code values for shear and bond are expressed in terms of the limit state of collapse. This limit state is considered in detail in the next chapter.

Bent-up bars for shear

It has been shown previously that when bars are no longer required as main reinforcement in the bottom of the beam they can sometimes be bent up to act as shear reinforcement, as well as provide additional main steel in the top of the beam at the support. The effective zone for bent-up bars is from the bend in the tension zone of the first bar to the bend in the compression zone of the last bar, as shown in figure 5.6.6. To ensure that the failure plane cuts at least one bar, the spacing should not exceed $0{\cdot}75d_1$. The analysis for bent-up bars is exactly the same as for inclined stirrups (see appendix C). The provision of minimum stirrups in all cases avoids any heavy local stresses which would occur if there were a sudden transition from (all) stirrups to (all) bent-up bars. Independent shear bars, such as the end bent-up bar in figure 5.6.6, may be used if necessary.

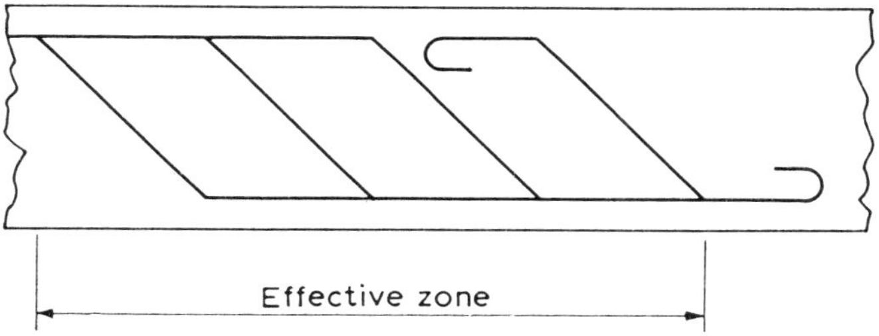

Fig 5.6.6 Effective zone for bent-up bars.

However, the cost of bending and placing such bars is not usually justified nowadays and additional vertical stirrups should be used instead. Notice that if θ is not greater than 45°, the presence of the main steel does not increase the shear resistance by contributing a horizontal force, since the horizontal component of the force in the shear steel itself is more than adequate. However, the Unified Code does not permit this possibility.

End anchorage—bent-up bars

Maximum tension occurs at the neutral axis in the case of tensile bars bent-up for shear, so that the bond length is measured from this point. Notice that this requirement may make it difficult adjacent to a support, as shown in figure 5.6.7. Vertical stirrups must sometimes be used entirely on their own in such cases. The bent-up bar can be brought nearer to the support by bending down the end of the bar into the tensile flange, but this can cause congestion of steel since tensile bars are also running through to the support.

End anchorage—stirrups

Stirrups are also fully stressed at the neutral axis. Hence the arms of the stirrups must possess adequate bond length beyond the neutral

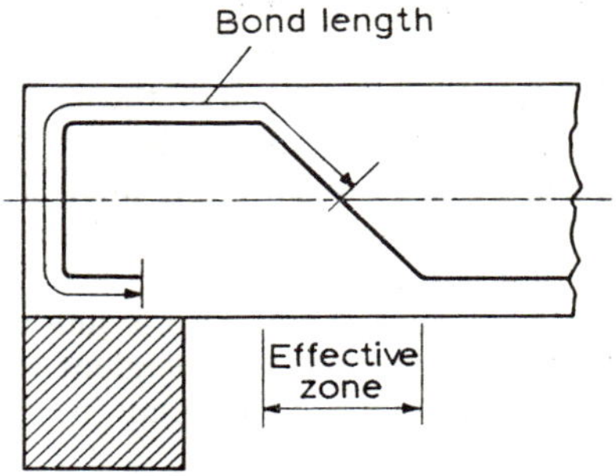

Fig 5.6.7 Development length for bent-up bar.

axis and it is not always satisfactory to simply hook the stirrups over longitudinal bars as in figures 5.6.8(*a*) and (*b*). If large diameter stirrups are used in a shallow beam then (*c*) may be necessary. Notice that (*b*) only provides additional bond length for one arm: it is the most commonly used stirrup, as compared with (*a*), because of its convenience in giving increased rigidity to the cage of reinforcement before concreting.

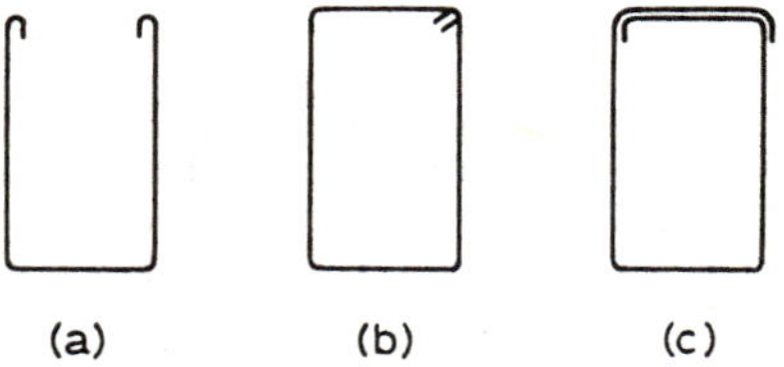

Fig 5.6.8 Various stirrups with differing bond lengths.

Concentrated loads on corbels

If a concentrated load is applied near to a support the concentrated load itself can inhibit failure on an inclined plane due to the vertical compressive stress which is present. The increased shear stresses permitted by the Unified Code are given in appendix C. The method of loading a member is considered further in the next section.

5.7 Failure of beams with shear reinforcement

It has been assumed that the main function of shear reinforcement is to accommodate the excessive diagonal tension which can be developed in the concrete. The way in which the formulae have been derived was based upon this principle. It is now necessary to study the actual behaviour of beams subjected to combined bending and shear in order to re-assess the justification of this basic assumption.

Stress redistribution

Redistribution of stress must take place once cracking occurs. If a shear crack develops as shown in figure 5.6.2(*a*) then the average shear stress across the remaining portion of the section, above point O, must be greater than that before the crack formed. The magnitude of this stress increase depends on the shape of the crack and the characteristics of the beam, with the bond stress adjacent to the cracked section reaching some maximum value.

Dowel action

Notice that the shearing force at the section gives rise to dowel action. The dowel force combined with the flexural tension force produces a resultant which may cause yielding of the main reinforcement.

Bond splitting

This dowel force can also cause "bond splitting" of the concrete at the level of the reinforcement, causing the bond strength between concrete and steel to be greatly reduced over an appreciable length. Unless the original bond lengths are considerably in excess of the minimum values considered adequate, premature flexural failure can occur because of bond failure accelerated by shear.

Method of load application

The usual method of loading a beam in the laboratory can give shear strengths that are extremely high compared with those for an actual structure. This is because the loads applied to the beam in practice are usually either uniformly distributed, or transmitted by secondary beams which *frame into* the member. A laboratory test,

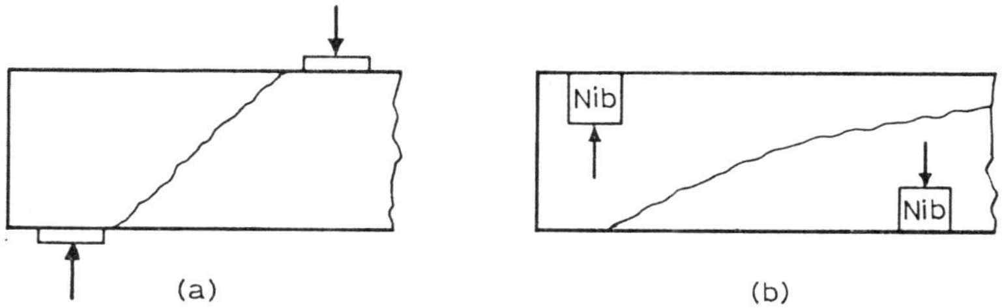

Fig 5.7.1 Different modes of failure due to different methods of load application.

however, usually applies one or two point loads via plates placed on the top surface of the beam, with supports pressing against the bottom surface. The latter method induces vertical compressive stresses which can greatly enhance the shear resistance of the beam, especially at the critical sections of short spans (see figure 5.7.1). The permissible shear stresses given in some of the earlier codes of practice were based on laboratory tests using point loads as in figure 5.7.1(*a*). They were therefore too high and some catastrophic failures occurred. The lesson to be learnt from these failures is that the testing techniques used in the laboratory should be reasonably representative of the conditions in practice.

Modes of failure

The various types of failure due to the combined action of shear and flexure are as follows:

Diagonal tension	(i) Shear-compression (ii) Constrained shear
Bending at a diagonal plane	(i) Shear-bond (ii) Shear-tension
Diagonal compression	(i) Web failure (ii) Support failure

Different authorities sometimes use different terms or attach different meanings to particular terms. However, the terms given above together with the meanings given below will be used here since they appear to give the most satisfactory description of the various modes of failure.

Diagonal tension

Failure due to diagonal tension can occur in two ways as follows:

(i) *Shear compression* The shear crack is inclined and is frequently a continuation of a normal bending crack. Failure eventually occurs with the extension of the crack becoming almost horizontal before turning up again into the compression zone as the concrete crushes in a manner similar to that in a flexural failure. A shear crack develops as an extension of a vertical crack which turns into an inclined crack near the neutral axis (see point 1 in figure 5.7.2), and propagation upwards into the zone of compression becomes flatter. The crack then tends to "crack back" with further increase in load, as shown by point 2. Propagation is usually gradual until some point (3) is reached, whereupon sudden failure occurs.

(ii) *Constrained shear* A diagonal crack is observed in the region of the neutral axis, and as the failing load is reached the crack suddenly extends up and through the compression zone with extreme violence. This type of failure occurs if concentrated loads are applied at the top (or bottom) of the beam on a short span as shown in figure 5.7.1(*a*). This form of failure is associated with very high shear stresses, which rarely occur in practice except for corbels, etc.

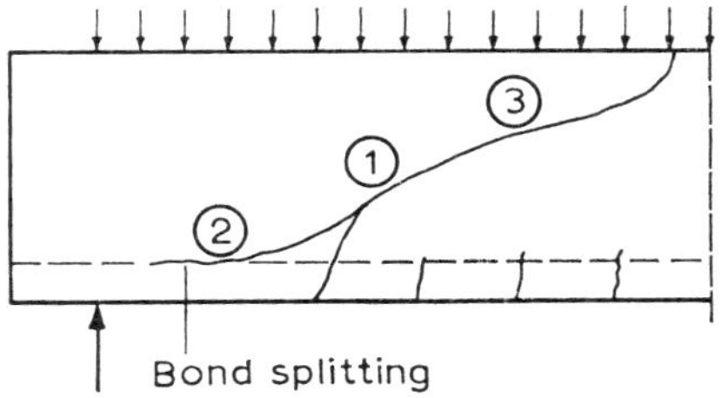

Fig 5.7.2 Typical shear-compression failure.

Bending at a diagonal plane

The diagonal crack widens due to the rotation of each portion about the point O shown in figure 5.6.2(*a*) until the steel which crosses the crack is no longer able to prevent a flexural failure at a diagonal plane. Flexural failure on a diagonal plane can occur in one of two ways:

(i) *Shear bond* Bond splitting occurs along the level of the reinforcement due to dowel action: the reduced effective bond length produces premature flexural failure due to slipping of the main reinforcement. Shear-bond failures can be avoided by providing nominal stirrups and adequate development lengths for the main reinforcement, as required by the Unified Code. Even very light stirrups can greatly offset the effect of reductions in bond due to splitting, by their own frictional contact force.

(ii) *Shear-tension* The main steel yields due to the combined effects of bending and dowel action. Calculations for bending at normal sections ensure that the areas of longitudinal reinforcement are satisfactory for the maximum moments. The provision of adequate development lengths for curtailed bars ensures that the flexural resistance in regions of high shear is slightly more than is necessary for pure flexure. Hence a shear-tension failure is unlikely before a flexural failure across a normal plane occurs elsewhere.

Diagonal compression

Local failure of the concrete can occur due to inclined compression as well as inclined tension. It is difficult to recognize as such since the outward manifestation of this type of failure is often inclined cracking in the same direction as before because the inclined compression acts at right angles to the inclined tension. This type of failure is likely to occur as follows:

(i) *Web failure* A T- or I-beam with a narrow web, suitably reinforced to take the inclined tension, can still fail due to crushing of the concrete under inclined compression. A limited number of tests carried out by Taylor [5.2] indicated that, for beams of a particular depth but varying cube strengths and web breadth, the nominal shear stress at the first sign of web crushing was approximately one-fifth of the cube strength. The Unified Code avoids such failures by stipulating that the shear stress should not exceed four times the value of q_c given in table C3. In general, the values of $4q_c$ are considerably less than $u_w/5\gamma_m$ and it is desirable to restrict the diagonal tension to the lower limit—i.e. $4q_c$—in order to avoid excessive cracking. However, for low cube strengths and high steel ratios, $u_w/5\gamma_m$ can be less than $4q_c$, so the lower value $(u_w/5\gamma_m)$ should be used.

(ii) *Support failure* Compression failure of the concrete can occur at the supports in short deep beams where considerable arching action can take place. This theory was first proposed by Faber

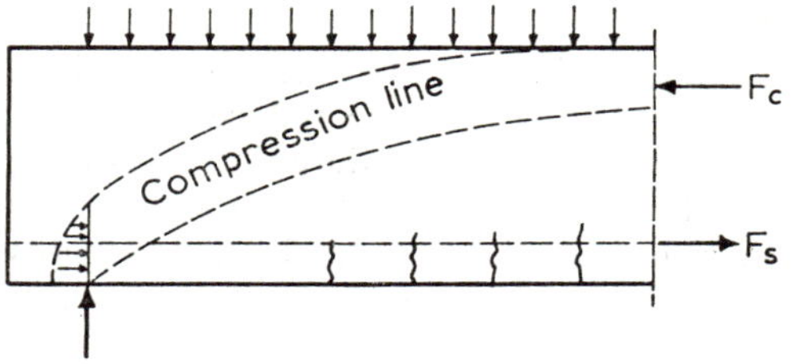

Fig 5.7.3 Possible compression failure due to arching action.

who assumed that the longitudinal reinforcement slips over its entire length to induce stresses as shown in figure 5.7.3. The compressive strength of the concrete should be checked in such cases, if necessary.

Discussion of shear failures

Shear-compression failures are the most important type to consider in design, so the earlier emphasis on diagonal tension is justified.

Provided that the designer is aware of the other possibilities, suitably prudent precautions can be taken by following the Unified Code requirements.

It is clear that although the actual behaviour of beams subjected to combined bending and shear is complex, these precautions are sufficient to eliminate failure modes other than shear-compression. The uncertainty in estimating shear compression failures is now not so very much greater than that for pure bending effects. However this larger uncertainty, together with the fact that shear failures can occur suddenly and with considerable violence, requires that the partial safety factor for material strength should still be greater for concrete in shear.

Comparison with CP114:1957

Comparison with CP114:1957 shows that the old code was relatively unsafe for beams without shear reinforcement, whereas for beams with shear reinforcement it was extravagant on steel, often requiring twice the amount that was really necessary. CP114:1957, before amendment, implied ultimate nominal shear stresses at failure of approximately 1·18 and 1·93 MN/m^2 for cube strengths of 20·7 and 41·4 MN/m^2, respectively, regardless of how the main steel was curtailed. Assuming that suitable values for r and γ_m are, say, 1% and 1·8 respectively, the Unified Code implies corresponding values of 1·08 and 1·53 MN/m^2 for $u = 22{\cdot}5$ and 45 MN/m^2 respectively. Although few failures occurred in actual structures, in the UK, this was probably due to the liberal use of nominal stirrups in detailing (which CP114 did not actually require) and a load factor which was intended to be at least 2. Users of the American Code were not generally so conservative in detailing and one or two catastrophic failures occurred before the ACI Code was amended.

REFERENCES

5.1 HUGHES, B. P. and ASH, J. E. Water gain and its effects on concrete. *Concrete* vol. 3 no. 12 Dec. 1969 pp 494–6.

5.2 TAYLOR, R. Some shear tests on reinforced concrete beams without shear reinforcement. *Magazine of Concrete Research* vol. 12 no. 36 Nov. 1960 pp 145–54.

Exercises

1. A doubly reinforced cantilever is 3 m long and carries a live point load of 20 kN at its free end. The beam measures 150 mm wide by 300 mm deep overall, and provides a dead load of 1·1 kN/m. The tension and compression bars are positioned at a distance of

50 mm from the tension and compression faces of the concrete respectively. Assume that $\gamma = 1{\cdot}6$ for live load and $1{\cdot}4$ for dead load.

(*a*) Calculate the maximum nominal shear stress developed under the design loads for the limit state of collapse. If q_c equals $1{\cdot}20$ MN/m², calculate the required pitch of two-leg vertical stirrups of 6-mm diameter steel with a characteristic strength of 250 MN/m². Is the calculated pitch necessarily satisfactory?

(*b*) Explain the meaning of the term "flexural bond strength". If the beam requires 4% tensile steel to carry the maximum bending moment, and 20-mm diameter bars are used, what flexural bond stress is developed? (UB2 1969, 40 min)

2. A beam of width 250 mm and effective depth 450 mm has a steel area of 2 500 mm². The section carries working load shear forces of 100 kN dead load and 150 kN live load. Determine the nominal shear stress if the partial safety factors are 1·4 for dead load and 1·6 for live load. If shear reinforcement is provided in the form of two-leg 12-mm diameter stirrups inclined at 45° to the horizontal, calculate the required pitch. If all the reinforcement consists of square twisted bars such that $f_{yt} = f_{yq} = 420$ MN/m², calculate the theoretical maximum "equivalent diameter" of bar which could be used if the maximum flexural bond stress is not to exceed 4·4 MN/m². Also calculate the anchorage length required for the stirrups.

3. A section of a reinforced concrete beam, 300 mm wide and of 600 mm effective depth, has a steel area of 1%. If the design shear force at this section is 450 kN, and the characteristic cube strength is 45 MN/m², calculate the required pitch of two-leg 12-mm diameter stirrups with a characteristic strength of 420 MN/m² arranged (*a*) at 45° and (*b*) vertically.

Chapter 6

Limit states of collapse

Plastic deformations occur in reinforced concrete members as they approach the limit state of collapse. Permanent deformations are produced and the basic assumption of elastic theory, that stress is proportional to strain, no longer applies. The distribution of moments throughout the structure, and the moment of resistance of each section, can only be predicted accurately by considering the actual conditions at collapse. Calculations for the limit states of collapse thus ensure an accurate margin of safety against a catastrophic failure due to overloading.

6.1 Introduction to collapse load analysis

The basic principles of limit state theory, including the general assumptions for structural analysis which apply to all limit states, have been outlined in chapter 1. Reinforced concrete as a structural medium has been introduced with special reference to elastic theory. It is now necessary to consider reinforced concrete with specific reference to the limit states of collapse. Although elastic theory has considerable application for the limit states of deflection and local damage, as discussed later, it is not valid for the limit states of collapse. Even for collapse, however, it will be seen that the assumed distribution of forces and moments *is* based on the values obtained from an elastic analysis.

General requirements at collapse

The structure must be designed so as not to collapse. Collapse of the structure, or part of the structure, could arise from rupture of one or more critical sections, from overturning, or from buckling due to elastic or plastic instability (including the effects of sway where appropriate). The design resistance to bending, shear, torsion and axial loads at every section should not be less than the appropriate value at that section produced by the most severe arrangements of design loads on the structure. For example, the maximum overturning moments due to the design loads should not exceed the counter-balancing moments. Hence notice that the partial safety factor for dead load given in appendix B for case II is as high as 1·4 where the load is assisting the overturning action, and only 0·9 where it is preventing the overturning action.

EXAMPLE 6.1.1

A continuous reinforced concrete beam, ABCD, rests on two supports B and C. BC = 6 m and cantilevers AB and CD = 2 m. The characteristic dead load (including self-weight) is 35 kN/m throughout and the characteristic imposed load is 15 kN/m throughout. Calculate the maximum and minimum span and support design moments for the limit state of collapse. Refer to appendix B.

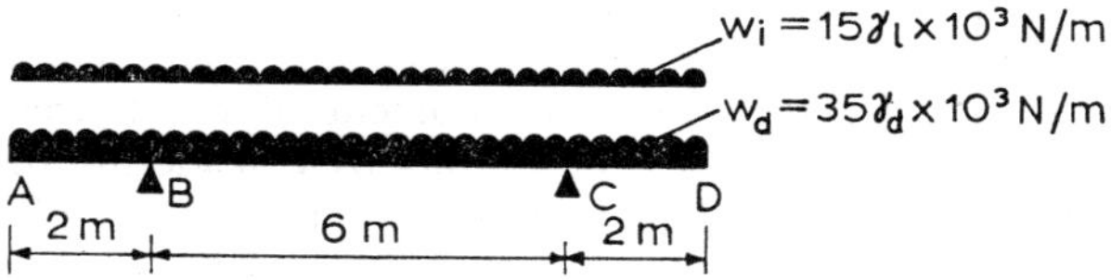

Fig 6.1.1 Example 6.1.1.

Solution

Span or cantilever loading From table B1, the partial safety factor γ_l for collapse under dead and imposed loading (case I) is 1·6 or 0 for imposed loading and 1·4 or 1·0 for dead loading. Hence the maximum value of w is

$$(1{\cdot}4 \times 35 + 1{\cdot}6 \times 15) \times 10^3 = 73 \times 10^3$$

and the minimum value is

$$1{\cdot}0 \times 35 \times 10^3 = 35 \times 10^3$$

Moments The maximum midspan moment is

$$(73 \times 6^2/8 - 35 \times 2^2/2) \times 10^3 = (328{\cdot}5 - 70) \times 10^3$$
$$= 258{\cdot}5 \text{ kNm} \qquad \text{(Answer)}$$

The minimum midspan moment is

$$(35 \times 6^2/8 - 73 \times 2^2/2) \times 10^3 = (157{\cdot}5 - 146) \times 10^3$$
$$= 11{\cdot}5 \text{ kNm} \qquad \text{(Answer)}$$

The maximum cantilever moment is 146 kNm. (Answer)

The minimum support moment is 70 kNm. (Answer)

EXAMPLE 6.1.2

Repeat example 6.1.1, if the characteristic imposed load is now only 10 kN/m for AB and BC and zero for CD.

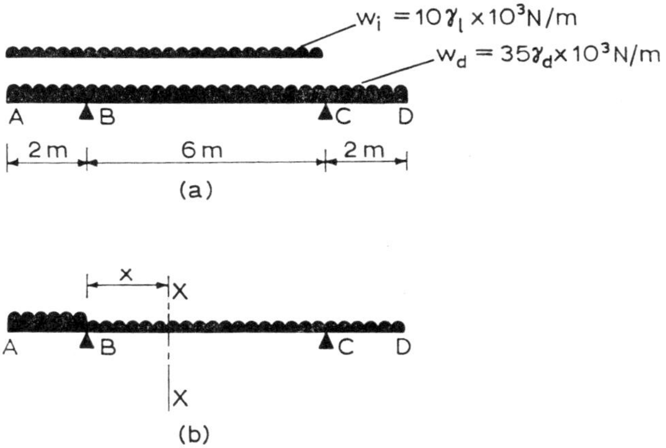

Fig 6.1.2 Example 6.1.2. (*a*) General loading conditions. (*b*) Loading for minimum span moments.

Solution For cantilever AB, the maximum value of w is

$$(1{\cdot}4 \times 35 + 1{\cdot}6 \times 10) \times 10^3 = 65 \times 10^3$$

The maximum moment (in magnitude) is

$$(65 \times 2^2/2) \times 10^3 = 130 \text{ kNm} \qquad \text{(Answer)}$$

and the minimum is

$$(35 \times 2^2/2) \times 10^3 = 70 \text{ kNm} \qquad \text{(Answer)}$$

(Notice that the maximum and minimum moments *algebraically* are -70 and -130 kNm).

For cantilever CD, the moment is 70 kNm in both cases. (Answer)
For span BC the maximum moment is

$$(65 \times 6^2/8 - 70) \times 10^3 = 222{\cdot}5 \text{ kNm} \qquad \text{(Answer)}$$

The maximum span moment for BC for the loading shown in figure 6.1.2(*b*) is no longer at midspan.

For span BC, the minimum (algebraic) moment is

$$R_B \times 6 = 65 \times 2 \times 10^3 \times 7 + 35 \times 8 \times 10^3 \times 2$$

$$R_B = (910 + 560) \times 10^3/6 = 245 \times 10^3$$

$$M_{XX} = 245 \times 10^3 x - 65 \times 10^3 \times 2 \times (x + 1) - 35 \times 10^3 \times x^2/2$$

$$= 10^3(115x - 130 - 17{\cdot}5x^2)$$

$$dM_{XX}/dx = 0 \quad \text{when} \quad 115 = 35x \quad \text{or} \quad x = 23/7$$

Therefore

$$M_{xx} = 10^3(115^2/35 - 115^2/2 \times 35 - 130) = 59 \times 10^3$$

which is positive and therefore does not induce reversed bending.

Stress-strain curves

Typical stress-strain curves for short-term loading of concrete in uniaxial compression, and steel in tension, are shown in figures 6.1.3 and 4. Typical stresses under working load conditions are contrasted

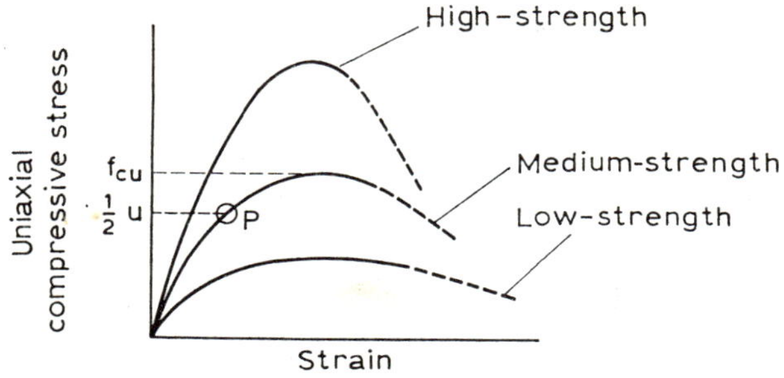

Fig 6.1.3 Typical stress-strain curves for concrete in compression.

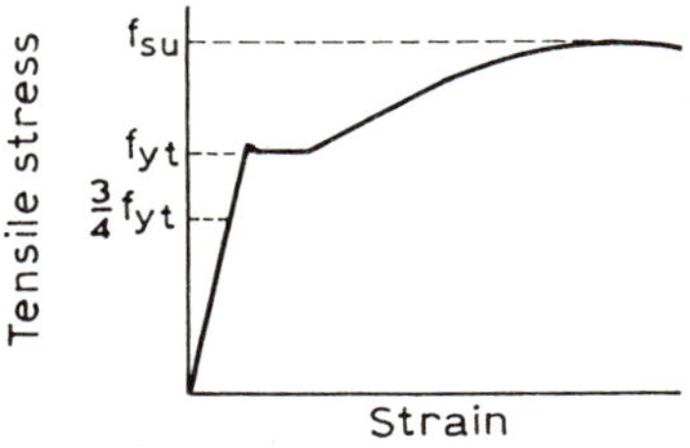

Fig 6.1.4 Typical stress-strain curve for steel in tension.

against typical stresses at collapse. Design stress-strain curves as given in the Unified Code are shown in figures D1 to 3 (appendix D). Notice that the design stress-strain curve for concrete relates to uniaxial tests. Hence it is assumed that the characteristic uniaxial strength, σ_k, is approximately $0{\cdot}78\, u_k$. The falling branch of the actual stress-strain curve in figure 6.1.3 is neglected, so the coefficient is reduced further to 0·67 to allow for this as shown in figure 6.1.5. Hence the design strength, $_*\sigma$, equals $0{\cdot}67u_k/\gamma_m = 0{\cdot}45u_k$ since $\gamma_m = 1{\cdot}5$.

Plastic theory

It is necessary to distinguish between two distinct aspects of plastic theory for reinforced concrete as follows:

(i) The increased resistance of a cross-section after its extreme fibre commences to yield.

(ii) The increased resistance of a redundant structure due to moment redistribution.

The ultimate resistances of individual sections of members are considered later in chapter 7. Limit theory and the analysis of framed structures for the limit state of collapse are discussed in the next section.

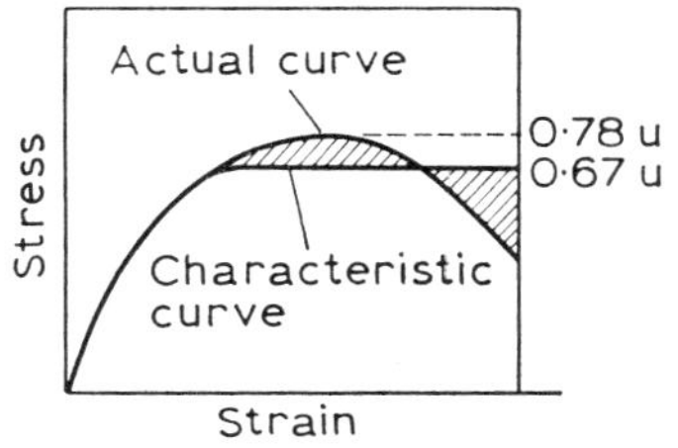

Fig 6.1.5 Fitting characteristic curve to actual stress-strain curve for concrete.

6.2 Limit theory

Limit theory refers to the changing distribution of moments which occurs throughout a structure as plastic hinges are formed. The ultimate strength of a structure can therefore be increased if redistribution of moments throughout the structure enables more sections to reach their ultimate capacity. Although the onset of yielding introduces large deflections, and should be avoided under working load conditions due to undesirable cracking and fatigue complications, the following implications of limit theory should be fully appreciated.

(i) A statistically indeterminate structure does not collapse when the first section yields.

(ii) A large reserve of strength usually exists between initial yielding and ultimate collapse of the structure.

Plastic hinge conception

The curvature of a member increases with the applied bending moment. Figure 6.2.1(a) shows a typical M-ϕ curve for an under-reinforced beam or slab, where ϕ is the change in slope over unit length of the member. The significant part of the M-ϕ curve for limit

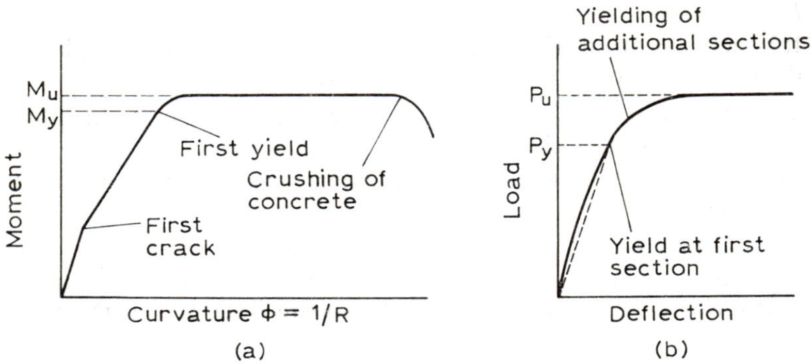

Fig 6.2.1 Typical moment-curvature and load-deflection diagrams. (*a*) Moment at a section. (*b*) Deflection of member.

design is that portion beyond the first yield moment, for which M_u remains substantially constant for a wide range of ϕ-values. This means that if a beam section develops a moment M_u it then behaves as a *plastic hinge*. It continues to resist the moment M_u, and further loading merely produces an increased curvature rather than an increased moment, so that other parts of the structure must then take the additional loading. The changes in moment produced elsewhere on the member due to the subsequent loading are just the same as if a real hinge existed at the plastic hinge point.

Collapse mechanism

The ultimate strength considered in limit theory is that which brings the structure to the verge of collapse, assuming perfect plastic hinge action. A member collapses under moment loading (other than by

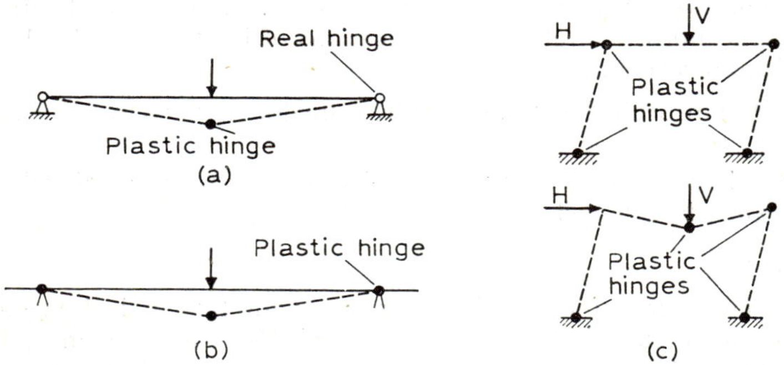

Fig 6.2.2 Typical collapse mechanisms. (*a*) Simple beam. (*b*) Continuous beam. (*c*) Two collapse mechanisms for a portal frame.

buckling) when there are sufficient plastic hinges to transform it into a mechanism. The required number of plastic hinges depends upon the original degree of redundancy, as shown in figure 6.2.2.

Moment redistribution

A study of working load conditions should usually involve an elastic analysis. Successive sections will normally pass the M_y value and reach the M_u value as increasing overloads are applied. Only a very special frame would have its maximum moments so equalized that all the plastic hinges required to form a mechanism would develop at any one loading stage and so lead to an immediate collapse. For any normal frame the collapse pattern can be radically different from the original elastic moment pattern, and can be established quite independenly of either the elastic or elasto-plastic stages.

Uniqueness theorem

The bending moment diagram which satisfies the following conditions is unique:

(i) Sufficient plastic hinges are formed to convert the structure to a mechanism.
(ii) The BMs are in equilibrium with the applied loads.
(iii) The ultimate moment is not exceeded at any point in the structure.

Upper bound (or limit) to load factor

Any arbitrary choice of hinge positions leads to an estimate of the collapse load which is greater than, or equal to, the actual. A structure which is on the verge of collapse must have developed a sufficient number of plastic hinges to change it into a mechanism. The true ultimate load for the structure can be no larger than the collapse load *for this particular mechanism.* It may be smaller for some other (more probable) mechanism. This approach therefore establishes an upper bound (or limit) on the value of the true load factor. If all possible mechanisms are investigated, the lowest collapse load is the true one. For this purpose it should be noted that the moments are all statically determinate when M_u values are used at each plastic hinge in a mechanism.

Lower bound (or limit) to load factor

Any arbitrary equilibrium state which nowhere causes a moment exceeding the ultimate leads to a collapse load which is less than, or equal to, the actual. Greenberg and Prager have shown that any moment diagram satisfying statics may be used as the basis of a lower-bound solution [6.1]. It is usually easier to develop an "arbitrary" moment diagram than it is to establish the elastic moment values. Different moment diagrams indicate different ultimate loads. Higher values of the lower bound are obtained when more and more sections for the assumed moment diagrams are at the M_u value. The moment diagram can therefore be "equalized" towards values establishing the true collapse load. The final solution can be checked by the upper-bound method above by assuming plastic hinges at the points of high moment.

Limitations

Limit theory assumes that

(i) Failure by shear or bond is prevented.

(ii) Sudden flexural compression failure is prevented.

The application of limit theory to normal reinforced concrete design has been retarded due to uncertainties concerning these assumptions. The M-ϕ curve for a concrete beam, unlike that for a mild steel I-beam, involves a gradual rather than a sudden change to plastic conditions. How is the diagonal tension and bond resistance affected by large deflections? What is the maximum permissible steel ratio consistent with an adequate plastic hinge formation? These complications have now been largely resolved (see, for example, reference 6.2). The Unified Code makes allowances for the rotational capacity of member sections and, in addition, has followed the lead of the Danish Code in permitting limit design for slabs for which the method is well suited. The application of limit theory to continuous structures is discussed in the next section.

EXAMPLE 6.2.1

According to elastic theory a single point load W at midspan of a uniform propped cantilever produces a moment of $3WL/16$ at the support and $5WL/32$ at the centre, where L is the span. Determine the additional load which can be carried assuming that the beam section has adequate rotational capacity at maximum moment and that the negative moment of resistance at the support is equal in magnitude to the positive moment of resistance at midspan.

Solution Let the ultimate moment of resistance of the beam section be M_r throughout.

The maximum load according to elastic theory is given when the maximum moment at any one section reaches the moment of resistance as shown by stage I. Hence

$$M_r = 3WL/16 \quad \text{or} \quad W = 16M_r/3L$$

If the support section is capable of yielding while still resisting its maximum moment of M_r, then the propped cantilever can sustain additional loading by subsequently acting as a simple beam. Collapse

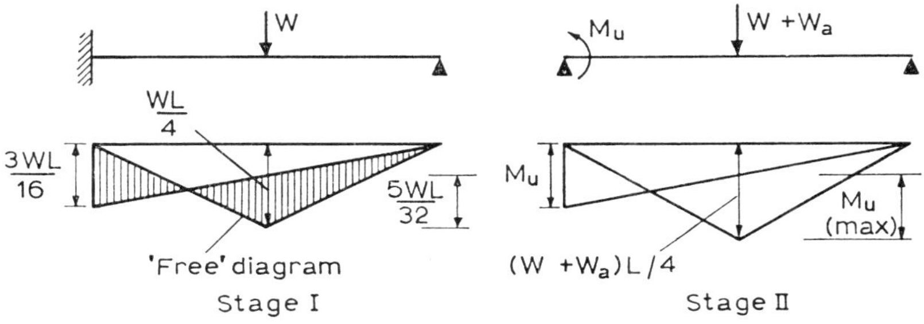

Fig 6.2.3 Example 6.2.1

finally occurs when the midspan section also reaches its maximum moment of resistance and the beam becomes a mechanism, i.e.

$$M_r = 5WL/32 + W_aL/4$$

where W_a is the additional load necessary to produce failure of the beam after the support section becomes a plastic hinge. But

$$M_r \text{ at midspan} = M_r \text{ at support} = 3WL/16$$

Hence

$$W_a = W/8 \qquad \text{(Answer)}$$

6.3 Analysis of continuous structures at collapse

The bending moments, axial loads and shear forces at sections in individual members of framed structures may be calculated by the application of elastic theory using the relative stiffnesses based on any of the sections described in section 1.8. The moments determined from an elastic analysis for any particular loading may then be redistributed by an amount not exceeding 30% of the maximum moment provided that equilibrium between the external forces and moments and the design loads is maintained, and provided that the rotational capacities of the members are adequate for shedding the excess moment at the sections concerned. In addition, the moment capacity provided at any section must nowhere be less than 70% of the undistributed moment. The ability of individual sections to deform plastically under any excess moment is considered later.

Collapse moments for continuous beams

The moments due to loading for continuous beam systems carried on simple supports, or for beams in framed structures, can be assessed in accordance with the above. Where such beams support uniformly distributed loads and are continuous over three or more spans which do not differ by more than 15% of the longest, the bending moments and shear forces which may be assumed according to the Unified Code, for load combination I only (appendix B), are given in tables 6.3.1 and 6.3.2 respectively.

Table 6.3.1 Simplified moments for continuous beams

W_v = **variable (design) load** (= $0{\cdot}4W_d + 1{\cdot}6W_i$)
W_d = **characteristic dead load**
W_i = **characteristic imposed load**
l = **effective span**

Type of load	Near middle of end span	Penultimate support	Middle of interior spans	Interior supports
Dead load	$+_*W_dl/12$	$-_*W_dl/10$	$+_*W_dl/24$	$-_*W_dl/12$
Variable load				
Maximum	$+W_vl/10$	$-W_vl/24$	$+W_vl/12$	$-W_vl/40$
Minimum	$-W_vl/50$	$-W_vl/9$	$-W_vl/24$	$-W_vl/9$

Redistribution of moments

The Unified Code requires that the redistribution of the moments must not exceed 30% of the maximum moment and that the moment capacity of any section must not be reduced to less than 70% of the maximum value for that section required by an elastic analysis. This latter requirement helps to ensure that no section is likely to suffer local damage at working loads, especially near points of contra-flexure as indicated by the bending moment diagram for the re-distributed moments.

No redistribution of moments is permitted for the moments given in table 6.3.1. More accurate determinations of the moments can be obtained using the coefficients given in appendix I and described in section 4.6. The curtailment of the reinforcement is carried out as indicated in section 5.4. The bending moment diagram given by table 6.3.1 can be used for this purpose, if desired, for approximately equal spans. The principle of redistributing moments to a certain value (e.g. 10%, 20% or 30%) according to the rotational capacity of the sections is illustrated in example 6.3.1.

Shear forces for continuous beams

The shear forces given in table 6.3.2 are satisfactory for most purposes, even when the use of coefficients is felt desirable for the bending moments.

Table 6.3.2 Simplified shear forces for continuous beams

Type of load	Outer support	Penultimate support: Outer side	Penultimate support: Inner side	Interior support
Dead load	$0{\cdot}4W_d$	$0{\cdot}6W_d$	$0{\cdot}5W_d$	$0{\cdot}5W_d$
Variable load	$0{\cdot}45W_v$	$0{\cdot}6W_v$	$0{\cdot}6W_v$	$0{\cdot}6W_v$

EXAMPLE 6.3.1

Determine the maximum moments at the midspans and the supports of a four-span beam ABCDE continuous over supports B, C and D. The spans are all equal. Consider both fixed and variable loads. Assume 10% maximum redistribution of moments and consider *either* alternate spans loaded *or* adjacent spans loaded for the variable load. Express the results in terms of coefficients (× load × span).

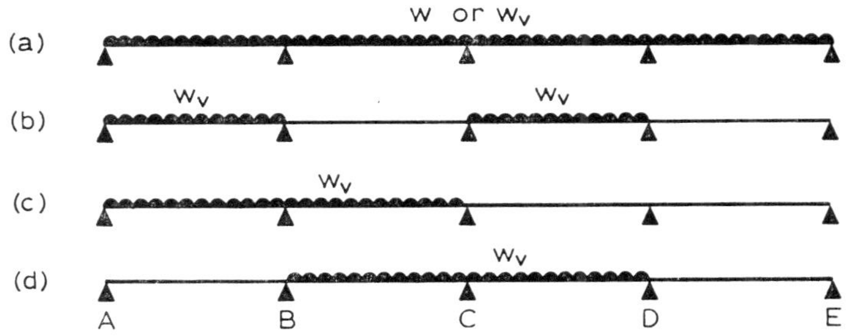

Fig 6.3.1 Example 6.3.1.

Solution The area A under the curve for the free bending moment diagram for a uniformly distributed loading w is given by

$$(2/3)(wL^2/8)L = wL^3/12$$

If L and I are common for all spans and $\bar{x} = L/2$ then the theorem of three moments given by equation 4.6.1 becomes

$$M_A + 4M_B + M_C = -\,3(A_1 + A_2)/L$$

This equation can be applied to adjacent spans in turn.

(*a*) Consider the fixed loading, w_d, with all spans loaded.

$$A_1 = A_2 = A_3 = A_4 = w_d L^3/12$$

Also

$$M_A = M_E = 0 \text{ (end supports)}$$

Hence

$$\begin{aligned} 4M_B + M_C \quad &= -\, w_d L^2/2 \\ M_B + 4M_C + M_D &= -\, w_d L^2/2 \\ M_C + 4M_D &= -\, w_d L^2/2 \end{aligned}$$

Now by observation of the above

$$4(M_B + 4M_C + M_D) - (4M_B + M_C) - (M_C + 4M_D) = 14M_C$$

Hence

$$14M_C = -\, w_d L^2 \quad \text{or} \quad M_C = -\, w_d L^2/14$$

Also

$$4M_B = -\, w_d L^2(7 - 1)/14 \quad \text{or} \quad M_B = -\, 3w_d L^2/28$$

and

$$M_D = -\, 3w_d L^2/28$$

For 10% relaxation of moments at supports,

$$M_C = -\, (0{\cdot}9/14)w_d L^2 = -\, 0{\cdot}064 w_d L^2$$

$$M_B = M_D = -\, 0{\cdot}096 w_d L^2 \qquad \text{(Answer)}$$

The corresponding midspan moments are as follows:

$$AB = DE = (0{\cdot}125 - 0{\cdot}048)w_d L^2 = 0{\cdot}077 w_d L^2 \qquad \text{(Answer)}$$

$$BC = CD = (0{\cdot}125 - 0{\cdot}080)w_d L^2 = 0{\cdot}045 w_d L^2 \qquad \text{(Answer)}$$

(*b*) Consider the variable (additional) loading, w_v, with alternate spans loaded. Applying equation 4.6.1 successively as before except for putting $A_2 = A_4 = 0$:

$$\begin{aligned} 4M_B + M_C \quad &= -\, w_v L^2/4 \\ M_B + 4M_C + M_D &= -\, w_v L^2/4 \\ M_C + 4M_D &= -\, w_v L^2/4 \end{aligned}$$

Therefore

$$M_C = -\, w_v L^2/28 = -\, 0{\cdot}036 w_v L^2$$

$$M_B = M_D = -\, 3w_v L^2/56 = -\, 0{\cdot}054 w_v L^2$$

Therefore the corresponding elastic midspan moments are as follows:

$$AB = (0{\cdot}125 - 0{\cdot}027)w_vL^2 = 0{\cdot}098w_vL^2$$

$$BC = (0 - 0{\cdot}045)w_vL^2 = -0{\cdot}045w_vL^2$$

$$CD = (0{\cdot}125 - 0{\cdot}045)w_vL^2 = 0{\cdot}080w_vL^2$$

$$DE = (0 - 0{\cdot}027)w_vL^2 = -0{\cdot}027w_vL^2$$

Therefore the elastic midspan moments (using symmetry of loading for spans BC and DE only loaded) are given by

$$AB = DE = (-0{\cdot}027 \text{ to } +0{\cdot}098)w_vL^2$$

$$BC = CD = (-0{\cdot}045 \text{ to } +0{\cdot}080)w_vL^2$$

The corresponding support moments are small. Hence the support moments need to be increased in order to reduce the span moments. That is, a relaxation of the span moments is assumed in this case since the midspan moments are relatively large. The most convenient approach is to increase the support moments by 10% and then check that the resulting midspan moments $\nless$ 90% of their unadjusted values. Thus the adjusted midspan moments are as follows:

$$AB = DE = (-0{\cdot}030 \text{ to } +0{\cdot}095)w_vL^2 \qquad \text{(Answer)}$$

$$BC = CD = (-0{\cdot}049 \text{ to } +0{\cdot}076)w_vL^2 \qquad \text{(Answer)}$$

The adjusted maximum values are $\nless 0{\cdot}9 \times$ (unadjusted values) and are therefore satisfactory.

(*c*) Consider variable loading w_v with adjacent spans AB and BC loaded. Applying equation 4.6.1 successively, as before, except for putting $A_3 = A_4 = 0$:

$$4M_B + M_C = -w_vL^2/2$$

$$M_B + 4M_C + M_D = -w_vL^2/4$$

$$M_C + 4M_D = 0$$

Therefore $14M_C = -w_vL^2(1 - \frac{1}{2})$

or

$$M_C = -w_vL^2/28 = -0{\cdot}036w_vL^2 \qquad \text{(Answer)}$$

$$4M_B = (1 - 14)w_vL^2/28 \quad \text{or} \quad M_B = -13w_vL^2/112$$

$$4M_D = +w_vL^2/28 \quad \text{or} \quad M_D = +0{\cdot}009w_vL^2 \qquad \text{(Answer)}$$

For 10% relaxation of moments at support B

$$M_B = -0{\cdot}105w_vL^2 \qquad \text{(Answer)}$$

Corresponding midspan moments are as follows:

$$\begin{aligned} AB &= (0{\cdot}125 - 0{\cdot}052)w_vL^2 = +0{\cdot}073w_vL^2 \\ BC &= (0{\cdot}125 - 0{\cdot}070)w_vL^2 = +0{\cdot}055w_vL^2 \\ CD &= (0 - 0{\cdot}014)w_vL^2 = -0{\cdot}014w_vL^2 \\ DE &= (0 + 0{\cdot}005)w_vL^2 = +0{\cdot}005w_vL^2 \end{aligned}$$

Therefore the range of midspan moments (using symmetry of loading for spans CD and DE only loaded) is given by

$$AB = DE = +0{\cdot}073w_vL^2 \qquad \text{(Answer)}$$

$$BC = CD = (-0{\cdot}014 \text{ to } +0{\cdot}055)w_vL^2 \qquad \text{(Answer)}$$

(*d*) Consider variable loading w_v with adjacent spans BC and CD loaded. Applying equation 4.6.1 successively, as before, except for putting $A_1 = A_4 = 0$:

$$\begin{aligned} 4M_B + M_C &= -w_vL^2/4 \\ M_B + 4M_C + M_D &= -w_vL^2/2 \\ M_C + 4M_D &= -w_vL^2/4 \end{aligned}$$

Therefore

$$14M_C = -w_vL^2(2 - \tfrac{1}{2}) \quad \text{or} \quad M_C = -3w_vL^2/28$$

$$4M_B = (3 - 7)w_vL^2/28 \quad \text{or} \quad M_B = M_D = -w_vL^2/28$$

For 10% relaxation of support moments:

$$M_B = M_D = -0{\cdot}032w_vL^2 \qquad \text{(Answer)}$$

$$M_C = -0{\cdot}096w_vL^2 \qquad \text{(Answer)}$$

The corresponding midspan moments are as follows:

$$AB = DE = (0 - 0{\cdot}016)w_vL^2 = -0{\cdot}016w_vL^2 \qquad \text{(Answer)}$$

$$BC = CD = (0{\cdot}125 - 0{\cdot}064)w_vL^2 = +0{\cdot}061w_vL^2 \qquad \text{(Answer)}$$

The worst moments for fixed and variable loading are therefore those shown in figure 6.3.2 *for the midspan and support sections.*

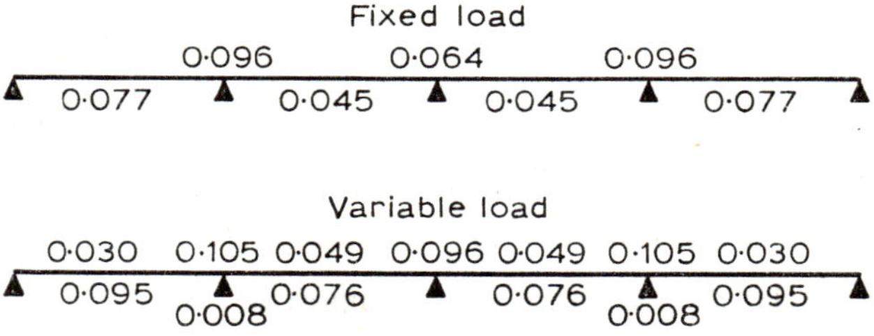

Fig 6.3.2 Example 6.3.1.

Notice that care must be taken to ensure that the different loading cases are not confused. Notice also that midspan sections were considered for simplicity. The moments for these may not equal the maximum span moments, particularly for end spans. For example, for fixed loading, maximum span moment occurs 0·404 L from end support for 10% relaxation and equals $0{\cdot}082wL^2$, which is appreciably greater than the midspan moment of $0{\cdot}077wL^2$ obtained above.

6.4 Economic aspects of moment redistribution

The Unified Code permits up to 30% redistribution of moments provided that the rotational capacity of individual sections is assured. However, assuming that the rotational capacity is satisfactory, how does one select the maximum value for the redistribution of moments in a particular instance? Is it better to design for the maximum 30% or for only 10%? This is a difficult question to answer since the choice depends mainly on the economics for a particular situation. The following example, however, indicates that it would rarely be desirable, even on economic considerations, to exceed the maximum redistribution value of 30%.

EXAMPLE 6.4.1

The "most economical" design for a continuous reinforced concrete beam could be considered to be given by either of the following alternatives:

(*a*) Adjustment of the support moments so that the area enclosed by the bending moment diagram and the axis (all areas considered positive) is a minimum.

(*b*) Adjustment of the support moments so that they are equal in magnitude to the maximum span moment.

Consider a uniformly distributed load w per unit length, acting on an encastré beam of span L. Determine the magnitude of the support moments which would provide the most economical beam according to each alternative. Hence calculate the percentage adjustment to the fixed end moments, as given by an elastic analysis, which is indicated in each case.

If economy is related to the quantity of steel can you explain why assumption (*a*) is an oversimplification which tends to overestimate the magnitude of the support moment for economical design? The Unified Code proposes that support moments may be reduced by up to 10% in general or up to 30% in certain circumstances. State, *very briefly*, the dangers which must be considered when trying to implement alternative (*b*). If maximum steel economy is in fact given by an adjustment which is intermediate between those represented by (*a*) and (*b*) do you think that a reduction in the elastic support moments of only 10% in the above encastré beam would be a very satisfactory compromise?

(UB3 1969, 45 min)

Solution

Free span moment $= wL^2/8$

End moments M_C and M_D for full fixity $= wL^2/12$

Hence let adjusted end moment $= kwL^2/12$

The maximum span moment for adjusted end moments is

$$wL^2/8 - kwL/12 = (1 - \tfrac{2}{3}k)wL^2/8$$

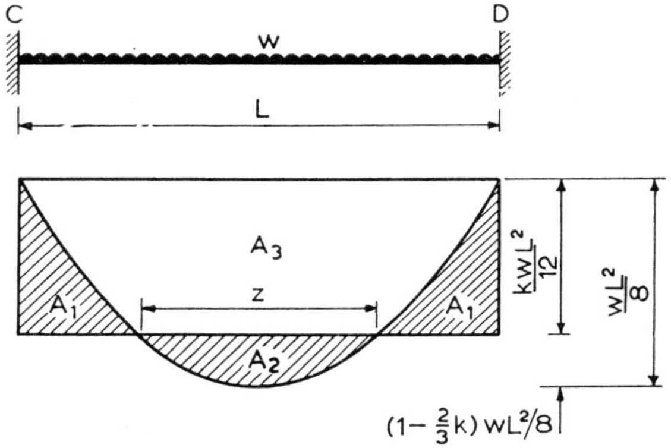

Fig 6.4.1 Example 6.4.1.

(*a*) The area enclosed between BMD and axis in figure 6.4.1 is $2A_1 + A_2$. The bending moment at a section distance x from support, M_x, is

$$\tfrac{1}{2}wx(L - x) - kwL^2/12$$

$M_x = 0$ when $x^2 - Lx + kL^2/6 = 0$, i.e. when

$$x = \tfrac{1}{2}L \pm \tfrac{1}{2}L\sqrt{(1 - \tfrac{2}{3}k)}$$

Let z = length of beam under sagging moment, hence

$$z = L\sqrt{(1 - \tfrac{2}{3}k)}$$

Now the area enclosed by a parabola is $\tfrac{2}{3} \times$ (enclosing rectangle).

Therefore

$$A_2 = \tfrac{2}{3}(1 - \tfrac{2}{3}k)wL^2z/8 = (wL^3/12)(1 - \tfrac{2}{3}k)^{3/2}$$

and

$$2A_1 = kwL^3/12 - (\tfrac{2}{3}wL^3/8 - A_2) = (wL^3/12)(k - 1) + A_2$$

i.e. the area A_3 forms a rectangle with $2A_1$ and a parabola with A_2. Therefore

$$\begin{aligned} 2A_1 + A_2 &= (wL^3/12)(k - 1) + 2A_2 \\ &= (wL^3/12)[k - 1 + 2(1 - \tfrac{2}{3}k)^{3/2}] \end{aligned}$$

$$\frac{d}{dk}(2A_1 + A_2) = 0 \quad \text{when} \quad 1 + 3(1 - \tfrac{2}{3}k)^{\frac{1}{2}}(-\tfrac{2}{3}) = 0$$

i.e. when

$$(1 - \tfrac{2}{3}k)^{\frac{1}{2}} = \tfrac{1}{2} \quad \text{or} \quad k = 9/8$$

Hence this criterion requires that the elastic end moment is *increased* by 12·5%. (Answer)

(*b*) For equal moments at supports and midspan

$$kwL^2/12 = \tfrac{1}{2} \times wL^2/8 \quad \text{or} \quad k = 3/4$$

i.e. this criterion requires that the elastic end moment is *decreased* by 25%. (Answer)

Alternative (*a*) does not allow for the fact that the excess support moment requires compression steel as well as additional tension steel. That is, the steel area per unit moment required for moments in excess of the balanced value is at least twice that for smaller moments.

Alternative (*b*) permits a "balanced" section to be used which is just adequate at both midspan and supports. The adjustment of −10% allowed by the Unified Code for all cases, and which does not introduce difficulties with the limit state of local damage, would seem to be a satisfactory compromise between the two extremes of +12·5% and −25%.

Alternative solution A simpler solution for alternative (*a*) is obtained as follows. The area of steel is assumed to vary linearly with bending moment. Hence the minimum volume of steel is obtained by equating the lengths of beam which must resist positive and negative bending moments *irrespective* of the actual form of the BMD. In figure 6.4.1, therefore, $z = 1/2$ since any vertical displacement of the axis from this position produces a net increase in $2A_1 + A_2$. Therefore

$$z = L/2 = L\sqrt{(1 - \tfrac{2}{3}k)} \quad \text{or} \quad k = 9/8 \text{ as before.}$$

REFERENCES

6.1 GREENBERG, H. J. and PRAGER, W. Limit design of beams and frames. *Transactions of the American Society of Civil Engineers* vol. 117 1952 pp 447–84.

6.2 CRANSTON, W. B. and CRACKNELL, J. A. *Tests on reinforced concrete frames. 2: Portal frames with fixed feet.* (Technical Report TRA 420, CACA, London 1969) p 35.

Exercise

1. A rectangular, reinforced concrete beam ABC is simply supported at A and C and is continuous over support B. Span AB = span BC = 9 m.

The uniformly distributed loadings and the partial safety factors for load respectively are as follows:

(Total) dead load: 8kN/m and 1·4

Live load: 12 kN/m and 1·6

(*a*) Calculate the design moments at the central support B and at midspans AB and BC assuming a maximum redistribution of moments from an elastic analysis of 10%.

(*b*) Design a suitable beam having a constant breadth of 300 mm and a constant depth throughout. You need only calculate the areas of main steel required at the support section.

Assume:

Characteristic cube strength = 22·5 MN/m^2

Characteristic steel strength f_y = 410 MN/m^2

Rectangular stress block for the concrete of 0·4u mean stress.

Partial safety factors for stress in steel = 1·15 (tension) and 1·15 + $f_y/2000$ (compression).

Redistribution of moments from an elastic analysis = 10%.

(UB3/H 1970, 45 min)

(*Note:* attempt part (*b*) only after studying chapter 8).

Chapter 7

Analysis of beam sections at collapse

Analysis concerns the determination of certain unknowns when the section (or member) is fully defined: that is, when the concrete dimensions, and the areas and positions of the steel, are known. The present chapter considers the determination of the collapse load for a given section subjected to pure bending.

7.1 Basic assumptions for individual sections

The basic assumptions which can be made when determining the collapse load for a member in bending are summarized below. It is important to visualize the section as carrying the collapse (or ultimate) load and about to fail.

(i) Plane sections remain plane in bending.

(ii) Good bond exists between concrete and steel. Hence the strain in the reinforcement, whether compressive or tensile, is given by the theoretical strain in the adjacent concrete.

(iii) The stress distribution in the concrete in flexural compression is given by the stress-strain curve for concrete in uniaxial compression.

(iv) The ultimate flexural compressive strain in the concrete at the outermost fibre provides a satisfactory failure criterion.

(v) The stress in the reinforcement is given by the appropriate stress-strain curve.

(vi) The tensile strength of the concrete is ignored.

Strain distribution

The basic assumption—already used in elastic theory—that plane sections remain plane after bending, is again made. Whilst the validity of this assumption has been verified by numerous investigators, the following limitations should be appreciated.

(i) *Imperfect bond* The linear strain distribution across a section refers to the *average* strain. Cracking occurs in the tension zone of a reinforced concrete beam at a series of localized points (since the bond strength is not infinite) so that rapid fluctuations of strain occur in both the concrete and the steel at the regions of discontinuity.

(ii) *"Deep" or "wide" beams* If the depth of a beam is more than about one-third of the span, the shear strains become significant compared with the normal strains due to bending. This results in a non-linear distribution of strain. Similarly, distortion of the plane section can also occur in beams or slabs where the applied bending moment is not uniformly distributed across the breadth of the section. Local distortions of the plane section occur in the region of concentrated loads or reactions.

Stress distribution

The stress at any point can be determined from the strain at that point and the appropriate stress-strain relation. The "appropriate" curves for concrete and steel are taken to be those in direct compression and direct tension respectively. The stress-strain curves for concrete in axial compression and pure bending have, in fact, been shown by Hognestad, Hanson and McHenry [7.1] to be similar.

The following additional assumptions affecting the stress distribution are also made.

(i) *Direct tensile stresses* Temperature and shrinkage stresses are negligible. If they are not, then separate calculations are necessary in order to take these effects into account.

(ii) *Tensile strength of concrete* All concrete in flexural tension is ignored. Calculations can be made taking the tensile strength of the concrete into account, but such refinements are rarely justified. Experimental evidence suggests that cracks can occur throughout the tensile zone and can even exist above the neutral axis if high temperature fluctuations or shrinkage effects occurred before the application of the load. Ignoring the tensile strength of the concrete partly compensates for (i) above.

Ultimate compressive strain

The compressive strain in the concrete at the extreme fibre, at collapse, is assumed to be given by the empirical value of Hognestad *et al.* or, more approximately, by a value of 3·5 millistrain.

7.2 Flexural moment at collapse

The area under the stress-strain curve for the concrete can be referred to as a *stress block*. The shape of the stress block for the compression side of the neutral axis of a beam varies according to the level of loading as shown in figure 7.2.1. The present section first describes these different shapes and then shows how the ultimate flexural moment can be related to the ultimate stress block by the introduction of suitable stress block factors. This approach is due to Hognestad, Hanson & McHenry [7.1].

Triangular stress block

It is assumed that plane sections of the beam remain plane after bending, hence the distribution of strain due to bending is linear as shown in figure 7.2.1(*a*). Initially, stress is approximately proportional to strain, hence the stress block is as shown in figure 7.2.1(*b*) for fairly small loads.

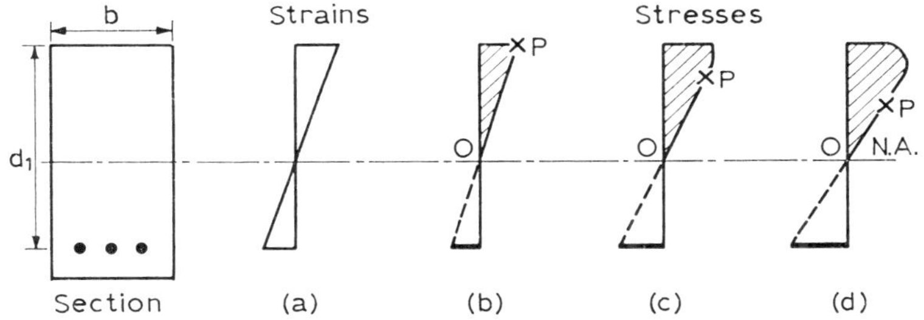

Fig 7.2.1 Strain and stress distributions for beam section subjected to an increasing moment.

Curved stress block

Higher loads produce a stress block as shown in figure 7.2.1(*c*). The strain is still proportional to the distance from the neutral axis, but since the stress-strain curve for concrete becomes curved at higher stresses, as shown in figure 6.1.3, this same curve should be plotted (to a suitable scale) as in figure 7.2.1(*c*) and (*d*).

Ultimate stress block

Just before failure the stress block becomes similar in shape to the entire stress-strain curve shown in figure 6.1.3. The point P in figure 6.1.3 is reproduced in figure 7.2.1 for comparison.

Stress block factors

Whatever the actual shape of the stress block, its area can be related by some factor K_1 to the area of the enclosing rectangle. Again, the position of the resultant concrete compressive force can be related by some factor K_2 to the depth of the neutral axis, d_n, as shown in figure 7.2.2. Similarly, let the maximum stress in the concrete be $K_3 f'_c$, where f'_c is the crushing strength of a 150 mm × 300 mm cylinder.

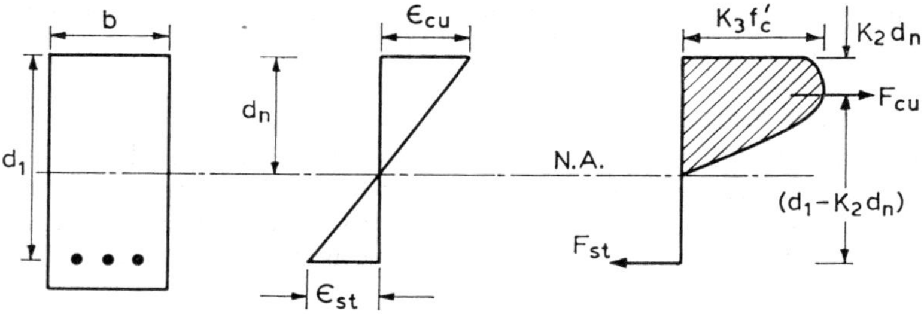

Fig 7.2.2 Strain distribution and stress block factors at collapse.

Longitudinal forces

The maximum compressive force in the concrete is given by

$$F_{cu} = K_1 K_3 f'_c b d_n$$

The tensile force in the steel is given by

$$F_{st} = f_{st} A_{st}$$

where f_{st} can be less than, equal to, or greater than f_{yt}. Equating the horizontal forces, F_{cu} and F_{st}, gives

$$d_n = f_{st} A_{st} / K_1 K_3 f'_c b \qquad (7.2.1)$$

Ultimate bending moment

The ultimate bending moment of the section is given by taking moments about the line of action of either the steel force or the concrete force:

$$M_{uc} = K_1 K_3 f'_c b d_n (d_1 - K_2 d_n) \qquad (7.2.2)$$

and

$$M_{ut} = f_{st} A_{st} (d_1 - K_2 d_n) \qquad (7.2.3)$$

Substituting for d_n from equation 7.2.1 in equation 7.2.3 and putting $M_{ut} = M_u$ gives

$$M_u = f_{st} A_{st} \left(d_1 - \frac{K_2 f_{st} A_{st}}{K_1 K_3 f'_c b} \right) \qquad (7.2.4)$$

Empirical values for the stress-block factors

Hognestad *et al.* found that age had very little effect upon the following empirical relations for the ultimate stress block. Other variables, such as the type and proportion of aggregate, were assumed to be negligible also although this may not be strictly correct.

$$K_1 K_3 = \frac{27 + 0{\cdot}35 f_c'}{22 + f_c'} \quad (f_c' \text{ in MN/m}^2 \text{ units})$$

$$K_2 = 0{\cdot}5 - f_c'/550$$

$$\varepsilon_{cu} = (4 - f_c'/45) \times 10^{-3}$$

Also, $f_c' = 0{\cdot}78u$ is a typical relation between the standard 150-mm diameter cylinder strength and the 150-mm cube strength.

7.3 Modes of failure and ultimate concrete strain

Although the ultimate flexural moment has been determined in terms of the ultimate stress block factors in equation 7.2.4, it is evident that its actual value cannot be calculated unless the steel stress f_{st} (and hence d_n) is known. It is therefore necessary to examine the alternative modes of failure which can occur.

Balanced section

A balanced section is one for which the steel yields just as the concrete reaches its ultimate strain. In other words, the beam fails simultaneously in flexural compression and tension. The position of the neutral axis in figure 7.2.1 has been shown, for convenience, to be constant at all stages of loading. While this may be approximately correct for a balanced design, it is not true in all cases.

Under-reinforced section

The steel becomes over-stressed and therefore yields. The strain increases from ε_{s1} to ε_{s2} without any increase in stress, hence the neutral axis rises and the concrete finally fails in compression as shown in figure 7.3.1(*b*). Although the final collapse occurs due to crushing of the compression zone, this is a secondary effect. The primary cause of failure is yielding of the tensile reinforcement, which allows the cracks in the beam to widen and progressively reduce the depth of the compression zone.

Over-reinforced section

The steel remains elastic while the concrete strains without increasing its moment (due to the descending portion of the concrete stress-strain curve). The steel strains elastically from ε_{s1} to ε_{s3} so that the neutral axis falls while the concrete rapidly strains to failure.

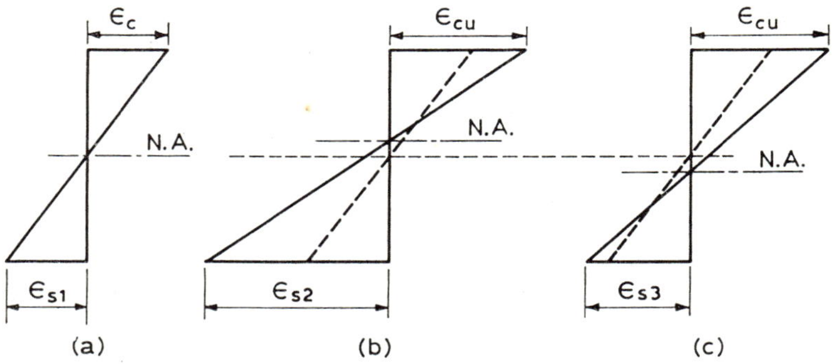

Fig 7.3.1 Variations in position of neutral axis for under- and over-reinforced sections.

Ultimate concrete strain

Regardless of the exact mode of failure, it is evident that the strain at the extreme concrete fibre reaches its ultimate value before flexural failure occurs. Hence in all cases

$$\varepsilon_{cu} = (4 - f_c'/45) \times 10^{-3}$$

as shown in figure 7.3.2.

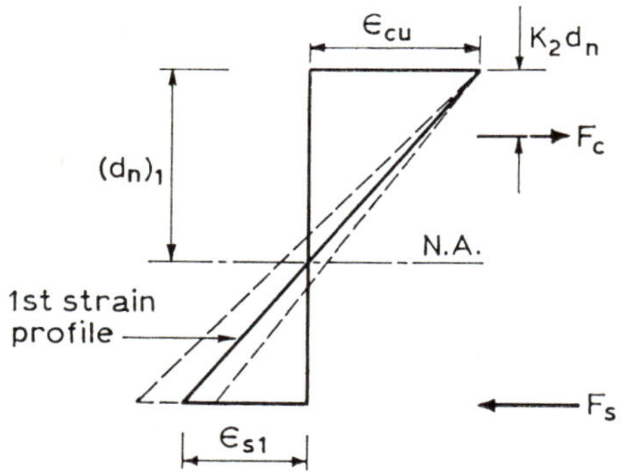

Fig 7.3.2 Tentative strain profiles for collapse.

7.4 Ultimate flexural strength analysis

If the ultimate concrete strain is eventually reached in all cases then one point on the actual strain profile for the concrete section at collapse is known. This fact can be utilized to determine the complete strain profile at failure and hence the ultimate flexural streng th.

Tentative strain profiles

If the correct strain profile at failure has a value at the extreme concrete fibre equal to the ultimate concrete strain, ε_{cu}, then the following trial and error method can be used. Draw any arbitrary strain profile using ε_{cu} as shown in figure 7.3.2. Calculate F_s from the tentative strain value ε_{s1} and the stress-strain curve for the steel. Calculate F_c from the expression $K_1K_3f_c'bd_n$. If the assumed profile is correct then F_s balances F_c. If not, repeat for different profiles until a balance is in fact obtained. The ultimate moment at collapse can then be calculated. This approach is basically simple but can be laborious. Consideration is now given, therefore, to making a reasonably sensible first estimate for d_n.

Neutral axis depth for balanced section

A balanced section is achieved if the steel yields just as the concrete reaches its ultimate strain. From the design curves:

(i) Ultimate concrete strain = 0·0035

(ii) Minimum yield strain in steel exceeds 0·002, and can be taken as approximately equal to ε_{cu} for convenience.

It follows from figure 7.4.1 and simple geometry that for a balanced section $d_n \simeq \frac{1}{2}d_1$. A more accurate, but less convenient, value would

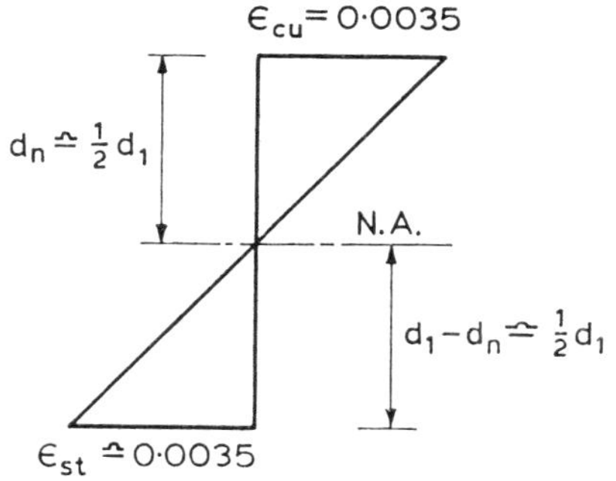

Fig 7.4.1 Strain profile at collapse for approximately balanced section.

be $d_n \simeq 0{\cdot}6d_1$. Putting $f_{st} = f_{yt}$ in equation 7.2.1 provides the equation

$$d_n = f_{yt}A_{st}/K_1K_3f_c'b$$

which is only true for a balanced section, i.e. $\frac{1}{2}d_1 < d_n < 0{\cdot}6d_1$ approximately. However, it can be used for any section in order to assess if it is over- or under-reinforced, as shown in the following procedure.

Recommended procedure for collapse-load analysis of beams

The collapse-load analysis of sections is a trial-and-error process. It is most conveniently carried out for sections in flexure as follows:

1. Determine a first approximation $(d_n)_1$ from

$$d_n = f_{yt}A_{st}/K_1K_3f'_cb$$

2. (i) If $(d_n)_1 < \frac{1}{2}d_1$ the beam is under-reinforced (i.e. A_{st} is less than balanced value).

   (ii) If $(d_n)_1 > \frac{1}{2}d_1$ the beam is over-reinforced (i.e. A_{st} exceeds balanced value).

3. (i) If under-reinforced, $(d_n)_1$ is low, since actual $f_{st} > f_{yt}$. A grossly under-reinforced beam has a value given by

$$(f_{su}/f_{yt})(d_n)_1$$

   Estimate a suitable value between these limits and check to see if forces balance for $(d_n)_2$.

   (ii) If over-reinforced, $(d_n)_1$ is high, since actual $f_{st} < f_{yt}$. If slightly high, choose a value intermediate between $(d_n)_1$ and $\frac{1}{2}d_1$. If very high, substitute

$$f_{st} = E_s\varepsilon_{st}$$

   in the equation for d_n to obtain

$$\varepsilon_{st} = Kd_n \quad \text{where}$$

$$K = K_1K_3f'_cb/A_{st}E_s$$

   From similar triangles for the strain diagram (see figure 7.4.1)

$$\varepsilon_{st}/\varepsilon_{cu} = (d_1 - d_n)/d_n$$

   or

$$(K/\varepsilon_{cu})d_n^2 + d_n - d_1 = 0$$

   and hence calculate $(d_n)_2$. See if forces balance for $(d_n)_2$.

4. If forces do not balance within 2% calculate

$$(d_n)_3 = (F_{st}/F_c)(d_n)_2$$

   (i) If under-reinforced, see if forces balance for $(d_n)_3$

   (ii) If over-reinforced, $(d_n)_3$ over-corrects, hence try mean of $(d_n)_2$ and $(d_n)_3$.

5. If forces still do not balance within 2% for $(d_n)_3$ estimate subsequent values for d_n by interpolation. This should be unnecessary, however, if reasonable estimates and slide-rule calculations have been made earlier.

6. Once the forces balance, and have therefore been determined correctly, the ultimate moment of resistance can be found by taking moments.

Compression steel

Consider the resultant concrete and steel forces in a balanced section of a doubly reinforced beam as shown in figure 7.4.2. The total tensile force in all the steel is given by

$$F_s = f_{st}A_{st} - f_{sc}A_{sc}$$

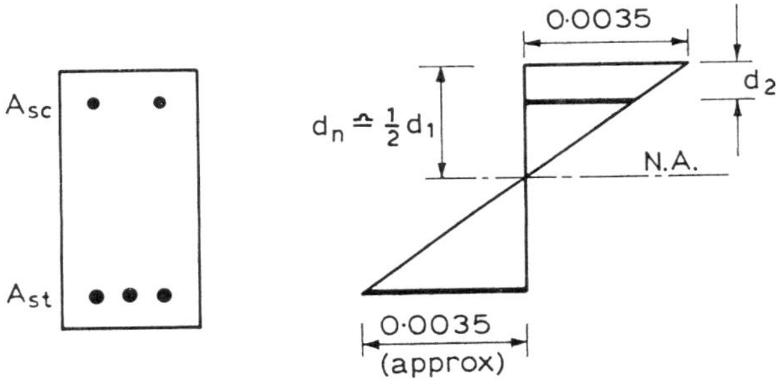

Fig 7.4.2 Strain profile for approximately balanced section of doubly reinforced beam.

If $d_2 \ngtr \frac{1}{4}d_1$ the compressive steel has a strain of 2 millistrain and f_{sc} equals f_{yc} which is rather less than f_{yt}. The total compressive force in the concrete, F_{cc}, is given by

$$F_{cc} = K_1K_3f'_c bd_n$$

Equating forces and rearranging gives

$$d_n = \frac{f_{st}A_{st} - f_{sc}A_{sc}}{K_1K_3f'_c b} \qquad (7.4.1)$$

The same general procedure given previously for singly reinforced beams can therefore be applied to doubly reinforced beams, provided that

$$\sum F_s \,(= F_{st} - F_{sc})$$

is substituted for F_{st}.

Notice that a compressive force has a positive sign when substituted into F_{sc} (since F_{sc} is specified as a compressive force) and a negative sign when substituted into F_s (in accordance with the general sign convention).

EXAMPLE 7.4.1

Calculate the collapse moment of resistance of the following beam, using the analytical approach of Hognestad, Hanson and McHenry:

Rectangular section 0·30 m deep × 0·15 m wide
Three 32-mm bars 0·20 m from compression face
Three 32-mm bars 0·25 m from compression face

The characteristic cube strength, u_k, is 37·5 MN/m². Assume that

$$f_c' = 0{\cdot}78u$$
$$K_1K_3 = (27 + 0{\cdot}35f_c')/(22 + f_c')$$
$$K_2 = 0{\cdot}5 - f_c'/550$$
$$\varepsilon_{cu} = (4 - f_c'/45) \times 10^{-3}$$

where f_c' is in MN/m² units.

The load-strain curve for 20-mm diameter bars is as shown in figure 7.4.3(*a*). The load-strain curve for 32-mm diameter bars is similar with the different ordinate scale as shown. Assume similar curves for compression up to 0·23% strain.

Solution

$$f_c' = 0{\cdot}78 \times 37{\cdot}5 \times 10^6 = 29{\cdot}2 \times 10^6$$
$$K_1K_3 = (27 + 0{\cdot}35 \times 29{\cdot}2)/(22 + 29{\cdot}2) = 0{\cdot}727$$
$$K_2 = 0{\cdot}5 - 29{\cdot}2/550 = 0{\cdot}447$$
$$\varepsilon_{cu} = (4 - 29{\cdot}2/45) \times 10^{-3} = 3{\cdot}35 \times 10^{-3}$$

Hence

$$K_1K_3f_c'b = 0{\cdot}727 \times 29{\cdot}2 \times 0{\cdot}15 \times 10^6 = 3{\cdot}18 \times 10^6$$

The force per 32-mm bar at "yield" (e.g. 2 millistrain) is 195 × 10³. If the steel is just yielding as section fails then

$$d_n = f_{yt}A_{st}/K_1K_3f_c'b = (6 \times 195/3{\cdot}18) \times 10^{3-6} = 0{\cdot}368$$

Try $(d_n)_1 = 0{\cdot}368$

For a balanced section

$$d_n = \tfrac{1}{2}d_1 = 0{\cdot}1125$$

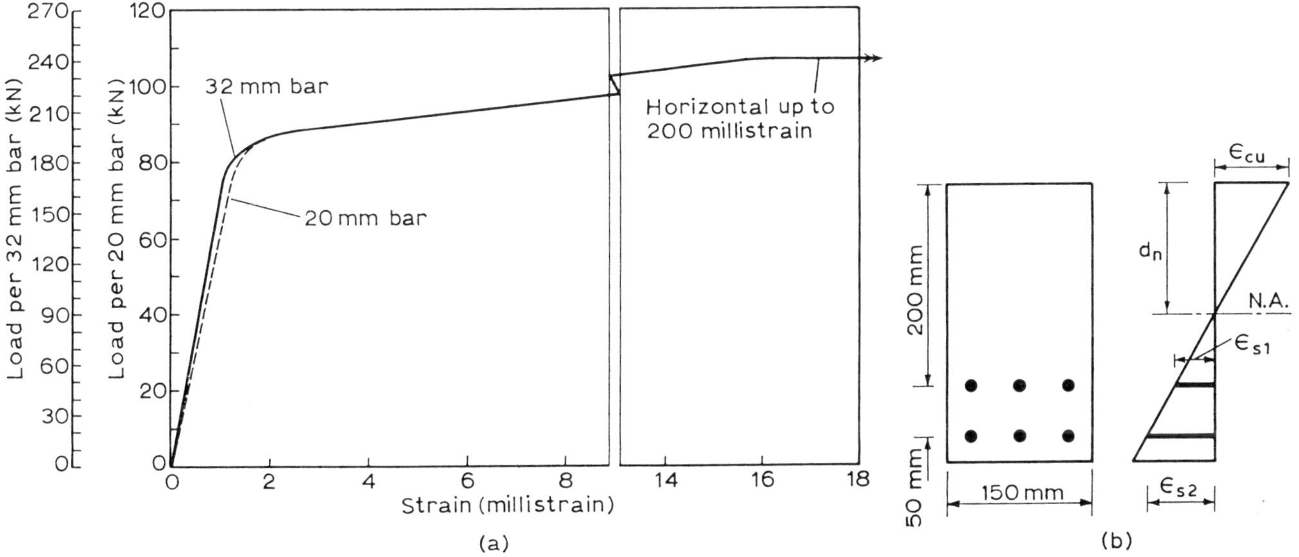

Fig 7.4.3 Example 7.4.1. (*a*) Load-strain curves for 32 mm and 20 mm bars. (*b*) Section and strain profile.

Since $(d_n)_1 \gg 0{\cdot}1125$, section is grossly over-reinforced and the steel is therefore behaving elastically. From the load-strain curve,

$$E_s = (160/\tfrac{1}{4}\pi \times 32^2 \times 1) \times 10^{3+6+3} = 200 \times 10^9$$

From similar triangles, the average steel strain is given by

$$\varepsilon_s = [(0{\cdot}225 - d_n)/d_n] \times 3{\cdot}35 \times 10^{-3}$$

For elastic steel strains:

$$F_{st} = \tfrac{1}{4}\pi \times 32^2 \times 6 \times 200 \times 10^{9-6}\varepsilon_s = 965 \times 10^6\varepsilon_s$$

But $F_{st} = F_{cu} = 3{\cdot}18 \times 10^6 d_n$. Therefore

$$(3{\cdot}18/965)d_n^2 + 3{\cdot}35 \times 10^{-3}(d_n - 0{\cdot}225) = 0$$

Hence

$$d_n = 0{\cdot}1895$$

Try $(d_n)_2 = 189{\cdot}5$ mm

$$\varepsilon_{s1} = (10{\cdot}5/189{\cdot}5) \times 3{\cdot}35 \times 10^{-3} = 0{\cdot}185 \times 10^{-3}$$

$$\varepsilon_{s2} = (60{\cdot}5/189{\cdot}5) \times 3{\cdot}35 \times 10^{-3} = 1{\cdot}07 \times 10^{-3}$$

Hence

$$\sum F_s = 3 \times (30 + 172) \times 10^3 = 606 \times 10^3$$
$$F_{cu} = 3{\cdot}18 \times 10^6 d_n = 603 \times 10^3$$

Therefore

$$\sum F_s = F_{cu} \quad \text{(within 2\%)}$$
$$K_2 d_n = 0{\cdot}447 \times 0{\cdot}1895 = 85 \times 10^{-3}$$

Therefore

$$M_u = [90(200-85) + 516(250-85)] \times 10^{3-3}$$
$$= 95{\cdot}5 \text{ kNm} \qquad \text{(Answer)}$$

T-beams

In the case of a T-beam, if d_n is less than the thickness of the compression flange, proceed as for a rectangular beam. This is the more usual case. If d_n exceeds the thickness of the compression flange an approximate solution only is possible. Divide up the compressive force as shown in figure 7.4.4 such that

$$F_c = F_{c1} - F_{c2}$$

F_{c1} = compressive force for an ordinary rectangular beam of breadth b.

F_{c2} = compressive force under the flange projections.

The stress-strain relation for the area below the soffit of the flange is often approximately linear, hence a suitable estimate for F_{c2} is given by

$$F_{c2} \geqslant \tfrac{1}{2}E_c\varepsilon_{c2}(b - b_r)(d_n - d_s)$$

where ε_{c2} is the concrete strain at the soffit of the flange at failure. Alternatively

$$F_{c2} \leqslant K_1K_3f_c'(b - b_r)(d_n - d_s)$$

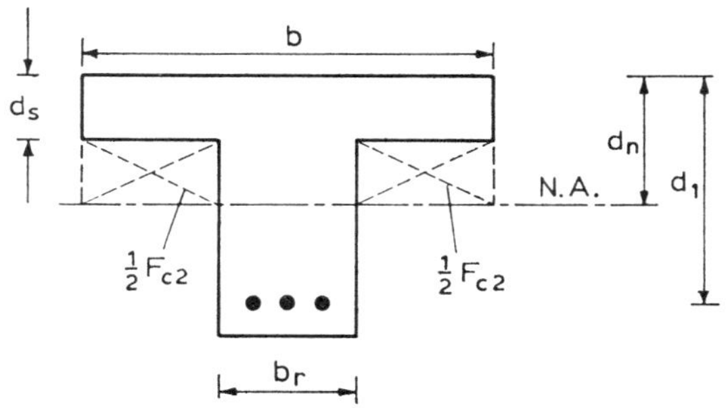

Fig 7.4.4 T-beam section.

EXAMPLE 7.4.2

Calculate the collapse moment of resistance of the following beam, using the empirical factors of Hognestad *et al.* and the load-strain curves given in example 7.4.1:

T-beam with flange 650 mm wide and 150 mm deep, overall depth 550 mm and web 150 mm wide, reinforced as follows:

Three 20-mm bars 400 mm from compression face
Three 20-mm bars 450 mm from compression face
Three 20-mm bars 500 mm from compression face

There are also four 10-mm bars (equivalent in area and strength to one 20-mm bar) 25 mm from top edge of flange. The concrete cylinder strength, f_c', is 30 MN/m^2.

Solution

$$f_c' = 30 \times 10^6$$
$$K_1K_3 = (27 + 10{\cdot}5)/(22 + 30) = 0{\cdot}721$$
$$K_2 = 0{\cdot}5 - 30/550 = 0{\cdot}445$$
$$\varepsilon_{cu} = (4 - 30/45) \times 10^{-3} = 3{\cdot}333 \times 10^{-3}$$
$$K_1K_3f_c'b = 0{\cdot}721 \times 30 \times 0{\cdot}65 \times 10^6 = 14{\cdot}06 \times 10^6$$

If $d_n < 150$ mm, beam can be considered as rectangular. The load for 20-mm diameter bar at "yield" (2 millistrain) is $86{\cdot}7 \times 10^3$.

$$f_{yt}(A_{st} - A_{sc})/K_1K_3f_c'b = [86{\cdot}7(9 - 1)/14{\cdot}06] \times 10^{3-6}$$
$$= 49{\cdot}3 \times 10^{-3}$$

Try $(d_n)_1 = 49{\cdot}3 \times 10^{-3}$

Since $\frac{1}{2}d_1 \simeq 0{\cdot}225$, section is grossly under-reinforced and is acting as a rectangular beam. The steel stress therefore reaches its ultimate strength value.

The maximum load per bar is 106·7 kN, and

$$(106{\cdot}7/86{\cdot}7)(d_n)_1 = 60{\cdot}6 \times 10^{-3}$$

Try $(d_n)_2 = 60{\cdot}6 \times 10^{-3}$

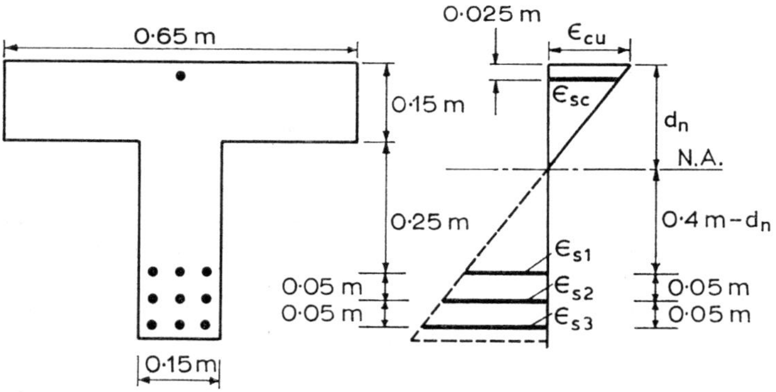

Fig 7.4.5 Example 7.4.2.

From figure 7.4.5.

$$\varepsilon_{s1} = (339{\cdot}4/60{\cdot}6) \times 3{\cdot}333 \times 10^{-3} = 18{\cdot}65 \times 10^{-3}$$

$\varepsilon_{s1} < \varepsilon_{s2} < \varepsilon_{s3}$ hence

$$\sum F_{st} = 9 \times 106{\cdot}7 \times 10^3 = 960 \times 10^3$$

$\varepsilon_{sc} = (35{\cdot}6/60{\cdot}6) \times 3{\cdot}333 \times 10^{-3} = 1{\cdot}955 \times 10^{-3}$ hence

$$F_{sc} = 86{\cdot}5 \times 10^{-3}$$

$$F_{cu} = 14{\cdot}06 \times 10^6 d_n = 852 \times 10^3$$

Therefore

$$\sum F_s = 873{\cdot}5 \times 10^3 \neq F_{cu}$$

and

$$(\sum F_s/F_{cu})(d_n)_2 = (873{\cdot}5/852) \times 60{\cdot}6 \times 10^{-3}$$
$$= 62 \times 10^{-3}$$

<u>Try $(d_n)_3 = 62 \times 10^{-3}$</u>

$$\varepsilon_{s1} = (338/62) \times 3{\cdot}333 \times 10^{-3} = 18{\cdot}18 \times 10^{-3}$$

$$\varepsilon_{s3} > \varepsilon_{s2} > \varepsilon_{s1}$$

so $\sum F_{st}$ equals 960×10^3 as before.

$$\varepsilon_{sc} = (37/62) \times 3{\cdot}333 \times 10^{-3} = 1{\cdot}99 \times 10^{-3}$$

Hence from the curve

$$F_{sc} = 86{\cdot}7 \times 10^3$$

$$F_{cu} = 14{\cdot}06 \times 10^6 d_n = 872 \times 10^3$$

Therefore $\sum F_s = 873{\cdot}3 \times 10^3$

Hence

$$\sum F_s = F_{cu} \quad \text{(within 2\%)}$$

$$K_2 d_n = 0{\cdot}445 \times 62 \times 10^{-3} = 27{\cdot}6 \times 10^{-3}$$

The depth of the resultant is

$$(872 \times 27{\cdot}6 + 86{\cdot}7 \times 25)/958{\cdot}7 \times 10^3 = 27{\cdot}4 \times 10^{-3}$$

Therefore $M_u = 959(450 - 27{\cdot}4) = 405 \times 10^3$

$= 405$ kNm (Answer)

REFERENCE

7.1 HOGNESTAD, E., HANSON, N. W., and McHENRY, D. Concrete stress distribution in ultimate strength design. *Journal of the American Concrete Institute*, vol. 27 no. 4 1955 pp 455–79.

Exercises

1. Sketch moment rotation curves for under- and over-reinforced beams subject to bending moments.*

Using the analytical method of Hognestad, Hanson and McHenry, determine the ultimate flexural moment of the beam shown in figure A. Is this beam under- or over-reinforced? The following information is given:

K_1K_3, K_2 and ε_{cu} as in exercise 2 below.

The appropriate stress-strain curve for the reinforcement is also given in figure A. The cylinder compressive strength is

$f'_c = 5\ 000\ \mathrm{lbf/in^2}\quad (= 34{\cdot}5\ \mathrm{MN/m^2})$

$1\ \mathrm{kip/in^2} = 6{\cdot}895\ \mathrm{MN/m^2}$

$1\ \mathrm{in} = 25{\cdot}4\ \mathrm{mm}$ (UB3/O 1968, 36 min)

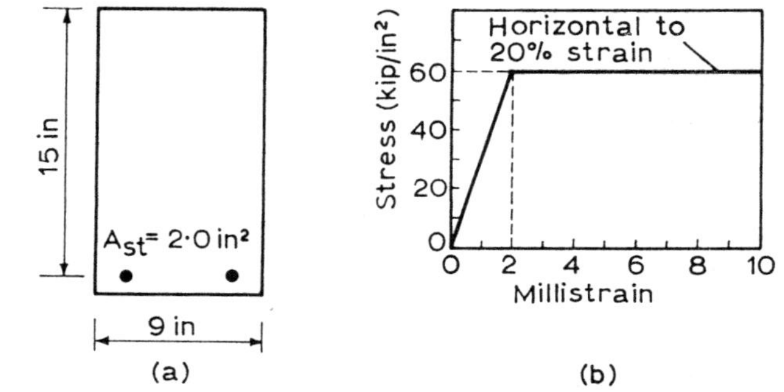

Ch. 7 Fig A Question 1. (*a*) Section (*b*) Stress-strain diagram.

2. (*a*) Determine by the method of Hognestad, Hanson and McHenry the ultimate bending moment of the beam shown in figure B. The tensile stress-strain curve for the reinforcement reaches a maximum of 420 MN/m².

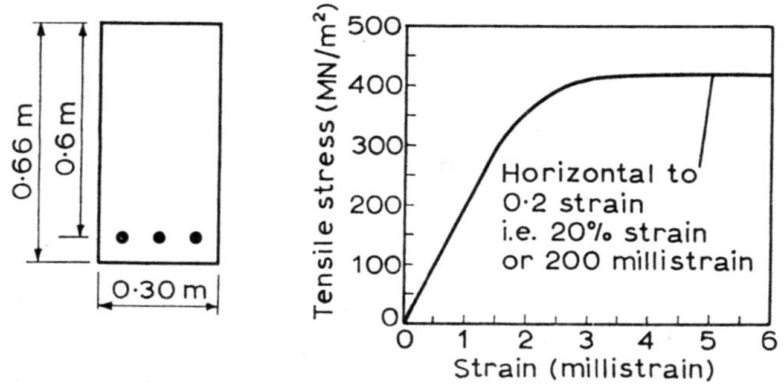

Ch. 7 Fig B Question 2.

The empirical factors (MN/m^2 units) according to the usual notation are:

$$f'_c = 0{\cdot}78u$$

$$K_1K_3 = (27 + 0{\cdot}35f'_c)/(22 + f'_c)$$

$$K_2 = 0{\cdot}5 - f'_c/550$$

$$\varepsilon_{cu} = (4 - f'_c/45) \times 10^{-3}$$

The concrete cube strength is 30 MN/m^2 and the reinforcement comprises three 32-mm diameter bars. Is this beam under- or over-reinforced?

(*b*)* Using the values of 420 MN/m^2 and 30 MN/m^2 as characteristic strengths of the steel and concrete respectively, determine the design strength of the section in bending by means of the design chart (figure E3). Why are the bending moment values different for (*a*) and (*b*)? (UB3/O 1969, 45 min)

* Refer to chapter 8 for first part of question 1 and for part (*b*) of question 2.

Chapter 8

Design of beams at collapse

The determination of a suitable section to resist a given moment is a design problem. It must not be confused with the analysis of a given section which is already fully defined.

A practical design approach can adopt one of two alternatives:

(i) Make additional (simplifying) assumptions. This approach was used in chapter 4 for elastic theory and will be considered first.

(ii) Analyse a large number of sections and use the information to draw up design charts. Once the underlying principles are fully understood and the appropriate charts are available designs can be carried out very rapidly. This is the normal method for use by an experienced designer.

8.1 General design considerations for collapse

Practical design must consider all the factors which affect the serviceability and safety of a structure. Limit state theory requires that each limit state is considered individually. This means that a section or member *designed* for one limit state is then checked by *analysing* the particular section for compliance with the requirements of other limit states. In theory it is possible to design in accordance with any limit state and then analyse in accordance with the remainder, making any adjustments that are necessary in the process. In practice it is desirable to try and reduce any subsequent adjustments for the other limit states to a minimum. The accumulated experience which is available for working load conditions greatly exceeds that which is available for structures which have failed. Hence it has been possible to evolve simple arbitrary rules which often ensure compliance with the limit states of local damage and deflection. The normal procedure will therefore be to design according to the limit states of collapse as well as complying with a number of arbitrary rules given in the Unified Code. The design will then be found, in general, to comply with the other limit states as appropriate, and subsequent adjustments or re-designs should usually be unnecessary. Design requirements for the limit state of collapse are now considered in detail.

Warning at failure

An under-reinforced beam undergoes extremely large deflections as the load approaches its ultimate value, since the steel yields and the neutral axis rises. Such a beam gives ample warning of approaching failure. An over-reinforced beam, on the other hand, fails at a relatively small deflection since the steel is still in the elastic range as the concrete crushes with explosive violence. The latter beam is therefore a potentially dangerous beam to use in any structure.

Moment-rotation curves

The moment-rotation characteristics of over- and under-reinforced concrete beams are as shown in figure 8.1.1. The over-reinforced beam has little plasticity and is typical of a brittle failure. This lack of plasticity means that if a redundant structure is constructed of such members, collapse occurs as soon as any single

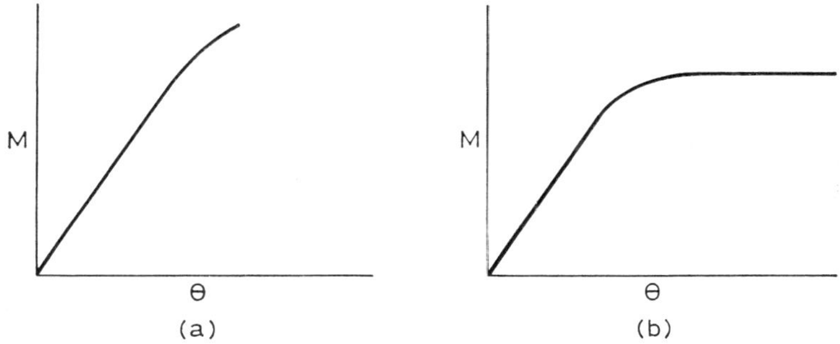

Fig 8.1.1 Moment rotation characteristics of (*a*) over- and (*b*) under-reinforced beams.

section reaches its ultimate moment. The under-reinforced beam, however, can undergo a very large deformation and still resist a moment very near the ultimate value. This therefore enables the full plastic strength of the structure to be realized.

Neutral axis position

The position of the neutral axis for a balanced section has been shown to be at approximately one-half of the depth of the beam. If it is arbitrarily stipulated that $d_n \ngtr \frac{1}{2}d_1$ then the steel strain at failure always exceeds the yield strain, and an under-reinforced (and therefore inherently safer) type of failure is assured. Furthermore, the Unified Code permits a decrease in the maximum elastic moments of up to 10% in this case: i.e. 10% redistribution of the bending moments. If d_n is made even smaller, the rotational capacity of the section is increased further. Up to 30% redistribution of moments throughout a structure is allowed if d_n is reduced to $0{\cdot}3d_1$.

Steel stress

The stress-strain curve for the steel is of the form shown in figure 6.1.4. If the steel *strain* at failure exceeds the yield strain then, for all but the greatly under-reinforced beams, the steel *stress* at failure equals the yield stress. The slightly higher stress value obtained for the relatively few beams where the steel strain becomes very much greater can reasonably be ignored, however, since such variations are on the safe side. This has already been done in the design curve shown in figure D2.

Concrete stress block

For design purposes it is convenient to assume a stress block shape which is as simple as possible, while still giving results which are near to the actual. Many different shapes have been suggested from time to time, such as an ellipse or a trapezium, but the one which formed the basis for earlier British and American Codes was the rectangular shape first proposed by C. S. Whitney in 1937 [8.1].

Comparison between theory and design

Reference to figure 8.1.2 indicates that the rectangular stress blocks have, in effect, given simple values to the constant factors in the

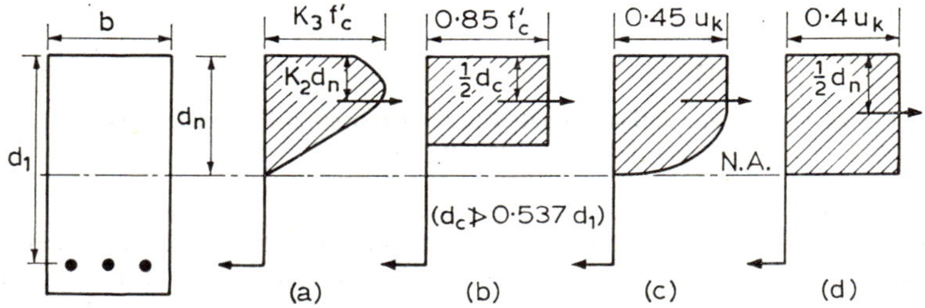

Fig 8.1.2 Stress blocks according to (*a*) Hognestad (*b*) Whitney (*c*) Unified Code and (*d*) Unified Code (simplified).

theory of Hognestad *et al.* For example, the simplified stress block assumed by the Unified Code is rectangular with an average stress value of $0{\cdot}4u_k$ and the depth of the resultant is simply $\frac{1}{2}d_n$. The ultimate moment values obtained are similar to Hognestad's more precise values *provided that the beams are under-reinforced.*

Unified Code requirements

The basic assumptions of the Unified Code when using the simplified design approach are as follows:

(i) The concrete compressive stress block at failure is rectangular, of depth d_n and average stress $0{\cdot}4u_k$.

(ii) The tension steel stress at failure is the upper design yield stress of $f_{yt}/1{\cdot}15$ for sections in pure flexure.
(iii) In all cases $d_n \ngtr \frac{1}{2}d_1$.
(iv) A redistribution of moments of up to 10% may be assumed in all cases.
(v) If a redistribution of moments of between 10 and 30% is assumed then

$$n_1 + (p'' - 10)/100 \ngtr \tfrac{1}{2} \quad \text{i.e.} \quad n_1 + p''/100 \ngtr 0{\cdot}6$$

where p'' is the percentage decrease in moment due to redistribution.

In general a redistribution of up to 30% can be assumed provided that equilibrium between the external forces and moments and the design loads is maintained. A maximum of 10% redistribution only can be assumed for structures having four or more storeys and which rely on the columns for stability.

8.2 Design of singly reinforced concrete beams

Once the design assumptions have been made, the moments of resistance can be determined by equating the longitudinal forces and taking moments about the lines of action of the steel and concrete forces.

Moment of resistance of steel

Using the simplified assumptions of the Unified Code and equating longitudinal forces then, from figure 8.2.1,

$$0{\cdot}4u_k b d_n = 0{\cdot}87 f_{yt} A_{st}$$

$$d_n \simeq 2{\cdot}2 f_{yt} A_{st}/u_k b \tag{8.2.1}$$

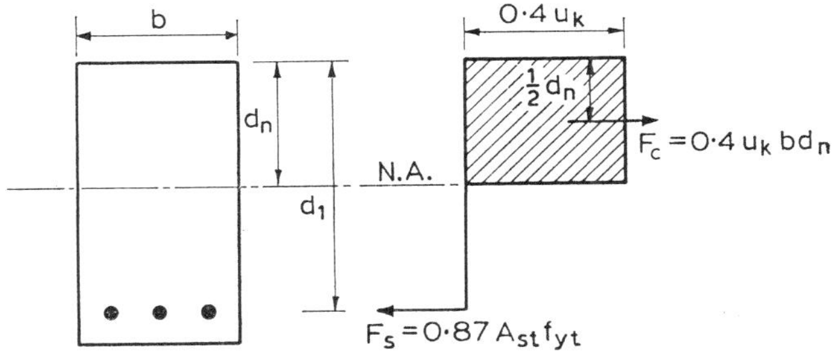

Fig 8.2.1 Simplified Unified Code assumptions for singly reinforced section.

Taking moments about the concrete force gives

$$M_{rt} = 0{\cdot}87 f_{yt} A_{st}(d_1 - \tfrac{1}{2}d_n)$$
$$= 0{\cdot}87 f_{yt} A_{st}(d_1 - 1{\cdot}1 f_{yt} A_{st}/u_k b) \qquad (8.2.2)$$

Since $A_{st} = rbd_1$ equation 8.2.2 can be expressed as

$$\frac{1{\cdot}15 M_{rt}}{bd_1{}^2 f_{yt}} = r\left(1 - \frac{1{\cdot}1 r f_{yt}}{u_k}\right) \qquad (8.2.3)$$

Equations 8.2.1 to 3 could have been obtained by substituting

$$f_{st} = 0{\cdot}87 f_{yt} \quad K_2 = \tfrac{1}{2} \quad K_1 K_3 f_c' = 0{\cdot}4 u_k \quad \text{and} \quad M_u = M_{rt}$$

in equations 7.2.1 and 7.2.4.

EXAMPLE 8.2.1

Calculate the collapse moment of resistance of the following beam, using the method given in the Unified Code.

Rectangular section 300 mm deep and 150 mm wide,
Four 10-mm bars 250 mm from compressive face
Four 10-mm bars 275 mm from compressive face

The minimum cube strength is 30 MN/m² and the characteristic yield stress for the steel is 420 MN/m² for tension.

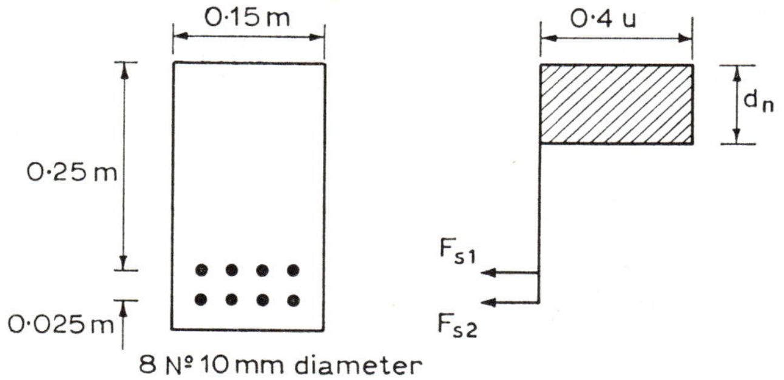

Fig 8.2.2 Example 8.2.1.

Is the assumed stress block satisfactory (i.e. is it on the safe side) if the accurate analysis of Hognestad *et al.* indicates an ultimate moment of resistance of 57·6 kNm? What is the percentage difference?

Solution

$$_{*}f_{yt} = 420 \times 10^6/1{\cdot}15 = 365 \times 10^6$$

Therefore

$$F_{s1} = F_{s2} = 4 \times \tfrac{1}{4}\pi \times 10^2 \times 365 \times 10^{-6+6} = 114{\cdot}7 \times 10^3$$

$$u_k = 30 \times 10^6$$

hence by equating forces and transposing

$$d_n = (2 \times 114{\cdot}7/0{\cdot}15 \times 0{\cdot}4 \times 30) \times 10^{3-6} = 127{\cdot}5 \times 10^{-3}$$

Therefore

$$M_r = 114{\cdot}7(250 + 275 - 127{\cdot}5) \times 10^{3-3} = 45{\cdot}6 \times 10^3$$

The percentage difference is

$$100 \times (45{\cdot}6 - 57{\cdot}6)/57{\cdot}6 = -20{\cdot}8\% \qquad \text{(Answer)}$$

(Notice that the partial safety factor of 1·15 gives a reduction of 1 − 0·87, i.e. 13%, and therefore accounts for most of the difference.)

Maximum steel requirement

The requirement that $d_n \ngtr \frac{1}{2}d_1$ can be rewritten by substituting for d_n from equation 8.2.1 to give

$$2{\cdot}2f_{yt}A_{st}/u_k b \ngtr \tfrac{1}{2}d_1$$

Since $A_{st} = rbd_1$:

$$r \ngtr u_k/4{\cdot}4f_{yt} \qquad (8.2.4)$$

The limitation concerning the neutral axis position is therefore directly equivalent to limiting the amount of steel that can be used. If f_c' equals $0{\cdot}78u$ then the condition becomes

$$r \ngtr 0{\cdot}292f_c'/f_{yt}$$

which compares with

$$r \ngtr 0{\cdot}40f_c'/f_{yt}$$

in the American Code for f_c' values less than 5000 lbf/in^2 (34·5 MN/m^2).

Balanced collapse moment

The balanced condition is theoretically given when failure occurs simultaneously in flexural tension and compression. For practical purposes, using the simplified design assumption of a rectangular stress block, the term refers to the beam containing sufficient reinforcement to give the maximum permissible value of d_n. The "balanced" collapse moment is therefore given by putting d_n equal to $\frac{1}{2}d_1$ and considering the moment of resistance of the concrete, i.e.:

$$M_{rc} = 0{\cdot}4u_k b(\tfrac{1}{2}d_1)(\tfrac{3}{4}d_1) = 0{\cdot}15u_k bd_1{}^2 \qquad (8.2.5)$$

Alternatively, this result could be obtained by substituting

$$r = u_k/4{\cdot}4f_{yt}$$

in equation 8.2.3.

Similarly, for cylinder strengths up to and including 5000 lbf/in^2 (34·5 MN/m^2) the American Code gives:

$$M_{rc} = 0{\cdot}306f_c' bd_1{}^2 \simeq 0{\cdot}24u_k bd_1{}^2$$

It is clear that the maximum tension-steel ratio permitted by the Unified Code ensures that all sections are appreciably under-reinforced even if only 10% moment redistribution is assumed. This is a deliberate policy to ensure that failures in bending are always due to flexural tension, and are therefore gradual.

Collapse design and theory

Table 8.2.1 compares the values obtained by using the design assumptions of the Unified Code with the analytical method of Hognestad *et al.* Notice that while both the line of action of the resultant and the average stress values of the Code differ appreciably

Table 8.2.1 Comparison between design and analysis for reinforcing steel with characteristic yield strength of 420 MN/m^2

Hognestad *et al.*		Unified Code (design chart)		Unified Code (simplified)	
Characteristic cube strength (MN/m^2)					
22·5	30	22·5	30	22·5	30
Average stress factor (0·78 K_1K_3 *for Hognestad* et al.)					
0·653	0·604	0·409	0·403	0·4	0·4
Depth factor of resultant (K_2 *for Hognestad* et al.)					
0·468	0·457	0·457	0·449	0·5	0·5
$M_u/bd_1{}^2$ $r = 1{\cdot}00\%$					
3·74	3·76	3·0	3·15	2·91	3·09
$M_u/bd_1{}^2$ $n_1 = 0{\cdot}5$					
5·64 (r=1·75%)	6·99 (r=2·16%)	3·5 (r=1·23%)	4·6 (r=1·64%)	3·37 (r=1·23%)	4·50 (r=1·64%)

from Hognestad's, the expression M_r/bd_1^2 (or M_u/bd_1^2) is not nearly so variable since the failure is primarily due to the steel: the smaller the steel ratio, the smaller the difference. The maximum error occurs at the maximum permissible values of r. However the error is on the safe side and is therefore very satisfactory.

EXAMPLE 8.2.2

Design a simply supported beam of 3·6 m span to carry a rolling point load of 36 kN (neglect dead weight). Assume that

$$u = 37{\cdot}5 \text{ MN/m}^2 \quad f_{yt} = 250 \text{ MN/m}^2$$

Use the partial safety factors given in appendix B. Assume a breadth of 150 mm and consider only the midspan section in bending.

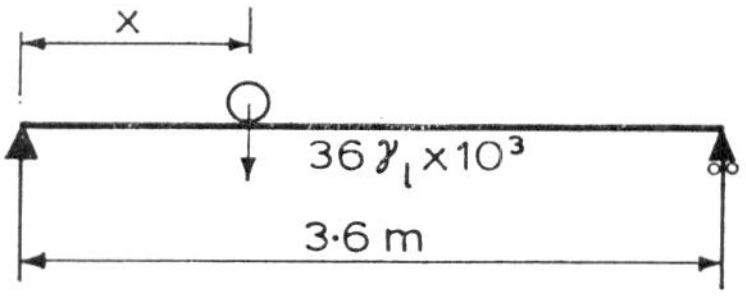

Fig 8.2.3 Example 8.2.2.

Solution

$$u = 37{\cdot}5 \text{ MN/m}^2$$

$$f_{yt} = 250 \text{ MN/m}^2$$

$_*M$ at midspan is

$$\tfrac{1}{4} \times 1{\cdot}6 \times 36 \times 10^3 \times 3{\cdot}6 = 51{\cdot}8 \times 10^3$$

Therefore

$$0{\cdot}15ubd_1^2 = 51{\cdot}8 \times 10^3$$

$$d_1 = \sqrt{\frac{51{\cdot}8 \times 10^{3-6}}{0{\cdot}15 \times 37{\cdot}5 \times 0{\cdot}15}} = 0{\cdot}248 = \text{(say) } 250 \text{ mm}$$

$$A_{st} \times 0{\cdot}87f_{yt} \times \tfrac{3}{4} \times 0{\cdot}250 = 51{\cdot}8 \times 10^3$$

$$A_{st} = 1{\cdot}27 \times 10^{-3} = 1270 \text{ mm}^2 \qquad \text{(Answer)}$$

8.3 Design of doubly reinforced rectangular beams

Doubly reinforced rectangular beams may be used for two reasons:

(i) The applied moment exceeds the maximum permissible for a singly reinforced beam, because the size of the beam is restricted. The moment in excess of $0{\cdot}15u_kbd_1^2$ must therefore be taken by compression steel and additional tension steel.

(ii) The applied moment does not exceed the maximum permissible for a singly reinforced beam, although compression steel is already present for other reasons, so the full compressive strength of the concrete is not utilized.

Additional longitudinal forces

Consider case (i) first. Let A'_{st} be the additional area of tension steel required, and A_{sc} the area of compression steel. Equating the additional longitudinal forces gives

$$A'_{st} \cdot {}_*f_{yt} = A_{sc} \cdot {}_*f_{yc}$$

or

$$A'_{st}f_{yt}/1{\cdot}15 = 2000A_{sc}f_{yc}/(2300 + f_{yc}) \qquad (8.3.1)$$

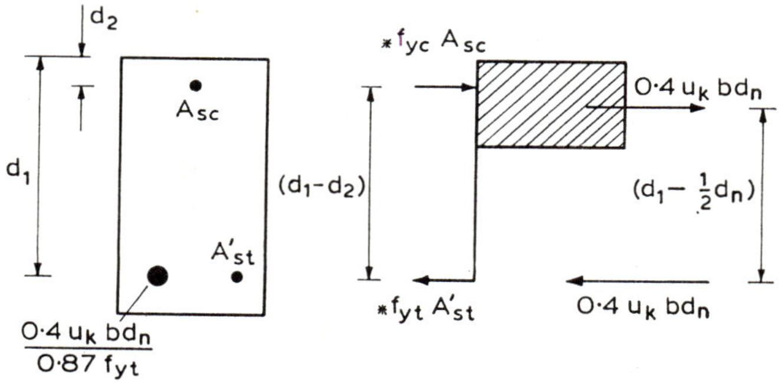

Fig 8.3.1 Simplified Unified Code assumptions for doubly reinforced section.

If the same steel is used in tension and compression then

$$f_{yt} = f_{yc} = f_y$$

where f_y is the characteristic yield strength in tension.

Notice that the term $A_{sc} \cdot {}_*f_{yc}$ should really be $A_{sc}({}_*f_{yc} - 0{\cdot}4u_k)$. However, the consequent effect is small and is allowed for since ${}_*f_{yc}$ is taken as a value which is considerably less than ${}_*f_{yt}$. Notice that f_{yc} must be in MN/m in equation 8.3.1.

Areas of steel

The area of compression steel is given by equation 8.3.1 and the total area of tension steel is given by

$$A_{st}f_{yt}/1{\cdot}15 = 0{\cdot}4u_kbd_n + 2000A_{sc}f_{yc}/(2300 + f_{yc}) \qquad (8.3.2)$$

The calculation of the additional steel for beams subjected to a high moment is therefore given simply by considering the *additional* moment to be resisted and using, in effect, the steel beam theory.

Collapse moment of resistance

The resistance moment is most conveniently given by taking moments about the tension steel. If the concrete strength is fully utilized then

$$M_r = 0{\cdot}15u_k bd_1^2 + 2000A_{sc}f_{yc}(d_1 - d_2)/(2300 + f_{yc}) \quad (8.3.3)$$

Minimum strain in compression steel

The foregoing equations assume that the strain in the compression steel reaches the minimum yield strain, i.e.:

$$d_2/d_1 \ngtr 0{\cdot}2.$$

If $d_2/d_1 > 0{\cdot}2$ then it is possible for the concrete to fail before the steel reaches its full yield value and equation 8.3.3 should not be used. In these cases the compressive stress in the steel should be obtained using assumptions (i) and (v) of section 7.1. That is, the simplified approach cannot be fully applied since it becomes necessary to use the full design stress-strain curve for the steel and not simply the design yield stress.

Beams under-reinforced in tension

Case (ii) is the less usual. However, if compression steel is present anyway, then economies in the tensile steel can be made if the presence of the compression steel is allowed for. It is no longer possible, of course, to use equation 8.3.3. If, as in case (ii), the

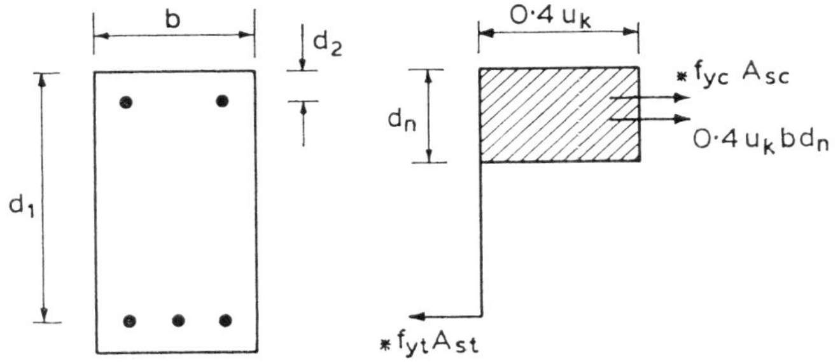

Fig 8.3.2 Doubly reinforced section which is under-reinforced.

compressive strength of the concrete is not fully utilized, then from figure 8.3.2

$$M_r = 0{\cdot}4u_k bd_n(d_1 - \tfrac{1}{2}d_n) + 2000A_{sc}f_{yc}(d_1 - d_2)/(2300 + f_{yc}) \quad (8.3.4)$$

The value of d_n can be found from equation 8.3.4 provided that the ratio d_2/d_1 is not greater than 0·2. Notice that the steel strain is more likely to be critical now that d_n is less than the balanced value. Once d_n is known, A_{st} can be obtained from equation 8.3.2.

EXAMPLE 8.3.1

Calculate the (design) collapse moment of resistance of the beam in example 8.2.1 if it also has four 10-mm bars 25 mm from the compression face. Assume that the design yield strength for the steel in compression is

$$2000f_y/(2300 + f_y)$$

where f_y is in MN/m².

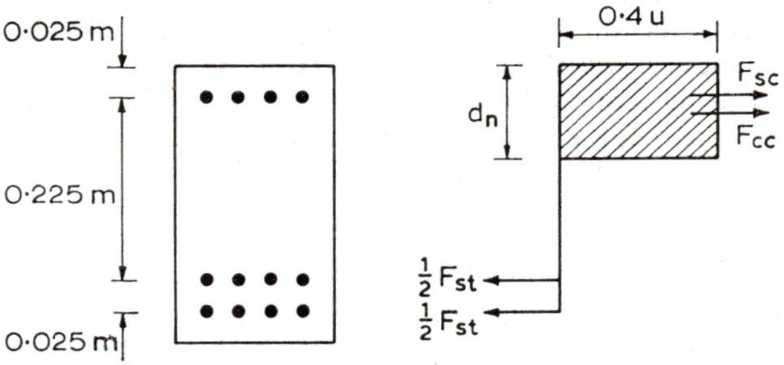

Fig 8.3.3 Example 8.3.1.

Solution (This is an analysis problem.)

$$*f_{yc} = (2000/2720) \times 420 \times 10^6 = 309 \times 10^6$$

Hence, since $d_2/d_1 \not> 0{\cdot}2$,

$$F_{sc} = \pi \times 10^2 \times 309 = 97{\cdot}0 \times 10^3$$

Equating forces:

$$0{\cdot}4u_k \times 0{\cdot}15d_n = (229{\cdot}4 - 97{\cdot}0) \times 10^3$$

or

$$d_n = 73{\cdot}6 \times 10^{-3} \quad \text{(i.e.} \gg d_2)$$

Taking moments about the compression steel:

$$M_r = 229{\cdot}4 \times 237{\cdot}5 \times 10^{3-3} - 0{\cdot}4 \times 30 \times 0{\cdot}15 \times 73{\cdot}6 \times (36{\cdot}8 - 25) \times 10^{6-3-3}$$

$$= 52{\cdot}9 \text{ kNm} \qquad \text{(Answer)}$$

EXAMPLE 8.3.2

Design a suitable section for the beam in example 6.1.1 if it is to have the same cross-section throughout. Take

$$u_k = 30 \text{ MN/m}^2 \quad \text{and} \quad f_{yt} = 420 \text{ MN/m}^2$$

and assume a breadth of 250 mm. Give the effective depth and the required areas of steel at both support and midspan sections. Assume a depth of 25 mm to any compression reinforcement.

Solution (This is a design problem)

(*a*) For balanced support section,

$$M_r = 0{\cdot}15ubd_1^2 = 0{\cdot}15 \times 30 \times 0{\cdot}25 \times 10^6 d_1^2$$
$$= 1{\cdot}125 \times 10^6 d_1^2$$

Hence

$$d_1 = \sqrt{\frac{146}{1{\cdot}125} \times 10^{3-6}} = 0{\cdot}36$$

(*b*) For balanced span section

$$d_1 = \sqrt{\left(\frac{258{\cdot}5}{1{\cdot}125} \times 10^{3-6}\right)} = 0{\cdot}48$$

A value nearer to the balanced span section is the best compromise. Hence let

$$d_1 = 0{\cdot}45$$

At support:

$$146 \times 10^3 = A_{st}(420 \times 10^6/1{\cdot}15)/(0{\cdot}45 - \tfrac{1}{2}d_n)$$

Equating forces

$$d_n = 0{\cdot}87 \times 420 A_{st} \times 10^{6-6}/0{\cdot}4 \times 30 \times 0{\cdot}25 = 123 A_{st}$$

Hence

$$61{\cdot}5 A_{st}^2 - 0{\cdot}45 A_{st} + 0{\cdot}40 \times 10^{-3} = 0$$

Therefore

$$A_{st} = \frac{0{\cdot}45 \pm \sqrt{(0{\cdot}45^2 - 4 \times 61{\cdot}5 \times 0{\cdot}40 \times 10^{-3})}}{123}$$
$$= (0{\cdot}45 \pm 0{\cdot}323)/123$$
$$= 0{\cdot}001032 \qquad \text{(positive root is inapplicable)}$$
$$= 1032 \text{ mm}^2 \qquad \text{(Answer)}$$

At midspan:

Balanced moment for section is

$$0{\cdot}15 \times 30 \times 0{\cdot}25 \times 0{\cdot}45^2 \times 10^6 = 0{\cdot}228 \times 10^6$$

A_{st} for balanced section is

$$0{\cdot}228 \times 10^{6-6}/0{\cdot}87 \times 420 \times 0{\cdot}75 \times 0{\cdot}45 = 1850 \times 10^{-6}$$

Additional moment is

$$(258{\cdot}5 - 228) \times 10^3 = 30{\cdot}5 \times 10^3$$

Therefore

$$A'_{st} = \frac{30{\cdot}5 \times 10^{3-6}}{0{\cdot}87 \times 420(0{\cdot}45 - 0{\cdot}025)} = 197 \times 10^{-6}$$

Therefore total tensile steel = 2047 mm^2 (Answer)

$$f_{yc} = 2000 \times 420 \times 10^6/(2300 + 420) = 309 \times 10^6$$

Therefore area of compressive steel is

$$A_{sc} = \frac{30{\cdot}5 \times 10^{3-6}}{0{\cdot}87 \times 309(0{\cdot}45 - 0{\cdot}025)}$$

$$= 267 \text{ mm}^2$$ (Answer)

8.4 Design curves for reinforced concrete beams

It has been shown that the simplified assumption of a rectangular concrete stress block can give reasonably accurate values of the collapse moment of resistance provided that over-reinforced sections are avoided. For repetitive and rapid design, however, it is convenient to refer to design charts. These can be based on the design stress-strain curves given in figures D1 to D3. They can therefore be slightly more accurate for under-reinforced sections and can, in addition, cover over-reinforced sections as well should these be required in certain circumstances. (Notice that an over-reinforced section *by design* may not be actually over-reinforced *in theory*, since a balanced section with n_1 equal to one half is really under-reinforced according to an accurate analysis.)

Variables in design

It is apparent from equation 8.2.3 that the unknowns when designing a singly reinforced beam are $bd_1{}^2$ and r. Hence for a given value of $bd_1{}^2$, rf_y (and hence r) can be evaluated from a general design chart as shown in figure E1 (appendix E) for given values of u_k and f_y. This chart covers all the main grades of concrete and steel and also shows the appropriate cut-off positions for neutral axis depths of 0·3, 0·4, 0·5 and $0{\cdot}6d_1$.

Design charts for doubly reinforced beams

A doubly reinforced beam can be designed by using figure E1 and the appropriate value of n_1 (e.g. 0·5) to determine M_r for the given values of u_k and f_y. A simple calculation for the *additional* moment to be resisted (see section 8.3) gives the required additional areas of tension and compression steel.

An alternative approach is to specify a value of d_2/d_1 and produce a separate design chart for each value of f_y and u_k. Typical examples of such charts are given in the Unified Code. Although very convenient in use the required number of design charts is considerable. Figure E2 and E3 are examples of such charts except that they have been made a little more general by giving a limited range of steel grades on one chart, e.g. figure E2 is suitable for $f_y = 410$, 420 or 425 MN/m^2.

EXAMPLE 8.4.1

Calculate the collapse moment of resistance of the beams in examples 8.2.1 and 8.3.1 using appendix E. What are the approximate percentage increases compared with the simplified approach?

Solution

$$u_k = 30 \times 10^6 \quad d_1 = 0{\cdot}2625$$

$$f_{yt} = 420 \times 10^6 \quad A_{st} = 100\pi \times 10^{-6} \times 2$$

$$A_{st}/bd_1 = 2 \times 100\pi \times 10^{-6+3}/0{\cdot}150 \times 262{\cdot}5 = 1{\cdot}596\%$$

(*a*) From figure E3 for $r = 1{\cdot}6\%$:

$$M_r = 4{\cdot}55 \times 150 \times 0{\cdot}2625^2 \times 10^{6-3} = 47{\cdot}1 \times 10^3 \quad \text{(Answer)}$$

The increase is

$$100 \times (47{\cdot}1 - 45{\cdot}6)/47{\cdot}1 = 3\% \quad \text{(Answer)}$$

(*b*) $A_{sc}/bd_1 = 0{\cdot}798\%$ i.e. $r_c \simeq 0{\cdot}8\%$

$$M_r = 5{\cdot}08 \times 150 \times 0{\cdot}2625^2 \times 10^{6-3} = 52{\cdot}4 \times 10^3 \quad \text{(Answer)}$$

The increase is

$$100 \times (52{\cdot}4 - 52{\cdot}3)/52{\cdot}3 = 0{\cdot}2\% \quad \text{(Answer)}$$

The reader should now repeat example 8.4.1 using figure E1.

Derivation of design charts

Design charts can be conveniently obtained by carrying out a series of calculations as follows:

1. Assume that the steel strains to some value ε_{st}. This gives a strain distribution as shown in figure 8.4.1(*a*). The values of both n_1 and ε_{sc} can be determined since a value of d_2/d_1 is selected for the latter.

2. The steel stresses can be determined from the design stress-strain curve for the steel, and figure 8.4.1(*b*) shows the concrete stress block.
3. Calculate the compressive force in the concrete for the given value of u_k. For example, if u_k is 30 MN/m² then

$$F_c = 0{\cdot}403 u_k b d_n$$

4. The concrete force equals the net tensile force in the steel. Hence for each percentage of compression reinforcement the value of A_{st}/bd_1 can be determined.

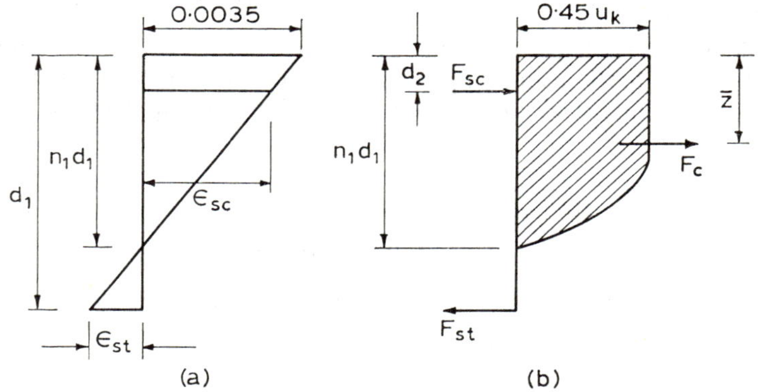

Fig 8.4.1 Stress block for deriving design charts.

5. Determine M_r/bd_1^2 by taking moments. This gives a point on the design curve for the particular value of A_{sc}/bd_1.
6. Repeat for other values of ε_{st} and compression steel ratios. This is most conveniently carried out using a digital computer programme.

EXAMPLE 8.4.2

Compare the values of the magnitude and position of the concrete force as given by the accurate and simplified methods of the Unified Code for a singly reinforced beam when

$$u_k = 30 \text{ MN/m}^2 \quad \text{and} \quad f_{yt} = 420 \text{ MN/m}^2$$

Solution The centre of the concrete compression force is determined by reference to figure D1 and figure 8.4.2 for which

$$0{\cdot}2\sqrt{(u_k)}/3{\cdot}5 = 0{\cdot}313$$

Hence for a unit width of beam:

$$F_{c1} = 0{\cdot}687d_n \times 0{\cdot}45u_k = 0{\cdot}309u_k d_n$$

$$F_{c2} = \tfrac{2}{3} \times 0{\cdot}313d_n \times 0{\cdot}45u_k = 0{\cdot}094u_k d_n$$

$$z_1 = 0{\cdot}343d_n$$

$$z_2 = (0{\cdot}687 + \tfrac{3}{8} \times 0{\cdot}313)d_n = 0{\cdot}804d_n$$

$$F_{c1} + F_{c2} = F_c = 0{\cdot}403u_k d_n$$

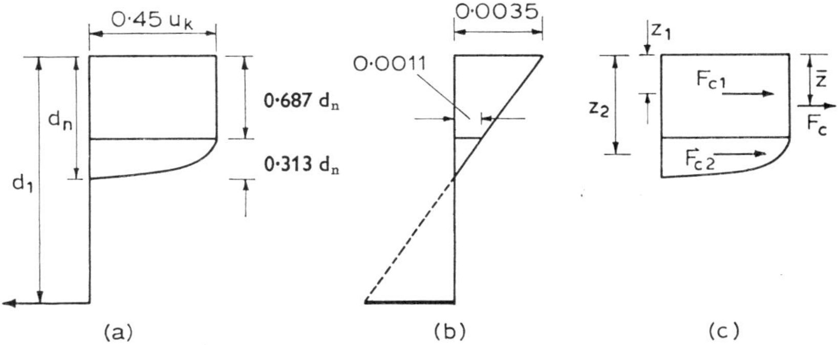

Fig 8.4.2 Example 8.4.2. (*a*) Concrete stress block. (*b*) Concrete strains. (*c*) Resultant concrete force.

This compares with $0{\cdot}4u_k d_n$ for simplified rectangular stress block.

(Answer)

$$F_c\bar{z} = F_{c1}z_1 + F_{c2}z_2$$

$$= (0{\cdot}309 \times 0{\cdot}343 + 0{\cdot}094 \times 0{\cdot}804)u_k d_n^2 = 0{\cdot}181u_k d_n^2$$

Hence

$$\bar{z} = 0{\cdot}449d_n.$$

This compares with $\bar{z} = 0{\cdot}5d_n$ for simplified rectangular stress block.

(Answer)

EXAMPLE 8.4.3

Determine the positions on the curve for $r_c = 0$ (i.e. a singly reinforced section) for the design chart given in figure E3, i.e.

$$u_k = 30 \text{ MN/m}^2 \quad f_y = 420 \text{ MN/m}^2 \quad d_2/d_1 = 0{\cdot}10$$

when the steel just reaches (*a*) lower yield and (*b*) upper yield.

Solution

(*a*) If steel has a stress of

$$0{\cdot}8f_{yt}/1{\cdot}15 = 292 \times 10^6$$

then steel strain equals $1{\cdot}46 \times 10^{-3}$ from figure D2, and the strain distribution is given by

$$n_1 = 3{\cdot}5/(3{\cdot}5 + 1{\cdot}46) = 0{\cdot}706$$

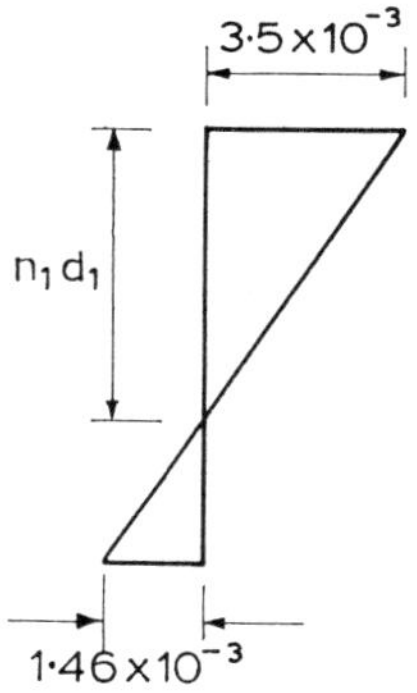

Fig 8.4.3 Example 8.4.3.

This point has therefore been excluded from figure E3 since n_1 exceeds 0·6. However the method of calculation is as follows:

The concrete force is

$$0{\cdot}403u_k bd_n = 8{\cdot}54bd_1$$

The steel force is

$$0{\cdot}8 \times 420 \times A_{st}/1{\cdot}15 = 292{\cdot}2A_{st}$$

Equating forces gives

$$A_{st}/bd_1 = 2{\cdot}92\%$$

Taking moments about the concrete force:

$$M_r = (0{\cdot}8f_y/1{\cdot}15)A_{st} \times (\text{lever arm}) = 292{\cdot}2A_{st}(d_1 - 0{\cdot}317d_1)$$

We know A_{st}/bd_1, hence

$$M_r/bd_1^2 = 5{\cdot}83 \qquad \text{(Answer)}$$

(*b*) At a steel stress of $f_y/1{\cdot}15$ the steel strain is 0·00383. From the strain diagram we obtain

$$n_1 = 0{\cdot}477$$

Using a technique similar to the one used before:

$$A_{st}/bd_1 = 1{\cdot}58\% \quad \text{and} \quad M_r/bd_1^2 = 4{\cdot}53 \qquad \text{(Answer)}$$

8.5 Design of T- and L-beams

The strength of the concrete in compression in a T- or L-beam is usually more than adequate since the breadth of the flange (up to $b_r + 12d_s$) is generally much greater than the breadth of the rib, b_r. The Unified Code therefore ignores the rib when calculating the resistance moments of the section.

Moments of resistance

Referring to figure 8.5.1 and taking moments about the tension steel gives

$$M_{rc} = 0{\cdot}4u_k b d_s(d_1 - \tfrac{1}{2}d_s) \qquad (8.5.1)$$

This expression is conservative provided that d_s is not taken to be greater than $\frac{1}{2}d_1$.

Taking moments about the compression force gives

$$M_{rt} = 0{\cdot}87 f_{yt} A_{st}(d_1 - \tfrac{1}{2}d_s) \qquad (8.5.2)$$

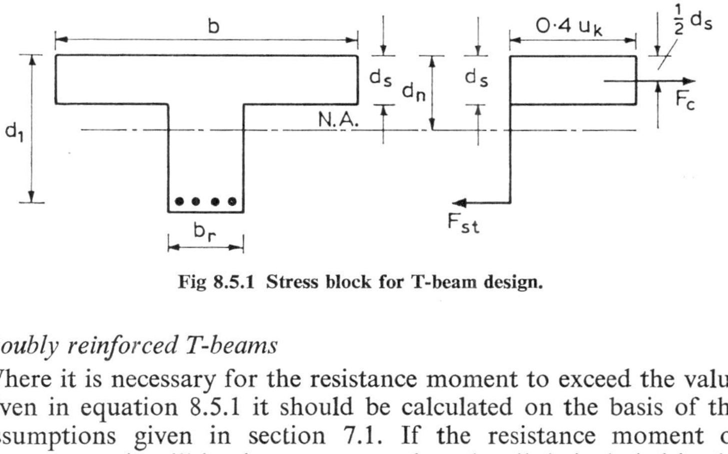

Fig 8.5.1 Stress block for T-beam design.

Doubly reinforced T-beams

Where it is necessary for the resistance moment to exceed the value given in equation 8.5.1 it should be calculated on the basis of the assumptions given in section 7.1. If the resistance moment of the concrete is still inadequate even when the rib is included in the more accurate calculations then the section must be doubly reinforced. However, it should be noted that the tension steel is usually very congested in such cases and it is far preferable to increase the beam size wherever possible.

EXAMPLE 8.5.1

Determine the moment of resistance of a singly reinforced T-beam using the assumptions of Whitney's stress-block and including the resistance of the rib.

Solution Equating longitudinal forces gives

$$A_{st}f_{yt} = 0{\cdot}85f_c'(b - b_r)d_s + 0{\cdot}85f_c'b_rd_c$$

Taking moments about the tensile steel gives

$$M_r = 0{\cdot}85f_c'(b - b_r)d_s(d_1 - \tfrac{1}{2}d_s) + 0{\cdot}85f_c'b_rd_c(d_1 - \tfrac{1}{2}d_c)$$

According to Whitney, maximum concrete strength is given by

$$d_c = 0{\cdot}537d_1$$

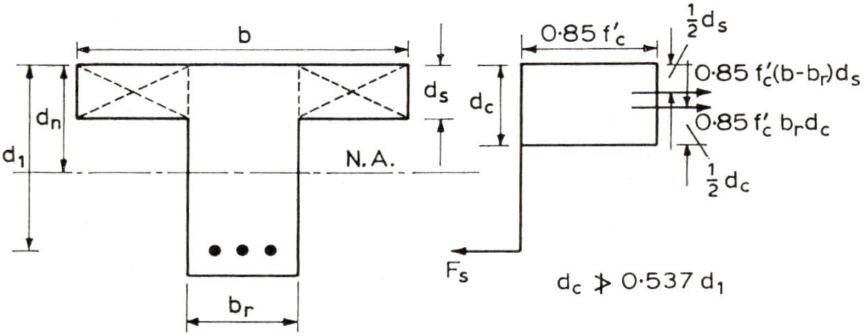

Fig 8.5.2 Example 8.5.1.

Hence

$$M_r = 0{\cdot}85f_c'[(b - b_r)d_s(d_1 - \tfrac{1}{2}d_s) + b_rd_1^2 \times 0{\cdot}537(1 - \tfrac{1}{2} \times 0{\cdot}537)]$$

$$= 0{\cdot}85f_c'bd_1^2[(1 - b_r/b)(d_s/d_1)(1 - d_s/2d_1) + 0{\cdot}393b_r/b]$$

$$= 0{\cdot}85f_c'bd_1^2K$$

A design table can be drawn up to give K for different values of the parameters b_r/b and d_1/d_s. CP114 used this design approach of utilizing the parameters b_r/b and d_1/d_s in conjunction with the old CP114 stress block.

REFERENCE

8.1 WHITNEY, C. S. Design of reinforced concrete members under flexure or combined flexure and direct compression. *Proc. A.C.I.* vol. 33 1937 pp 483–98.

Exercises

1. A simply supported beam spans 16 ft (4·88 m). It is subjected to a uniformly distributed dead load (including self-weight) of 300 lbf/ft (4·37 kN/m) and an imposed loading of 500 lbf/ft (7·29 kN/m). The

midspan section is 8 in (203 mm) wide × 10 in (254 mm) in effective depth and is reinforced with three 7/8-in (22·2-mm) diameter mild steel bars. Is it adequate for the above loading according to (*a*) limit state of collapse and (*b*) ultimate load theory as given in CP114:1965?

(*a*) For the limit state of collapse assume a rectangular stress block with a mean stress of $0{\cdot}4u$, where u is the characteristic cube strength = 4000 lbf/in^2 (27·6 MN/m^2). The characteristic strength of the steel = 40000 lbf/in^2 (276 MN/m^2). Assume the following partial safety factors:

(i) Loads
 Dead load 1·4
 Imposed load 1·6

(ii) Material strength
 Steel 1·15
 Concrete: included in factor of 0·4

(*b*) For ultimate load theory according to CP114:1965 assume a rectangular stress block with a mean stress of $4u/9$ where:
u = works cube strength = 4000 lbf/in^2 (27·6 MN/m^2).
The yield strength of the steel = 40000 lbf/in^2 (276 MN/m^2).
Take the (overall) load factor to be 1·8.

(UB3/H 1968, 36 min)

2. Refer to question 2, chapter 3 and complete part (*b*); refer to question 1, chapter 6, and complete part (*b*).

3. State briefly what you understand by "limit state theory".

A doubly reinforced cantilever is 3 m long and carries a live point load of 20 kN at its free end. The beam measures 150 mm wide by 300 mm deep overall, and provides a dead load of 1·1 kN/m. Calculate, using the rectangular stress block method, the areas of tension and compression steel required, if the bars are positioned at a distance of 50 mm from the tension and compression faces of the concrete respectively. Assume:

$u_w = 45{\cdot}0$ MN/m^2 $f_y = 420$ MN/m^2
$\gamma_l = 1{\cdot}6$ for live load
$= 1{\cdot}4$ for dead load

The maximum steel stress is $0{\cdot}87f_y$ in tension and

$$2000f_y/(2300 + f_y)$$

in compression.

The mean concrete stress in the rectangular stress block is $0{\cdot}40u_w$.

(UB2 1969, 40 min)

Chapter 9

Columns

Vertical members with heights much greater than their breadths are known as *columns*. They are used in buildings to support floor beams, in towers to support elevated tanks, and in trestle piers, etc. A strut need not be vertical but is still designed as a column. Piles and walls can also act as columns. RC columns have the advantage, in general, of acting as *short* columns: that is, columns which do not have a tendency to buckle. Special consideration is necessary for *slender* (or *long*) columns, for which additional bending effects become significant.

9.1 General considerations for columns and walls

A column may be considered to carry either (i) a direct (i.e. axial) load, or (ii) an eccentric load (i.e axial load combined with a bending moment). The stresses in an axially loaded column are purely compressive, whereas those for an eccentric load include bending stresses. Bending moments can be present due, for example, to beams framing into columns, or to wind loading.

Short columns

A short column as defined by the Unified Code is one whose effective height does not exceed twelve times its least lateral dimension. Short columns fail primarily by compression of the concrete or tension of the steel, because buckling effects can be ignored.

Column types

There are two general types of column as follows:

(i) Rectangular
(ii) Round or octagonal.

Each type has main reinforcement consisting of longitudinal bars. The secondary reinforcement usually consists of transverse links or ties (i.e. stirrups) although spiral or helical reinforcement can also be used in octagonal or round columns. Octagonal and round columns require more complicated formwork and have therefore tended to be used less frequently. However, the introduction of fibre-glass shuttering, which is virtually indestructible for round columns and gives a "glazed" finish to the concrete, greatly reduces the cost of repetitive work and can make round columns comparable in cost with rectangular columns.

Columns resisting axial loads are normally made square or circular, while columns subjected to eccentric loading are preferably constructed with unequal sides, the longer side being arranged to provide increased resistance to the bending moment.

Longitudinal reinforcement

The minimum bar diameter is 12 mm since the bar must be sufficiently rigid to stand up straight in the column boxes during fixing and concreting The total cross-sectional area of the longitudinal steel must not be less than $0{\cdot}15\,{}_*W/f_y$ and not more than 6% of the gross cross-sectional area of the column (see appendix C). If the column is cast horizontally the value can be increased to not more than 8% since the steel congestion is not quite so critical, but the value at laps must not exceed 10% in both cases.

Transverse reinforcement

Transverse reinforcement is essential in order to resist the tendency of the longitudinal bars to buckle outwards and split the concrete. In rectangular columns it consists of links (also called ties or binders), which are similar in shape to stirrups in beams. The Code (see appendix C) governs the diameter and spacing of the ties, and every longitudinal bar must be restrained in two directions by the links.

Continuous columns

Columns which continue up through a floor must be proportioned with due regard to the beams framing into the column. Column bars can either pass outside the beam bars, or bend inwards to miss the beam bars, to lap onto the bars for the next storey.

Restrained (or braced) columns

A column is assumed to be restrained in a given direction if lateral stability of the structure is provided by walls or other suitable bracing designed to resist all lateral forces in that direction. A column is unrestrained if the lateral stability of the structure in a given direction is provided by the columns alone. Hence a wall is restrained if lateral stability perpendicular to the wall is provided by cross-walls.

In-situ concrete walls

In-situ walls may be necessary if moderate structural strength is required, as in walls providing lateral stability to a structure, retaining walls etc. They are generally more expensive, if used above ground level, than brickwork cavity walls, precast concrete block walls or precast concrete wall panels. They are also often inferior with respect to both durability in industrial atmospheres and moisture penetration. The limit state of local damage is important for in-situ concrete walls—because of shrinkage and thermal movement (see section 11.3)—as well as the limit state of collapse. The limit state of deflection need not be checked for plain walls or, normally, for reinforced walls.

Precast concrete walls

Precast concrete walls can consist of either precast concrete blocks or precast concrete wall panels. Precast concrete wall panels are frequently used in conjunction with the many industrialized building systems which are available. It is essential to check the stability of a system of walls as a whole as well as satisfying the limit states for the individual panels. The reader should refer to section 8.3 of the proposed Code (*Large panel construction*).

Reinforced concrete walls

A reinforced concrete wall is a wall which acts as a vertical load-bearing member, whose length between vertical edges is at least four times its thickness, and which has steel ratios (on gross area) of not less than 0·4% and 0·2% for vertical and horizontal steel respectively. RC walls can be designed as narrow columns in most respects but there is the additional consideration that increases in collapse load can be allowed where the ratio of (clear height between supports) to (length of wall) is less than 1·5. These increases are as for a "plain" concrete wall with fully compacted concrete having u_w not less than 22·5 MN/m² (see table C1). The shear strength and the limit state of deflection need not normally be checked for reinforced concrete walls.

"Plain" concrete load-bearing walls

A "plain" concrete wall has less than 0·4% vertical steel and need never have transverse links. It can either be in-situ or precast. Wall panels are in general restrained or continuous along two or more edges. Connections are assumed to be either hinged or fully continuous (see section 8.3.5.4 of the Code for flexural continuity of horizontal connection), hence partial continuity is not considered. The primary and additional moments are considered as for columns but the total bending moment rarely exceeds $0{\cdot}33Wt$, where W is the direct load on the wall and t is the thickness of the wall. Sections of wall and slab are considered as being equivalent to part of a framework if there is a continuity between walls and adjoining slabs. For hinged connections the primary moments are given by the load acting at an eccentricity of not less than $0{\cdot}05t$. Floor loads, etc., should be assumed to act on the wall at the outer third point of the area of contact. The reaction at the base of a braced wall due to such a load can be considered to act through the centre line of the area of contact between the bottom edge of the wall and the supporting member.

Analysis of plain walls

Plain concrete walls can be considered as columns which are loaded either axially or with bending. Sections can be analysed according to the expression given in appendix C. The eccentricity of loading in the plane of the wall should *either* not exceed $\frac{1}{3}$ of wall length *or* vertical reinforcement should be provided ($r' \nless 0{\cdot}3\%$) for length of wall which is in vertical tension. Eccentricity of load perpendicular to an unbraced wall is given by allowing for the eccentricities of all vertical loads and the overturning moments produced by any lateral loads above that level. For a braced wall the eccentricity of the resultant load immediately above the lateral load is zero. When considering cracked sections the thickness of the compression zone alone should be assumed to carry the load.

Shear strength of plain walls

The shear strength of plain concrete walls should be considered at positions of maximum shear for the various combinations of design loads for the limit state of collapse. Shear strengths to resist forces both normal to, and in the plane of, the walls are given in appendix C (addendum).

9.2 Elastic analysis of axially loaded short columns

An axial load is such that its resultant acts at the centroid of the section. The safe axial load for a column reinforced with longitudinal bars and transverse reinforcement depends upon the sum of the resistances of the concrete and steel. In the following, the axial load is first determined by means of an elastic analysis; this is then considered in the light of actual test results and shown to give misleading values.

Elastic analysis

Elastic theory assumes that the strains in the concrete and steel are equal. The permissible concrete stress in uniform compression, p_{cc}, needs to be lower than that for flexure, p_{cb}, since the latter occurs at the extreme fibre only. For conventional elastic theory p_{cc} is approximately $\frac{1}{4}u$. The concrete achieves its permissible value before the steel hence, referring to figure 9.2.1 and putting p_{cc} equal to f_{cc},

$$P = p_{cc}A_{cc} + mp_{cc}A_{sc}$$
$$= p_{cc}A_c + (m - 1)p_{cc}A_{sc} \qquad (9.2.1)$$

where A_{cc} = actual (or net) concrete area
A_c = gross concrete area (= bd)

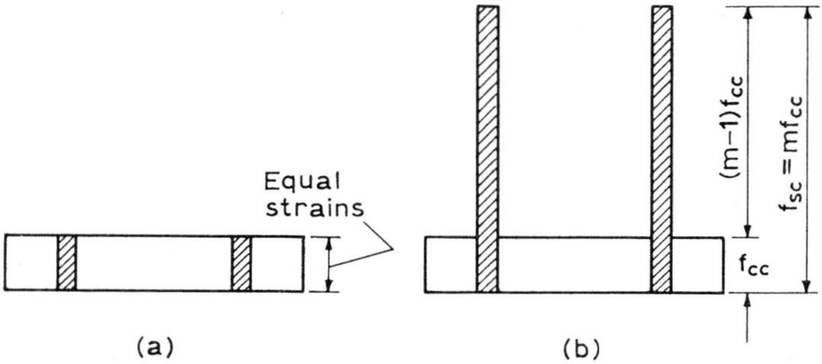

Fig 9.2.1 Elastic strains (*a*), and stresses (*b*), for column.

Significance of the modular ratio

It has been shown already that concrete is not an elastic material and that the concrete modulus varies with rate, intensity and duration of loading. The concrete modulus is defined, for convenience, in terms of the total strain (elastic strain plus creep strain) so that its value is time-dependent. Hence the modular ratio is also time-dependent since only the steel is elastic up to the working load. The modular ratio for long-term loading has been assumed, for convenience, to be twice the short-term value although it can be larger than this. In the case of beams it was clear that variations in modular ratio only affect, directly, the lever arm. Such variations therefore have a secondary effect only upon the moment of resistance of the section. The resistance of a column section, however, is directly affected by changes in modular ratio, as shown by equation 9.2.1. Hence although large uncertainties could be tolerated in the modular ratio value for beams this is not possible for columns.

Stress redistribution in columns

An axially loaded column subjected to a constant load undergoes a gradual but significant redistribution of stress with time: the concrete stress decreases while the steel stress increases. Figure 9.2.2(*a*) shows a typical concrete strain-time curve where a load is applied suddenly at time $t = 0$, maintained constant, and then suddenly released at time $t = t_1$.

Effect of sudden sustained load

The instantaneous elastic concrete strain, ε_0, equals the initial steel strain. The concrete immediately commences to deform inelastically with time whereas the steel remains elastic. Part of the load initially

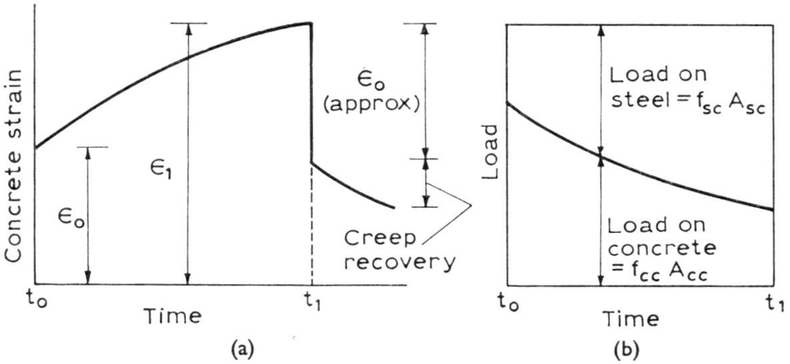

Fig 9.2.2 Stress redistribution in columns.

carried by the concrete is therefore transferred to the steel as shown in figure 9.2.2(*b*). The increase in the load carried by the steel causes the latter to deform elastically, so that the column as a composite unit also strains with time. Notice that the apparent modular ratio at time t is very different to that at time t_0 since the stress, f_{cc}, decreases with time although the strain increases with time.

Effect of sudden removal of load

If the load is removed quickly both the steel and the concrete undergo an instantaneous elastic recovery. However since the load carried by the steel has increased, while that for the concrete has decreased, the steel tries to recover much more than the concrete. The steel therefore retains some residual compression while the concrete becomes stressed *in tension* since the external load is now zero. This tension can cause cracking of the concrete.

Plastic theory for columns

The foregoing shows that even at moderate column loads the stresses in the concrete and steel are not necessarily given by the elastic theory. Considerable plastic flow or creep occurs in the constantly stressed concrete. Tests show that failure of a column occurs when both the concrete and the steel have reached their maximum load-carrying capacities. This limitation of conventional elastic theory was one of the first to be generally recognized. Equation 9.2.1 was discarded by codes of practice in favour of an equation for ultimate load conditions even when elastic theory for beams was still retained as the main design method.

9.3 Axially loaded short columns at collapse

It has been shown that conventional elastic analysis is unsatisfactory for reinforced concrete columns. The limit state of collapse, on the other hand, can be designed for very satisfactorily. It is again important, however, to distinguish between the similarities and differences which exist when considering columns as compared with beams.

Axial collapse load

The design stress for concrete in bending is $0{\cdot}45u_k$ (see appendix D). For compression over the entire section however, rather than at the extreme fibre only, it is necessary to assume a lower value. The Unified Code uses the value of $0{\cdot}40u$ which coincides with the value for the rectangular stress block assumed for sections in flexure. The partial safety factor for the compression reinforcement is also increased from $(2300 + f_y)/2000$ to $1{\cdot}5$ for the reasons given below. The Unified Code therefore gives the following expression for the design load:

$$_*W = 0{\cdot}4u_kA_c + 0{\cdot}67f_yA_{sc} \tag{9.3.1}$$

where A_c = area of concrete
A_{sc} = area of compression steel
f_y = characteristic (tensile) strength of compression steel.

Load-contraction characteristics

Brittle compression failures in over-reinforced beams can be avoided by limiting the depth of the neutral axis, and hence the tension-steel ratio. For axially loaded columns, however, the compression is high throughout the entire cross-section and no tension is present. An alternative approach is therefore necessary. The load-contraction curve for an axially loaded column can become approximately "flat-topped" if the load taken by the compression steel is capable of increasing beyond the design value after the concrete reaches its ultimate strain. That is, the compression steel can compensate for the decreasing proportion of the total load taken by the concrete at very high strains. The Unified Code does this by specifying a minimum steel area of $0{\cdot}15\,{}_*W/f_y$ (see appendix C) and increasing the partial safety factor for the compression steel to 1·5 (i.e. the latter allows for the change from $0{\cdot}45u$ to $0{\cdot}40u$). This ensures that the steel can carry at least 15% of the total axial load.

Restrained short columns with low moments

If a restrained short column supports an approximately symmetrical arrangement of beams and imposed load it may be designed without special calculations for the actual moments if the safety factors are effectively increased for the axial load as follows:

$$_*W = 0{\cdot}35u_kA_c + 0{\cdot}60A_{sc}f_y \tag{9.3.2}$$

Equation 9.3.2 may be used provided that

(i) h_a/d and $h_t/t \not> 12$
(ii) Beams carry a uniformly distributed imposed load
(iii) Beam spans differ by $\not> 15\%$

Wind loads do not affect restrained columns, hence the column load is calculated from combination I for dead and imposed loads only (see table B1).

9.4 Column footings

Suitable footings are necessary to distribute the loads from columns and walls onto the subsoil. The main types of footing are as follows:

(i) *Isolated footing*, usually square or rectangular, supporting an individual column.
(ii) *Combined footing*, supporting two or more columns.
(iii) *Strip footing*, supporting a continuous wall.

Any of these footings may be directly supported on the soil or, possibly, on piles. Under very poor soil conditions it is sometimes necessary to support the entire structure on a single mat or slab known as a *raft foundation*.

Critical stresses

The following stresses must be considered in footings:

(i) Bearing pressure between column and top of footing
(ii) Bearing pressure between soil and base of footing
(iii) Diagonal tension stress for concrete
(iv) Punching shear stress for concrete
(v) Bending stresses for concrete and steel
(vi) Bond stress between concrete and steel

Concrete bearing stress

The load from the column steel must be transferred into the footing by means of dowel (or "starter") bars. If the grade of concrete in the column is much higher than that for the footing then additional dowel bars may be necessary. The bearing stress on the footing concrete may be increased from $0{\cdot}4u$ to $0{\cdot}67u$ if the column is centrally placed on less than one-third the area of the footing (i.e. the central core of concrete in the base is effectively confined).

Soil bearing pressure

The permissible pressure on the ground determines the minimum overall area of the footing. The soil pressure is usually considered uniform for axial loading and trapezoidal or triangular for eccentric loadings (i.e. pressure varies linearly across the base). Notice that according to soil mechanics theory the pressure distribution is rarely uniform, so that the calculated shear and bending stresses in the concrete may not be those for the worst case. No redistribution of bending moments should be considered. It is therefore preferable to be conservative when designing bases, and to consider the possibility of reversed stresses in the case of combined footings. The permissible soil pressure must exceed the combined effects of column load, self-weight of footing and any overburden weight. It is usually convenient to consider the "net soil pressure", given by the column load divided by the footing area, and to compare this with the permissible pressure less the self-weight and overburden pressure.

Diagonal tension

The critical sections for shear along a vertical section producing diagonal tension should be taken across the full width of the base at a distance from the column faces equal to the effective depth of the base. In the case of rectangular footings, the vertical section extending across the full width of the base should be considered to resist a shearing force equal to the pressure on the area on the remote side of the section, as shown in figure 9.4.1.

Punching shear

The critical positions for punching shear along a vertical section should be taken at a distance from the column faces equal to $\frac{1}{2}d_1$. Hence the length of the perimeter CCCC in figure 9.4.1 resists a shearing force due to the base area on the remote sides of section CCCC. Stresses due to punching shear should be considered as for concentrated loads on slabs (see appendix C).

Bending stresses

The critical section for bending in the base should be taken at the face of the column or wall, the moment being given by the forces over the entire area on one side of the section. The reinforcement provided to resist these bending moments should, in general, be

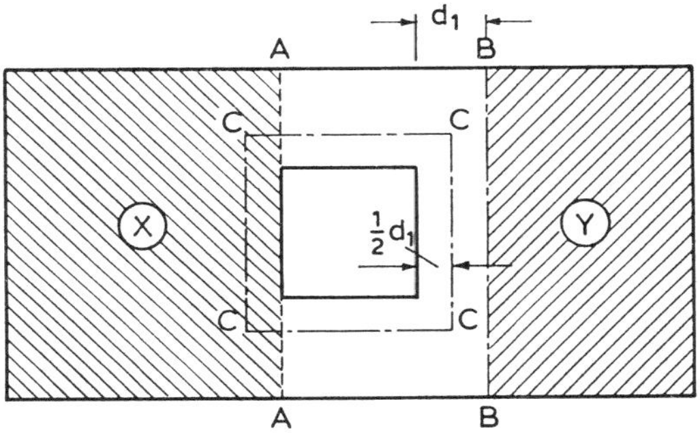

Fig 9.4.1 Critical sections for bending (AA) and shear (BB and CCCC) for rectangular footing.

The moment due to shaded area X is used for bending and bond at face AA, and the shear force due to shaded area Y acts on section BB.

distributed across the full width of the section. In rectangular bases, however, the reinforcement across the short span should be more closely spaced near the column. An amount

$$[2/(R + 1)] \times \text{total area of steel}$$

should be spread over a central band of width equal to the shorter dimension of the base. R is the ratio of longer side to shorter side (i.e. greater than unity). The remainder of the steel should be spread evenly over the outer bands.

Bond stress

The critical section for local bond stress is usually the same section as the critical section for bending moment, i.e. at the face of the column or wall. A lower bond stress is sometimes used in two-way footings, since it is reasoned that cracks parallel to the bars may actually lie along these bars and lower their resistance to bond, due to bending in the other direction. Bond stresses should also be considered for sections at which the depth changes or at which reinforcement is curtailed.

Eccentric loads

Variable moments are frequently applied to footings, so that eccentric loading occurs and a trapezoidal soil pressure results.

This modifies the magnitude of the design moment and shear, but not the general design procedure.

Strip footing

The design of wall footings is simpler than that of column footings since bending occurs in one direction only. Because of the uniform nature of a wall loading, it is sufficient to consider a typical one-metre length of wall. Plain concrete is often suitable for light loadings.

Footings on rock

The high bearing pressures possible on rock often necessitate purely nominal footings. It should be noted, however, that good rock is usually stiffer in compression than is reinforced concrete of normal proportions in bending, hence the latter cannot effectively distribute the loading. A rock footing requires an extra-large depth for shear stiffness, and little if any reinforcement for tension on its bottom face.

Combined footing

A single footing carrying two or more columns is known as a combined footing. They are particularly useful where there is:

(i) close proximity of two or more columns,

(ii) an exterior column which cannot be made symmetrical because of site boundary or other limitations.

The centroid of the combined footing is made to coincide with the resultant of the individual column loads. The cantilever projections on three sides of each column in a two column footing could therefore act very similarly to an isolated footing. It is usual, however, to

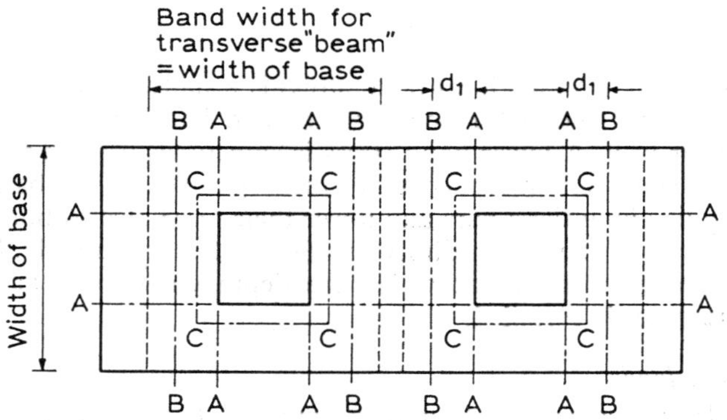

Fig 9.4.2 Critical sections for a combined footing.

consider strips of slab under the column as acting as transverse beams. There should, therefore, be a concentration of transverse steel for a band-width equal to the short-side dimension of the base under the columns, as shown in figure 9.4.2.

Piled foundations

Piles are often necessary on poor ground, and can be considered as concentrated reactions on the footing. The designer must, however, allow for the fact that the location of piles can vary by as much as 300 mm or so from the intended position, due to difficulties in driving.

9.5 Analysis of column sections at collapse

The two main types of flexural failure that occur in beams can also occur in eccentrically loaded short columns. They are:

(i) *Compression failure,* due to primary crushing of the concrete.

(ii) *Tension failure,* due to yielding of the steel and subsequent crushing of the concrete.

In either case, however, it is clear that the attainment of an ultimate concrete strain in the extreme fibre can be taken as the failure criterion for the column. The basic approach for the accurate analysis of column sections can therefore follow that of Hognestad, Hanson and McHenry [7.1] for sections in pure flexure.

Equilibrium conditions

For sections subjected to pure flexure the sum of the internal forces is zero. In the case of columns, however, the sum of the internal forces must equal the applied axial load. Apart from this essential difference in the equilibrium requirements the approach can be very similar. Tentative strain profiles with the strain at one extreme fibre equal to the ultimate concrete strain are again considered, as shown in figure 9.5.1.

Known eccentricity of load

The basic problem for columns is to determine the collapse load when that load acts at a known eccentricity. For equilibrium of longitudinal forces,

$$W = F_c - \sum_{1}^{n} F_s$$

$$= K_1 K_3 f_c' b d_n - A_{s1} f_{s1} - A_{s2} f_{s2} - A_{s3} f_{s3} - A_{s4} f_{s4} \quad (9.5.1)$$

Notice that f_{s3} and f_{s4} are compressive, and therefore negative for the neutral axis position shown in figure 9.5.1 when using the general sign convention (see appendix Y).

Also, for equilibrium of moments,

$$We = F_c(\tfrac{1}{2}d - K_2 d_n) - \sum_{1}^{n} F_s y$$

$$= K_1 K_3 f_c' b d_n(\tfrac{1}{2}d - K_2 d_n) - A_{s1} f_{s1} y_1 - etc. \qquad (9.5.2)$$

Note: Equation 9.5.2 has general applicability but it should be remembered that if e is made negative, $(\frac{1}{2}d - K_2 d_n)$ would have to be considered negative when measured in the negative y-direction.

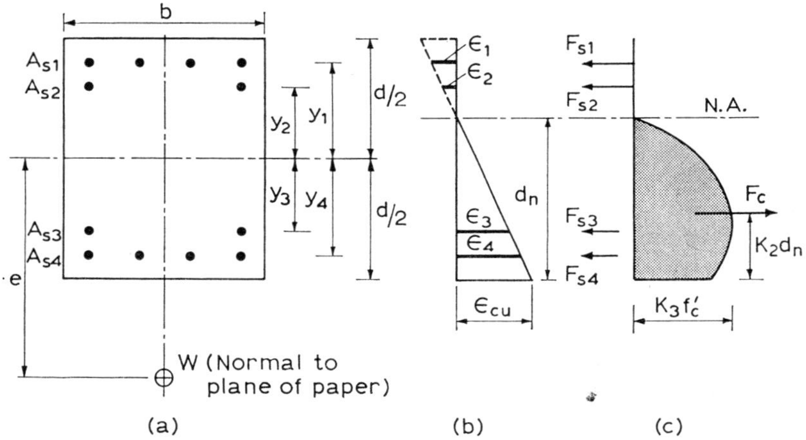

Fig 9.5.1 Column section at collapse. (*a*) Section. (*b*) Strains. (*c*) Forces.

Strain profiles

The stresses f_{s1}, f_{s2} etc. must be compatible with the strains ε_1, ε_2, etc., and due attention must be paid to the signs of both f_s and y. Notice, for instance, that *either* f_s *or* y must be negative in equation 9.5.2 so that all terms are additive. As for beams, a trial-and-adjustment procedure is used to find the strain distribution at failure. The essential requirement for columns is that the two values of W which must be made to coincide are obtained by considering forces and moments respectively: that is, the values of W obtained from equations 9.5.1 and 9.5.2 must be equal to within about 2%. A suitable solution is achieved most conveniently simply by inspection followed by trial-and-error adjustments to the value of d_n.

EXAMPLE 9.5.1

Calculate the collapse load for a 0·5 m × 0·5 m column using the method of Hognestad *et al.* when the load acts 75 mm outside the midpoint of one edge as shown. The column is reinforced with six 32-mm diameter bars of load-strain characteristics in tension and

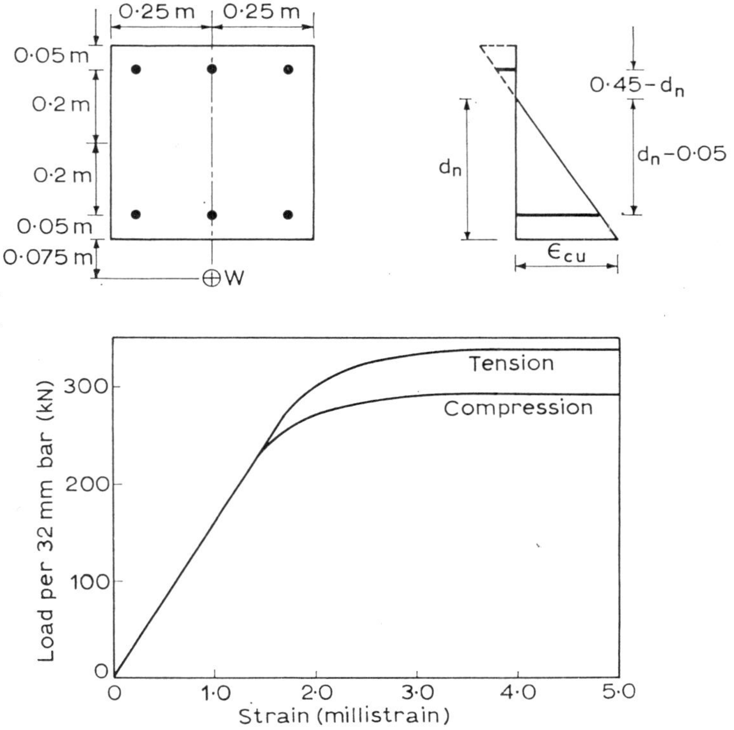

Fig 9.5.2 Example 9.5.1.

compression as shown. The cube crushing strength of the concrete is 37·5 MN/m². Assume that f_c' is equal to $0{\cdot}78u_k$ and refer to section 7.2 for the values of K_1K_3, K_2 and ε_{cu}.

Solution

$$f_c' = 0{\cdot}78 \times 37{\cdot}5 \times 10^6 = 29{\cdot}2 \times 10^6$$
$$K_1K_3 = (27 + 0{\cdot}35 \times 29{\cdot}2)/(22 + 29{\cdot}2) = 0{\cdot}731$$
$$K_2 = 0{\cdot}5 - 29{\cdot}2/550 = 0{\cdot}447$$
$$\varepsilon_{cu} = (4 - 29{\cdot}2/45) \times 10^{-3} = 3{\cdot}35 \times 10^{-3}$$

Eccentricity of load is fairly large hence try

$$f_{st} = f_{sc} = f_{yc}$$

i.e. load per bar is 292×10^3. Hence

$$W = F_{cc} + 0 \qquad (1)$$

and

$$0{\cdot}325W = F_{cc}(\tfrac{1}{2}d - K_2 d_n) - \sum F_s y \qquad (2)$$

$$F_{cc} = K_1 K_3 f_c' b d_n$$

$$= 0{\cdot}731 \times 29{\cdot}2 \times 0{\cdot}5 \times 10^6 d_n = 10{\cdot}67 \times 10^6 d_n$$

Therefore from (1) and (2)

$$F_{cc}[0{\cdot}325 - (0{\cdot}25 - 0{\cdot}447d_n)] = 6 \times 292 \times 0{\cdot}2 \times 10^3$$

or

$$0{\cdot}447d_n^2 + 0{\cdot}075d_n - 32{\cdot}8 \times 10^{-3} = 0$$

$$d_n = +0{\cdot}179$$

If $(d_n)_1 = 0{\cdot}179$ then $f_{st} > f_{sc}$; hence increase d_n.

Try $(d_n)_2 = 0{\cdot}20$

$$\varepsilon_{st} = \frac{0{\cdot}45 - 0{\cdot}20}{0{\cdot}20} \times 3{\cdot}35 \times 10^{-3} = 4{\cdot}19 \times 10^{-3}$$

$$\varepsilon_{sc} = \frac{0{\cdot}20 - 0{\cdot}05}{0{\cdot}20} \times 3{\cdot}35 \times 10^{-3} = 2{\cdot}51 \times 10^{-3}$$

Hence

$$\sum F_{st} = 3 \times 337{\cdot}5 \times 10^3 = 1013 \times 10^3$$

and

$$\sum F_{sc} = 3 \times 285 \times 10^3 = 855 \times 10^3$$

$$F_{cc} = 10{\cdot}67 \times 0{\cdot}20 \times 10^6 = 1{\cdot}13 \times 10^6$$

Therefore

$$W = F_{cc} + \sum F_{sc} - \sum F_{st} = 1{\cdot}97 \times 10^6$$

The external moment is

$$0{\cdot}325 \times 1{\cdot}97 \times 10^6 = 640 \times 10^3$$

The internal moment about the centre line is

$$1868 \times 10^3 \times 0{\cdot}2 + 2130 \times 10^3(0{\cdot}25 - 0{\cdot}447 \times 0{\cdot}20)$$

$$= 715{\cdot}7 \times 10^3 \neq \text{external moment}$$

Now

$$(715{\cdot}7/640)(d_n)_2 = 0{\cdot}224$$

but since the internal moment also increases with d_n use a slight overcorrection.

Try $(d_n)_3 = 0{\cdot}225$

This gives $W = 2{\cdot}25 \times 10^6$

Internal moment = external moment = 731×10^3 (Answer)

9.6 Columns in continuous frames

Floor loadings produce bending and deflections in beams and slabs which, due to the stiffness of the connections, induce deflections in the columns also. These bending moments must be evaluated before the sections can be designed. Internal columns which support an arrangement of beams and loading which is approximately symmetrical can be designed as indicated in section 9.3. However, external columns and internal columns which are not symmetrical must be designed to take the appropriate bending moments in conjunction with the axial load. Transverse loads, such as wind pressure, can also produce bending moments in (unrestrained) columns.

Structural stiffness

The stiffness of a member is given by dividing the second moment of area of the section by the length of the member, provided that the cross-section is constant throughout the length of the member. It is usual to consider the concrete section and to ignore the reinforcement, but any of the three sections listed in appendix B may be used. The breadth of the flange for T-beams is that assumed for midspan calculations. For members which are assumed to be fully fixed at both ends the length of a member is taken, for convenience, as the distance between centres of supports.

Primary moments

The moments caused by the out-of-balance load on the members framing into the column are called *primary* moments. For columns in framed structures the moments due to vertical loads on the beams may be calculated by simple moment-distribution procedures. It may be assumed that the far ends of members meeting at the joint under consideration are fixed, except that for single-bay structures the beam stiffness should be halved. The moments in the columns are therefore as shown in table 9.6.1. K_l, K_u and K_b are the stiffnesses of the lower column, upper column and beam respectively. ΣK_b is the sum of the stiffnesses of the beams on each side of an internal column.

Table 9.6.1 Moments in external and internal columns

	External columns		Internal columns
	Single-bay	Multi-bay	
	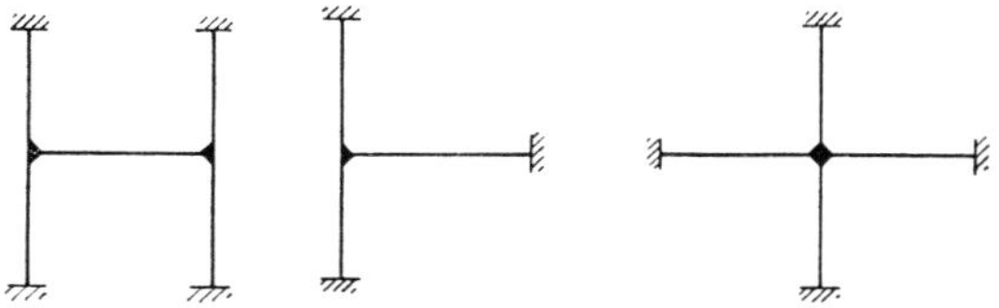		
Moment at foot of upper column	$\dfrac{K_u M_e}{K_l+K_u+\frac{1}{2}K_b}$	$\dfrac{K_u M_e}{K_l+K_u+K_b}$	$\dfrac{K_u \Delta M_e}{K_l+K_u+\Sigma K_b}$
Moment at head of lower column	$\dfrac{K_l M_e}{K_l+K_u+\frac{1}{2}K_b}$	$\dfrac{K_l M_e}{K_l+K_u+K_b}$	$\dfrac{K_l \Delta M_e}{K_l+K_u+\Sigma K_b}$

M_e is the end moment for a beam framing into an external column, assuming full fixity at both ends. ΔM_e is the difference in end moments for beams on either side of an internal column.

Restrained short columns with applied moments

Where a column is subjected to significant bending moments about one or both axes, due to the nature of the structure and the design load pattern, the cross-section should be designed to withstand an axial load W together with end moments M_{ex} and M_{ey}. When bending is about one axis only, M_{ex} or M_{ey} as appropriate should be put equal to zero. In framed structures where the columns are arranged on a rectangular grid, loading patterns giving biaxial bending need not be considered except for corner columns.

EXAMPLE 9.6.1

A three-bay three-storey framed building is shown in elevation in figure 9.6.1. The characteristic dead and imposed loadings for the first floor beams are 14 kN/m and 20 kN/m respectively uniformly distributed. Calculate the maximum design end moments at the feet of columns KG and LH and at the heads of columns GC and HD if the columns are restrained.

All external columns measure 300 mm × 250 mm and internal columns are 250 mm square. Beam EFGH measures 250 mm × 500 mm overall.

Solution Clear distance between vertical restraints is

$$H = 3{\cdot}3 - 0{\cdot}5 = 2{\cdot}8$$

Hence

$$H/D = 2{\cdot}8/0{\cdot}25 = 11{\cdot}2 \quad \text{i.e.} \quad < 12$$

$$w_d = 14 \text{ kN/m} \quad \text{and} \quad w_i = 20 \text{ kN/m}$$

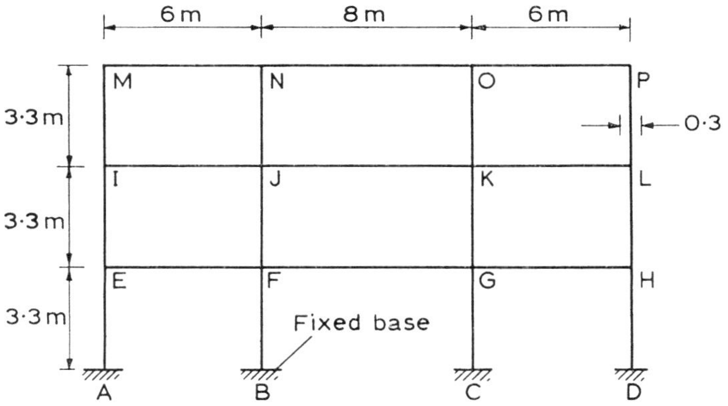

Fig 9.6.1 Example 9.6.1.

External column:

The maximum design end moment is given by

$$1{\cdot}4w_d + 1{\cdot}6w_i = (19{\cdot}6 + 32)\text{kN/m}$$

Therefore maximum M_e is

$$wl^2/12 = 51{\cdot}6 \times 6^2 \times 10^3/12 = 154{\cdot}8 \times 10^3$$

Internal column:

The minimum design end moment is given by

$$1{\cdot}0w_d = 14 \text{ kN/m}$$

The maximum out-of-balance moment is

$$\Delta M_e = (51{\cdot}6 \times 8^2 - 14 \times 6^2) \times 10^3/12 = 233 \times 10^3$$

Second moments of area

For external column

$$I = bd^3/12 = 0{\cdot}25 \times 0{\cdot}3^3/12 = 562 \times 10^{-6}$$

Similarly, I for internal column $= 326 \times 10^{-6}$

I for beam $= 2\ 605 \times 10^{-6}$

Stiffnesses

The relative stiffnesses are based on distances between centres. For external columns (assumed fixed)

$$K_u = E \times 562 \times 10^{-6}/3{\cdot}3 = 170{\cdot}4 \times 10^{-6}E = K_l$$

For the end beam

$$K_b = E \times 2605 \times 10^{-6}/6 = 434{\cdot}5 \times 10^{-6}E$$

Therefore the maximum design moment at foot of column LH is

$$\frac{K_u M_e}{K_u + K_l + K_b} = \frac{170{\cdot}4 \times 154{\cdot}8 \times 10^3}{340{\cdot}8 + 434{\cdot}5}$$

$$= 34{\cdot}0 \text{ kNm} \qquad \text{(Answer)}$$

Similarly, at head of column HD

$$\text{moment} = 34{\cdot}0 \text{ kNm} \qquad \text{(Answer)}$$

For the internal beam

$$K_b = E \times 2605 \times 10^{-6}/8 = 326 \times 10^{-6}E$$

$$\sum K_b = (326 + 434{\cdot}5) \times 10^{-6}E = 760{\cdot}5 \times 10^{-6}E$$

$$K_u = K_l = 326 \times 10^{-6}E/3{\cdot}3 = 98{\cdot}8 \times 10^{-6}E$$

at node G. Therefore the maximum design moment at foot of column KG is

$$\frac{K_u \Delta M_e}{K_u + K_l + \sum K_b} = \frac{98{\cdot}8}{958{\cdot}1} \times 233 \times 10^3$$

$$= 24{\cdot}0 \text{ kNm} \qquad \text{(Answer)}$$

Similarly, at head of column GC,

$$\text{moment} = 24{\cdot}0 \text{ kNm} \qquad \text{(Answer)}$$

Unrestrained short columns

Unrestrained short columns may be designed as above for restrained columns except that all three load combinations from appendix B must be considered since wind loads must now be allowed for. The moments due to the lateral forces may be assessed from the assumption that there are points of contraflexure at the centres of beams and columns.

9.7 Design of column sections at collapse

The assumptions made by the Unified Code for sections subjected to flexure combined with axial load are exactly the same as for sections subjected to pure flexure. The stresses in the concrete in compression are therefore given by either of the methods explained in chapter 8; that is, either from the design stress-strain curves or from the simplified approach assuming a rectangular stress block of $0{\cdot}4u$ mean stress for the concrete compression zone.

Simplified approach for columns

When using the simplified approach for sections in pure flexure it could be assumed that all the tension steel was at the upper yield stress and that in general the compression steel also reached its yield value. In the case of columns this is not the case. The steel can have a stress which can vary, for different eccentricities, from yield in compression to yield in tension. When determining the depth to the neutral axis for the concrete, therefore, it is necessary to assume values for the stresses in the steel near each face. Having obtained a value for d_n it is then necessary to check that the assumed steel stresses were reasonable. If not, then it is necessary to recalculate d_n assuming stresses in the steel which are more compatible with the position of the neutral axis. The design stress-strain curves in figures D2 and D3 should be used. The method is illustrated in example 9.7.1, which emphasizes the unsuitability of the approach.

EXAMPLE 9.7.1

Calculate the design collapse load for the column in example 9.5.1 according to the simplified assumptions of the Unified Code if the characteristic yield strength of the bars in tension, f_y, is 420 MN/m². (The design yield strength in compression, $_*f_{yc}$, is

$$2000f_y/(2300 + f_y)$$

if f_y is in MN/m²).

Solution Assume that the stresses are at the yield values for a first assessment; *but remember to check later that this is valid.* That is, assume that

$$f_{sc} = {_*f_{yc}} = (2000 \times 420/2720) \times 10^6 = 309 \times 10^6$$

and

$$f_{st} = {_*f_{yt}} = 0{\cdot}87 \times 420 \times 10^6 = 365 \times 10^6$$

Hence

$$F_{sc} = 3 \times \tfrac{1}{4}\pi \times 32^2 \times 10^{-6} f_{sc} = 2{\cdot}41 f_{sc} \times 10^{-3} = 746 \times 10^3$$

$$F_{st} = 2{\cdot}41 f_{st} \times 10^{-3} = 882 \times 10^3$$

Therefore

$$W = 0{\cdot}4ubd_n - (F_{st} - F_{sc})$$

$$= 0{\cdot}4 \times 37{\cdot}5 \times 0{\cdot}5d_n \times 10^6 - 136 \times 10^3$$

$$= 7{\cdot}5 \times 10^6 d_n - 136 \times 10^3$$

and

$$0{\cdot}325W = 7{\cdot}5 \times 10^6 d_n(0{\cdot}25 - \tfrac{1}{2}d_n) + (746 + 882) \times 10^3 \times 0{\cdot}2$$

Solving the last two equations and simplifying:

$$d_n^2 + 0{\cdot}15d_n - 98{\cdot}6 \times 10^{-3} = 0 \quad \text{hence} \quad d_n = 0{\cdot}248$$

Try $(d_n)_1 = 0{\cdot}248$

Then

$$\varepsilon_{st} = (0{\cdot}202/0{\cdot}248) \times 3{\cdot}5 \times 10^{-3} = 2{\cdot}85 \times 10^{-3}$$

From the design curve in figure D2,

$$\varepsilon_{st} \text{ at upper yield} = 0{\cdot}002 + (365/200) \times 10^{-3} = 3{\cdot}825 \times 10^{-3}$$

$$\varepsilon_{st} \text{ at lower yield} = (0{\cdot}8 \times 365/200) \times 10^{-3} = 1{\cdot}46 \times 10^{-3}$$

Hence a strain ε_{st} of $2{\cdot}85 \times 10^{-3}$ is insufficient for attaining the upper design yield stress and would provide a stress of

$$365\left(0{\cdot}8 + \frac{2{\cdot}85 - 1{\cdot}46}{3{\cdot}825 - 1{\cdot}46} \times 0{\cdot}2\right) \times 10^6 = 335 \times 10^6$$

Hence error in $(F_{st})_1$ is

$$(335 - 365)(F_{st})_1/365 = -(30/365)(F_{st})_1$$

The assumed value of f_{sc}, i.e. equal to ${}_*f_{yc}$, is correct. Hence

$$\text{error in } (W)_1 = -(30/365)(F_{st}/W)_1$$

where

$$(W)_1 = 7{\cdot}5 \times 248 - 136 = 1724$$

Try $(d_n)_2 = 0{\cdot}24$

$$\varepsilon_{st} = [(0{\cdot}45 - 0{\cdot}24)/0{\cdot}24] \times 3{\cdot}5 \times 10^{-3} = 3{\cdot}06 \times 10^{-3}$$

This is between the strains for upper and lower yield. Hence from the stress-strain curve,

$$f_{st} = 365\left(0{\cdot}8 + \frac{3{\cdot}06 - 1{\cdot}46}{3{\cdot}825 - 1{\cdot}46} \times 0{\cdot}2\right) \times 10^6 = 341 \times 10^6$$

Therefore

$$F_{st} = 2{\cdot}41 \times 341 \times 10^3 = 823 \times 10^3$$

$$F_{sc} = 746 \times 10^3 \quad \text{as before}$$

Therefore W (for $\sum$ internal forces) is

$$7{\cdot}5 \times 10^6 d_n - (823 - 746) \times 10^3 = 1{\cdot}72 \times 10^6$$

Also

$$0{\cdot}325W = 7{\cdot}5d_n(0{\cdot}25 - \tfrac{1}{2}d_n) \times 10^6 + (823 + 746) \times 10^3 \times 0{\cdot}2$$

$$= 548 \times 10^3$$

or W(for $\sum$ external forces) is

$$548 \times 10^3/0{\cdot}325 = 1{\cdot}69 \times 10^6$$

$$= W(\text{for} \textstyle\sum \text{internal forces}) \quad (\text{within } 2\%)$$

Hence

$$W = 1{\cdot}71 \text{ MN} \qquad \text{(Answer)}$$

Notice that the simplified assumptions for columns did not produce a satisfactory design method in this case since it was not possible to ascertain the stress in the tensile steel without degenerating into a trial-and-error approach. It is therefore little simpler than an accurate analysis approach. The design of column sections is therefore carried out far more conveniently by means of design charts based on the design stress-strain curves for both concrete and steel. Notice also that the collapse load value of 1·71 MN represents a reduction of 24% in the value of Hognestad *et al.* This is due mainly to γ_l giving reductions of 13%, 13%+ and 33% for tensile steel, compressive steel and concrete respectively.

Design curves

The axial load is the maximum which a particular section can carry. Hence, for symmetrically reinforced rectangular columns, for example, there is a horizontal portion for the design curves, as shown in figure E4 (e.g. see portion AB). Once the load given by the design stress-strain curves for concrete and steel for an eccentric load becomes less than

$$W_{axial} = 0{\cdot}4uA_c + 0{\cdot}67f_yA_{sc}$$

then the design curves decrease linearly with increasing eccentricity (e.g. BC). As the eccentricity becomes even greater some tension occurs in the section and the curve becomes non-linear (e.g. CD). At high eccentricities the tension steel reaches the lower yield stress and is given by a distinct change in slope of the curves (e.g. point D). At very high eccentricities the tension steel reaches the upper yield stress (e.g. point E) and failure occurs simultaneously in compression and tension. This represents the "balanced" condition for a column section. Further decreases in axial load result in a decrease in the moment that can be resisted once tension failures occur (e.g. beyond E).

EXAMPLE 9.7.2

Calculate the design collapse load for the column in examples 9.5.1 and 9.7.1 using the design chart which is based upon the full assumptions of the Unified Code. Assume as before that

$$f_y = 420 \text{ MN/m}^2 \quad \text{and} \quad u_k = 37{\cdot}5 \text{ MN/m}^2$$

Solution

$$u_k = 37{\cdot}5 \times 10^6 \quad f_y = 420 \times 10^6$$

For figure E4

$$r' = 6 \times \tfrac{1}{4}\pi \times 32^2 \times 10^{-6}/0{\cdot}5 \times 0{\cdot}5 = 1{\cdot}93\%$$

and

$$r'f_y = 8{\cdot}11,$$

$$d_1/d = 0{\cdot}45/0{\cdot}50 = 0{\cdot}90$$

$$e = 0{\cdot}325 \quad \text{and} \quad e/d = 0{\cdot}65$$

Now

$$\frac{M_r/bd^2}{W/bd} = M_r/Wd = e/d = 0{\cdot}65$$

Hence drawing the line for $e/d = 0{\cdot}65$ on figure E4 gives

$$W_r/bd = 7 \quad \text{or}$$

$${}_*W = 1{\cdot}75 \text{ MN} \qquad \text{(Answer)}$$

Notice that this load is 2% greater than the value obtained using the simplified assumption in example 9.7.1.

Redistribution of moments for columns

Notice that columns should be designed for the full moment arising from an elastic frame analysis. This is the case regardless of what is done with the beams. This is not incompatible with the general requirements for sections since in columns generally the neutral axis depth exceeds $0{\cdot}5d_n$.

Biaxial bending

If biaxial bending occurs and the axial load, W, exceeds $0{\cdot}2uA_c$, the section may be designed such that

$$1/W \nless 1/W_x + 1/W_y - 1/W_o \qquad (9.7.1)$$

where W_x = calculated load capacity for moment M_x only
W_y = calculated load capacity for moment M_y only
W_o = calculated load capacity for axial load only
M_x, M_y = moments about major and minor axes respectively.

In general, corner columns only need be considered for biaxial bending.

9.8 Design of long columns at collapse

The moments induced by the lateral deflection of a loaded column become significant if the column is long: that is, if the effective height exceeds 12 times the least lateral dimension of the column. These moments are termed *additional* moments and affect the centre moments in the restrained long columns. For unrestrained long columns the additional moments are included in the expressions for the total end moments. Accidental eccentricities for long columns arising from constructional errors produce a further effect which is significant if the *primary* moment at the end of the column is small. These accidental eccentricities can be allowed for by assuming that the primary moment should not be taken to be less than the appropriate minimum value given in the proposed Unified Code.

Limit state of buckling

The CEB-FIP recommendations for buckling in compression and bending advise a buckling check for columns with a slenderness greater than 35. The slenderness is the ratio of buckling length to radius of gyration of the concrete section. The slenderness should never exceed 140 for normal aggregate concretes or 80 for lightweight aggregate concretes. Uncertainties concerning the point of application of the load are allowed for either by means of a behaviour coefficient γ_c (equal to 1·2 and 1·4 for normal aggregate and lightweight concrete respectively), or by introducing an additional eccentricity, in the most unfavourable direction, of $d/30$ or 20 mm, whichever is greater (where d is the total depth of the section irrespective of shape; e.g. the diameter, for a circular section). The

Unified Code proposals given below adopt the principle embodied in the latter recommendation, since the various expressions ensure a certain minimum eccentricity of loading.

Restrained long columns

The load combination of W, M_{ex} and M_{ey} should be considered as for short columns except that it is important that

$$M_{ex} \nless 0{\cdot}05Wd \quad \text{and} \quad M_{ey} \nless 0{\cdot}05Wt$$

to allow for accidental eccentricities (t and d are the smaller and greater lateral dimensions of the column respectively). In addition, however, the load combination of W and total centre moments M_{cx} and M_{cy} should also be considered. The values of M_{cx} and M_{cy} for each type of loading are as follows:

(1) *Restrained columns bent about minor axis* ($M_{cx} = 0$)

$$M_{cy} \ngtr \mu M_{ey} + (Wt + \mu M_{ey})(h_t/t)^2/2500$$

and

$$\ngtr \mu M_{ey} + 2Wt(h_t/t)^2/2500$$

where μM_{ey} is the primary moment at centre of column about minor axis (μ must not be less than a half, to allow for accidental eccentricities).

(2) *Restrained columns bent about major axis*

(*a*) In all cases,

$$M_{cx} \ngtr \mu M_{ex} + (Wd + \mu M_{ex})(h_d/d)^2/2500$$

and

$$\ngtr \mu M_{ex} + 2Wd(h_d/d)^2/2500$$

$$M_{cy} = Wt(h_t/t)^2/2500$$

(*b*) If $d \ngtr 2t$, centre moments may be calculated either from (*a*) above or from

$$M_{cx} \ngtr \mu M_{ex} + (Wd + \mu M_{ex})(h_d/t)^2/2500$$

and

$$\ngtr \mu M_{ex} + (2Wd)(h_d/t)^2/2500$$

$$M_{cy} = 0$$

where μM_{ex} is the primary moment at centre of column about major axis (μ not less than $0{\cdot}5$).

(3) *Restrained columns bent about both axes*

M_{cx} as for (2*a*)

M_{cy} as for (1)

Biaxial bending need only be considered, in general, for corner columns as indicated previously.

Unrestrained long columns

The three load combinations from appendix B (i.e. including wind loads) should be considered. The end moments used in design for long columns must be the sum of the *primary* moment and an *additional* moment. These *total* end moments, $\sum M_x$ and $\sum M_y$, are equal in magnitude to the total central moments, M_{cx} and M_{cy}, given above for restrained long columns provided that μ is put equal to unity in all cases. For example, for an unrestrained column bent about minor axis:

$$\sum M_y \not> M_{ey} + (Wt + M_{ey})(ht/t)^2/2500$$

and

$$\not> M_{ey} + (2Wt)(ht/t)^2/2500$$

$$\sum M_x = 0$$

Exercises

1. (*a*) Draw a typical stress-strain curve for concrete under uniaxial compression. Explain why elastic theory is usually justified for short-term working loads, and show how the non-linear behaviour of concrete is taken into account in collapse theory.

(*b*) A short concrete column has overall dimensions of 250 mm by 250 mm. It is reinforced by four 25-mm diameter bars in the corners of the column fixed so as to allow 25 mm cover on each face. Calculate the maximum axial live load that the column could carry under working load conditions (i) by elastic theory (ii) by collapse theory. Assume:

$$u_w = 45{\cdot}0 \text{ MN/m}^2 \quad f_y = 420 \text{ MN/m}^2$$

For elastic analysis:

$$p_{cc} = \tfrac{1}{4}u_w \quad p_{sc} = \tfrac{1}{2}f_y \quad m = 13{\cdot}2$$

For collapse analysis:

$$\gamma_l = 1{\cdot}6$$

The maximum steel stress is $0{\cdot}67f_y$ and the mean concrete stress in the rectangular stress block is $0{\cdot}40u_w$ (UB2 1969, 40 min)

2. A mass concrete wall of weight w per unit volume is of height H and width $H/2$. It is subjected to a uniform horizontal pressure of p per unit area. Find p if the criterion of no tension is just satisfied at the base section. It is decided to increase the height of the wall by qlacing blocks centrally on top as shown in figure A. If the blocks

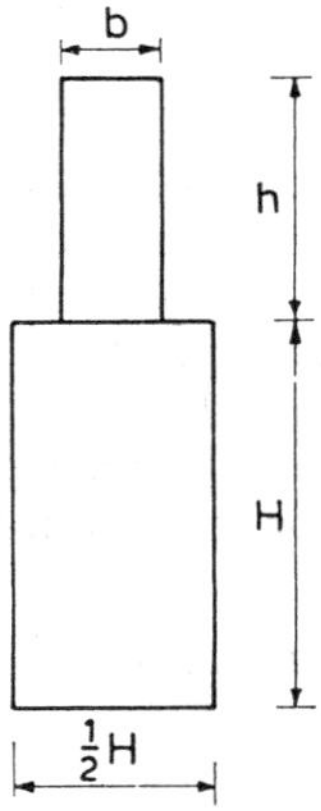

Ch. 9 Fig A Question 2.

weigh $4w$ per unit volume and the same pressure p acts over the total height, find the dimensions b and h of the blocks to satisfy the two criteria:

(*a*) a factor of safety of 1·6 against the block sliding on the wall (coefficient of friction between block and wall is 0·1)

(*b*) no tension on the base section. (UL2 1969, 36 min)

3. The characteristic cube strength of the concrete for the column analysed in example 9.5.1 was found to be only 30 MN/m^2 when tested at 28 days due to the use of cement which had deteriorated following prolonged exposure to damp air. Determine the collapse load for the column by an accurate analysis according to the method of Hognestad *et al.* What is the percentage reduction in collapse load for this 20% fall in concrete cube strength?

4. Repeat question 3 using the simplified assumptions of the Unified Code if the characteristic yield strength of the bars in tension is 420 MN/m^2. (Refer to example 9.7.1 for the collapse load of the column when u equals 37·5 MN/m^2).

5. Figure B shows part of a multi-storey framed building. The columns BE and ED are 0·25 × 0·25 metres square, and the beams AE and EC are 0·2 m wide by 0·4 m deep. Determine the reinforcement required for the column ED by considering the direct loads

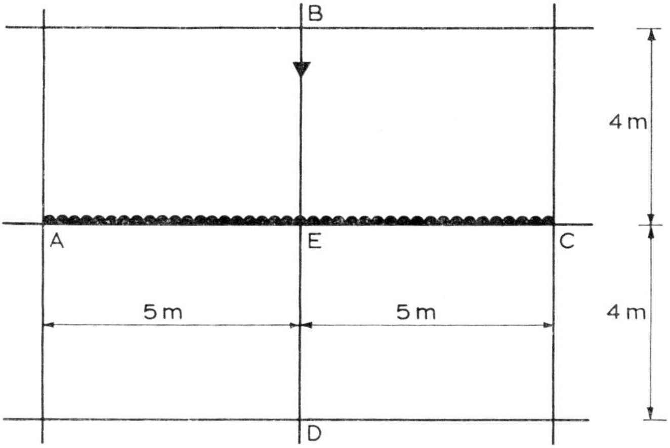

Ch. 9 Fig B Question 5.
The characteristic loads acting on column BE from the floors above equal 300 kN (live) and 200 kN (dead). The characteristic loads acting on beam AEC equal 30 kN/m (dead) and 120 kN/m (live).

down column BE and the loads on the floor beams AE and EC. Neglect the self-weight of the columns and assume that the frame is laterally stable. You may use the design chart for

$$u_w = 37{\cdot}5 \text{ MN/m}^2, f_y = 420 \text{ MN/m}^2 \text{ and } d_1/d = 0{\cdot}90$$

(UB3/0 1969, 45 min)

6. Calculate the ultimate load of an 18-in (457-mm) square column, using the method of Hognestad *et al.*, when the load acts 2 in (50·8 mm) outside the midpoint of one edge as shown in figure C. (1 kip is 1 000 lbf, or 4448 N). The column is reinforced with eight

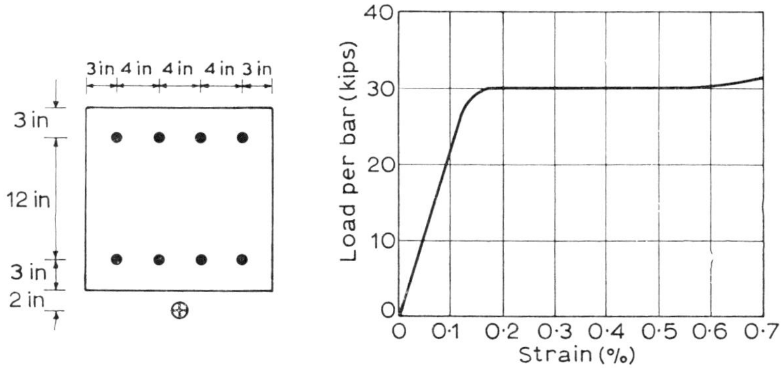

Ch. 9 Fig C Question 6.

1-in (25·4-mm) diameter mild steel bars, each bar having a load-strain curve as shown. The cube crushing strength of the concrete, u, is 6000 lbf/in^2 (41·3 MN/m^2). Assume the following values for the standard factors K_1K_3 and K_2, and the ultimate compressive concrete strain, ε_{cu} (f_c' is in lbf/in^2 units and equals $0{\cdot}78u$):

$$K_1K_3 = (3900 + 0{\cdot}35f_c')/(3200 + f_c')$$

$$K_2 \times 10^3 = 500 - f_c'/80$$

$$\varepsilon_{cu} \times 10^3 = 4 - f_c'/6500$$

(UB3/H 1968, 36 min)

Chapter 10

Limit states of deflection

Excessive deflections and deformations can impair the appearance and the efficiency of a structure and cause discomfort or alarm to the occupants or users. Visual offence and reduced efficiency can occur due to excessive slopes, pond formation, etc., and disturbance can occur due to dynamic effects under variable load. The most severe consequences, however, are usually due to local damage of the finishes and non-load-bearing members even though the limit state of local damage is not exceeded for the structure itself. For example, excessive deflections can cause cracking and possible separation of plaster finishes, crushing of partition walls, or cracking of glazing units.

The most important limit states of deflection and deformation are therefore as follows:

(i) Excessive deformability of the structure causing structural loads to be applied to non-load-bearing members.

(ii) Excessive deformation of the structure itself causing excessive slopes, etc.

(iii) Incompatible deformation of the structure, under the influence of shrinkage, creep and variable loads, with neighbouring load-bearing members (e.g. horizontal cracks in brickwork at the corners of slabs which are built in).

(iv) Excessive vibration under the effects of moving loads.

10.1 Deflections and design values

The maximum deflections which are permitted by the Unified Code under normal working loads are given, usually in terms of span or height, and are summarized in appendix B. Experience has indicated that deflections are likely to be satisfactory if certain limiting span/effective-depth ratios are not exceeded. A suitable alternative procedure has therefore been evolved which is summarized in appendix C and is discussed later. A basic understanding of the effects of deflections is, however, a first essential. The increased slenderness which is now possible for members designed for the limit state of collapse has made the consideration of deflections more important than in earlier codes.

Overall sag

The overall sag of floors, roofs and other horizontal members, given by the long-term deflection as measured below the level of the supports, is restricted in general to span/250. This ensures both satisfactory appearance and efficiency of the structure (e.g. satisfactory drainage from roofs, verticality of equipment placed on floors, etc.). In special cases deflection can be increased to span/200 if it can be shown that the increased deflection is unlikely to cause difficulties.

Pre-camber

The harmful effects of large deflections can be reduced by the provision of an initial camber. Such provision must be restricted, however, otherwise a satisfactory minimum stiffness may not be provided and dynamic effects under variable loads could become objectionable. The pre-camber of horizontal members should not generally exceed span/250.

Damage to finishes

The acceptable deflection for partitions and finishes depends on their ability to absorb or resist strain. The Code specifies that the additional deflection after the application of finishes must not exceed either span/350 or 20 mm, unless provision is made for the movement. The limit of 20 mm is clearly necessary and becomes operative for spans over 7 m. These limits could be considered high compared with those of earlier codes. However, it is shown later that calculations tend to overestimate the deflection and so compensate for this fact. In any case the Code emphasizes that suitable adjustments may be necessary in certain cases. For example, deflections must be reduced if very brittle finishes or partitions are used. Designers must therefore use their discretion; either in selecting more restrictive limits, or in exceeding the normal limits if they can justify such action.

Lateral deflections

The lateral deflections can be readily restricted to height/500, which ensures virtually complete freedom from difficulties with glazing panels and opening lights, etc. The lateral deflection of a frame structure need not be checked when the columns comply with the slenderness requirements given in appendix C.

Design values for deflection

The limit states of deflection are concerned with the probable maximum loading under normal working conditions throughout the life of the structure. Hence the characteristic loads can often be used directly since, in general, the partial safety factor for load is unity as shown in table B1. The properties of the materials are assumed to be those associated with the design strengths. The partial safety factors for material strength are also unity and the short-term modulus for concrete is related directly to the cube strength by the relation

$$E_c = 4500\sqrt{u} \quad (\text{MN/m}^2 \text{ units}).$$

10.2 Elastic deflections

The lateral deformation of an elastic member depends upon its flexural rigidity, *EI*. Several methods are available for determining elastic deflections and are adequately described in standard textbooks on structures. The direct method—considering the equation of a beam subjected to pure flexure—is discussed here, as well as approximations which can be made for complex loadings and members with non-uniform sections.

Bending of elastic beams

Simple elastic theory for the bending of beams assumes small deflections and obtains the following relation:

$$-EI\frac{d^2y}{dx^2} = M_x \tag{10.2.1}$$

where M_x is the bending moment at distance x from the origin, and y is measured downwards in the normal way as shown in figure 10.2.1.

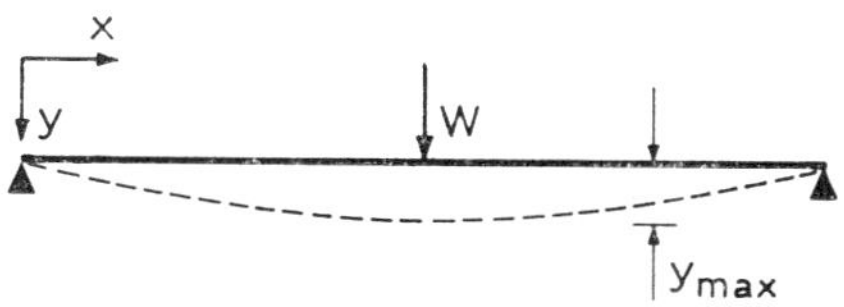

Fig 10.2.1 Elastic deflection of simple beam.

Shape of deflected beam

Integrating equation 10.2.1 once with respect to the horizontal distance x gives the slope, dy/dx, at any position. Integrating twice gives an expression for the vertical deflection y in terms of x. Two constants of integration are introduced which are determined by reference to points with known conditions. For example, the slope at midspan for a symmetrical beam is zero.

EXAMPLE 10.2.1

A uniform beam AB of span L and constant flexural rigidity EI is subjected throughout its length to a uniformly distributed load of w per unit length. Determine the maximum central deflection if it is (*a*) simply supported and (*b*) fixed ended.

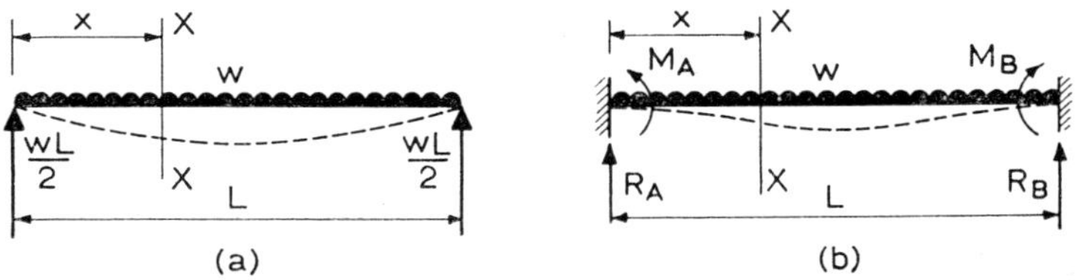

Fig 10.2.2 Example 10.2.1.

Solution

(*a*) $M_x = \frac{1}{2}w(Lx - x^2)$

Substituting in equation 10.2.1 and integrating gives

$$-EI\frac{dy}{dx} = \frac{w}{2}\left(\frac{Lx^2}{2} - \frac{x^3}{3}\right) + A$$

The slope is 0 at $x = L/2$, hence

$A = -wL^3/24$. Integrating again gives

$$-EIy = \frac{w}{2}\left(\frac{Lx^3}{6} - \frac{x^4}{12}\right) - \frac{wL^3}{24}x + B$$

$y = 0$ when $x = 0$ or L, hence $B = 0$. Thus the maximum deflection (at $x = L/2$) is given by

$EIy = 5wL^4/384$

$y = 5wL^4/384EI$ (Answer)

(*b*) By symmetry

$M_A = M_B$ and $R_A = R_B = \frac{1}{2}wL$

$M_x = \frac{1}{2}w(Lx - x^2) - M_A$

Hence

$$-EI\frac{dy}{dx} = \frac{w}{2}\left(\frac{Lx^2}{2} - \frac{x^3}{3}\right) - M_Ax + A$$

Slope $= 0$ at $x = 0$ or L (fixed ends). Hence $A = 0$ and

$M_AL = \frac{1}{2}wL^3(\frac{1}{2} - \frac{1}{3})$ or $M_A = wL^2/12$

Integrating again gives

$$-EIy = \frac{w}{2}\left(\frac{Lx^3}{6} - \frac{x^4}{12}\right) - \frac{wL^2x^2}{24} + B$$

$y = 0$ when $x = 0$ or L, hence $B = 0$. Thus the maximum deflection (at $x = L/2$) is given by

$$EIy = wL^4/384$$

$$y = wL^4/384EI \qquad \text{(Answer)}$$

Curvature

The curvature, ϕ, is the reciprocal of the radius, i.e. $1/R$, and is given approximately by $-d^2y/dx^2$. The simple elastic beam formula, equation 10.2.1, is given in terms of R by

$$M/I = E/R$$

Hence

$$\phi = -\frac{d^2y}{dx^2} = \frac{1}{R} = \frac{M}{EI} \qquad (10.2.2)$$

Integrating M/EI twice w.r.t. x gives an expression for which the power of x is raised by two. It is therefore possible to substitute a particular value of ϕ in the final expression for maximum deflection, y_{max}, for a uniform rectangular beam, in order to obtain an expression of the form

$$y_{max} = kL^2\phi \qquad (10.2.3)$$

where k = a constant given by the laws of mechanics applied to a beam of constant cross section
L = effective span of beam
ϕ = curvature of beam due to loading at a suitable section along the member (usually point of maximum sagging moment).

This expression is given in the Unified Code.

EXAMPLE 10.2.2

Determine the value of k in equation 10.2.3 for the beams in example 10.2.1 for ϕ_{max} equal to the value of ϕ at midspan.

Solution

(*a*) $\phi_{max} = M_{max}/EI = wL^2/8EI$

Therefore

$$y_{max} = 5wL^4/384EI = (5/48)L^2\phi_{max}$$

$$k = 5/48 \qquad \text{(Answer)}$$

(b) $\phi_{max} = M_{max}/EI = wL^2/24EI$

Therefore

$$y_{max} = wL^4/384EI = (1/16)L^2\phi_{max}$$

$$k = 1/16 \qquad \text{(Answer)}$$

(Notice that the value of k depends on the position at which ϕ is calculated).

Central deflection

The values of k which are appropriate for the central deflection of a fully elastic beam due to a variety of loading conditions have been tabulated by Beeby & Miles [10.1] and are included in appendix K. If it is assumed that the maximum deflection is never significantly different from the central deflection then it is possible to compute the deflection under more complex loading arrangements by obtaining values of k by superposition. It should be noted that the foregoing assumption is reasonable for beams but not cantilevers. The actual values of the maximum moments concerned must be carefully considered when using this approach, however, as shown below.

EXAMPLE 10.2.3

Determine the central deflection for the fixed-ended beam subjected to a uniformly distributed load considered in example 10.2.1 by suitable superposition of the data in appendix K.

Solution The fixed end moment of $wL^2/12$ must first be related to the maximum moment for the basic loading (conveniently taken as the uniformly distributed load in this case) which is $wL^2/8$. Hence for fixed end moments of $-M$ at each end,

$$-M = wL^2/12 = \tfrac{2}{3}M_{free}$$

From appendix K,

$$k = (5/48 - \tfrac{2}{3} \times \tfrac{1}{8}) \times M_{free}/M_{midspan}$$

$$M_{midspan} = wL^2/8 - wL^2/12 = wL^2/24$$

Hence

$$k = (1/48) \times (24/8) = 1/16$$

(which agrees with value previously obtained in example 10.2.2) and

$$y_{midspan} = kL^2\phi_{midspan} = (L^2/16) \times (wL^2/24EI)$$

$$= wL^4/384EI \qquad \text{(Answer)}$$

Significance of support section
Equation 10.2.3 for elastic deflections takes no account of variation in section properties, I_e or I_o, along the span. The support and midspan sections, for example, usually contain varying amounts of reinforcement. A rigorous analysis and the method given in an appendix to the Unified Code are discussed elsewhere [10.1]. However, the use of values for these terms based on the section at midspan has been found to be satisfactory for both simple and continuous beams. It should be noted that for continuous beams the appropriate loading condition for the limit state of deflection is when the full live load is applied to the span under consideration and both adjacent spans are unloaded. This reduces the support moments to a minimum so that the effect of the support section is in general small and little accuracy is lost if the deflection is calculated on the basis of the midspan section only. The use of a numerical mean of the second moment of area of the support and midspan sections should therefore be avoided. The Code states that the curvature at midspan should be considered, except for cantilevers and continuous members where the support moments exceed *three times* the midspan moment, in which cases ϕ should be calculated at the face of the support. The effective span, l, of a cantilever should be taken as the distance from the free end to the face of support plus ten times the diameter of the largest longitudinal tension bar at support, and the possibility of overall rotation at the support should also be considered.

l for simple beams is as for general requirements of other limit states

l for continuous beams equals the clear distance between faces of supports.

10.3 Short-term deflections

Elastic theory assumes linear behaviour between stress and strain even at high loads. The concrete in reinforced concrete members, however, cracks at relatively low tensile stresses so that the load deflection curve tends to be bilinear. The short-term deflection is therefore determined according to whether or not the member is cracked or uncracked, from the short-term moduli associated with the design strengths of the materials.

Load-deflection curves
The load-deflection behaviour of reinforced concrete members is best considered in terms of a typical experimental curve as shown in figure 10.3.1. Although linear behaviour occurs initially the first

crack produces a distinct change in slope which is accentuated by the incidence of further cracks. Yielding of the reinforcement for an under-reinforced beam causes the curve to become horizontal at a

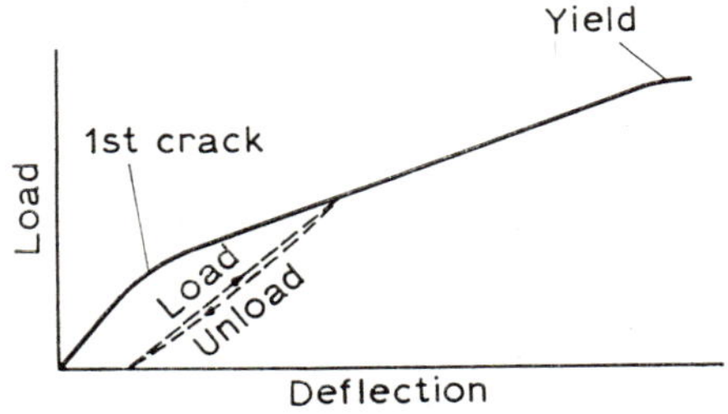

Fig 10.3.1 Typical experimental load-deflection curve (after reference 10.1).

much higher load as shown. If the member is unloaded at some load, W_1, the deflections measured during unloading and reloading cycles would follow the broken lines as shown.

Moment-curvature curves

The moment-curvature curves for an individual section are similar in form to the load-deflection curves since

$$M \propto W$$

and from equation 10.2.3

$$\phi \propto W$$

The main difference is that the moment-curvature relation concerns a single section, so the sudden change in slope for a cracked section is more marked for this curve than for the load-deflection curve,

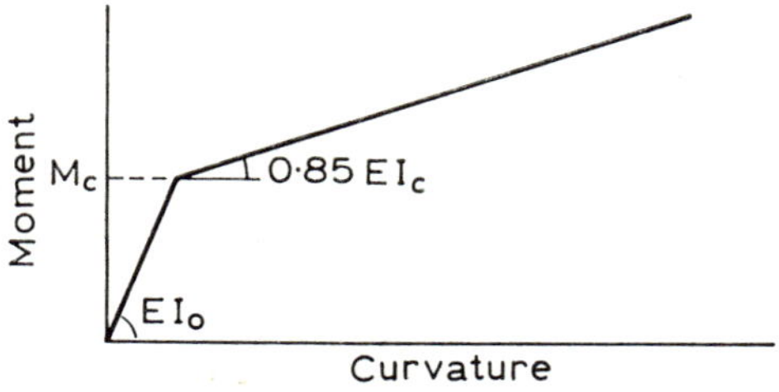

Fig 10.3.2 Bilinear moment-curvature relation.

which is affected by the cracking of sections along the full length of the beam. The idealized bilinear moment-curvature relation shown in figure 10.3.2 can therefore represent the actual moment curvature relation very closely. The slope for the uncracked section is EI_o while that for the cracked section is assumed to be $0{\cdot}85EI_c$ [10.1].

Curvature at uncracked section

The curvature at an uncracked section for short-term loading is given by

$$\phi = M/E_cI_o \tag{10.3.1}$$

where E_c = concrete modulus for short-term loading
I_o = second moment of area of uncracked section (based on *either* concrete section *or* gross concrete and steel section) for maximum sagging moment
M = moment at the section ($\not> M_c$)

Curvature at cracked section

The curvature at a cracked section for short-term loading is given by the Unified Code as

$$\phi = \frac{M - M_c}{0{\cdot}85E_cI_e} + \frac{M_c}{E_cI_o} \tag{10.3.2}$$

where I_e = second moment of area of cracked section, i.e. the equivalent (or transformed) sectionf or which concrete in tension is ignored
M_c = effective cracking moment given by

$$f_{ct}I_o/(d - d_n)$$

f_{ct} = flexural tensile strength of concrete at 28 days
$d - d_n$ = distance from tension face to the neutral axis of the concrete or gross section as appropriate.

Calculation of short-term deflections

The above bilinear relation enables the curvature at each section along a member to be determined and hence, by integration, the deflection at any point. The moment and corresponding curvature diagrams for a simple beam subjected to two-point loading are shown in figure 10.3.3. The value of k in equation 10.2.3 is governed by the shape of the curvature diagram. Hence it varies slightly from the elastic value which is given by the broken lines in figure 10.3.3. However, a very satisfactory approximation is obtained by calculating the maximum curvature using the *bilinear* relation in figure 10.3.2 and then assuming a *linear* relation between zero and this value. The values of k are then the same as those for a fully elastic beam.

The Code has modified equation 10.3.1 by substituting I_e for I_o so that the expression

$$\phi = M/E_cI_e$$

then overestimates the curvature. The only merit of this substitution is that the modified equation can be used for either cracked or uncracked sections, although equation 10.3.2 is still more accurate for cracked sections.

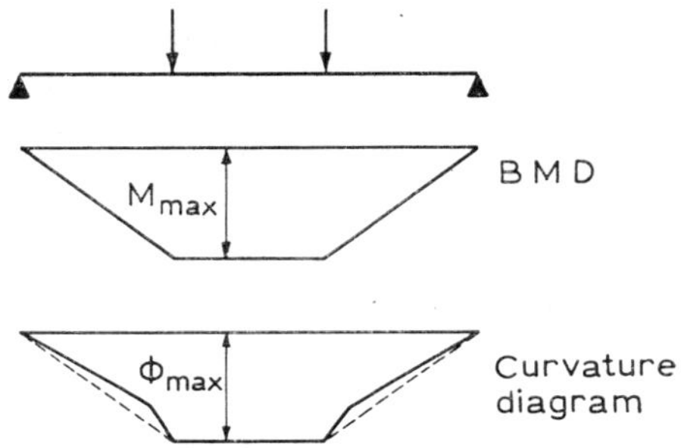

Fig 10.3.3 Bending moment and curvature diagrams for simple beam under two-point loading.

10.4 Long-term deflections

Accurate determination of the long-term deflection involves separate considerations of the actual creep and shrinkage coefficients of the concrete which is to be used. However, if the expected shrinkage and creep strains are not excessive (i.e. not abnormal) then the total long-term deflection is assumed to be the sum of the short-term deflection due to all loads and the additional deflection due to creep and shrinkage. This additional deflection is assessed by multiplying the short-term deflection by suitable factors which are given in the code.

Shrinkage effects

The amount of shrinkage depends upon the curing conditions, size and shape of member, mix proportions and type of aggregate used. If concrete is unrestrained and not reinforced it can shrink uniformly. However, the movement of the concrete is restrained if reinforcement is present, in such a way that the steel becomes compressed and the concrete is subjected to tension, as shown in figure 11.3.1 (chapter 11). In the case of a singly reinforced beam there is restraint on the tension side only; thus a strain gradient is imposed on the section which causes the beam to deflect. The addition of compression steel to the section will restrain the compression zone also and so tend to reduce the curvature. Curvature due to shrinkage therefore depends upon the relative quantities of compression and tension steel.

Creep effects

The effect of creep is to increase the compressive strain in the concrete by an amount $C_t\varepsilon_{ct}$ where C_t is the ratio of the creep strain

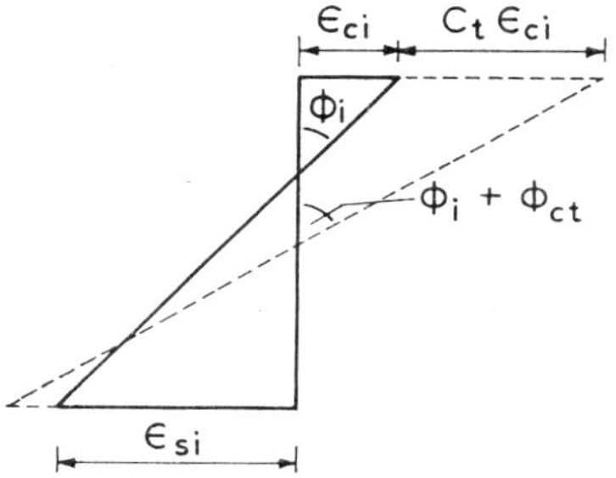

Fig 10.4.1 Short- and long-term strain profiles for singly reinforced beam.

to the short-term strain, ε_{ci}. The strain in the tension steel also increases by an amount which is governed by the reduction in the lever arm as shown in figure 10.4.1 for a singly reinforced beam. The curvature of the section increases by an amount ϕ_{ct} and hence the deflection increases by the factor $(\phi_i + \phi_{ct})/\phi_i$ where ϕ_i is the short-term curvature.

Additional deflections

If the ultimate shrinkage strain, ε_{sh}, does not exceed 0·6 millistrain, and neither does the ratio, C_t, of creep strain to elastic strain exceed 3·5, then the additional deflections may be assumed not to exceed that part of the short-term deflection which is due to the permanent load multiplied by the appropriate factor given in table 10.4.1.

Table 10.4.1 Additional deflection as a multiple of short-term deflection

	AGE AT LOADING			
	14 days	28 days	90 days	1 year
$A_{sc}=0$	2·0	1·8	1·5	1·3
$A_{sc}=\frac{1}{2}A_{st}$	1·2	1·1	0·9	0·7
$A_{sc}=A_{st}$	0·8	0·7	0·5	0·3

Reductions in additional deflection outstanding

When a time interval elapses between the member becoming self-supporting and the erection of non-load-bearing partitions and finishes, a proportion of the additional deflection due to the self-weight of the member will take place during this time interval. The outstanding percentage of the additional deflection due to self-weight should be assumed to be not less than the values given in table 10.4.2.

Table 10.4.2 Additional deflections outstanding after various time intervals

Time interval	14 days	28 days	90 days	6 mths	12 mths	36 mths
Percentage additional deflection outstanding	70	60	40	30	20	10

EXAMPLE 10.4.1

A uniform rectangular beam, 0·15 m wide and of 0·3 m effective depth, carries a uniformly distributed dead load (including self-weight) of 2·67 kN/m after 14 days, and a uniformly distributed imposed load of 4 kN/m after 90 days. Calculate the maximum deflection for a simple span of 3 m if the area of steel at midspan is 226 mm². Assume that the ratio of creep strain to elastic strain $\not> 3{\cdot}5$, shrinkage strain $\not> 600$ microstrain, and that 40% of imposed load is permanent. $u = 22{\cdot}5$ MN/m², $f_{yt} = 250$ MN/m², $E_s = 200$ GN/m², $E_c = 4500\sqrt{u}$ (MN/m² units) and $A_{sc} = 0$ at midspan.

Solution $n_1 = 0{\cdot}251$ and I_e for short-term loading $= 117 \times 10^{-6}$, by the same working as in example 3.3.1.
Midspan moment for design loads is

$$(1{\cdot}0 \times 2{\cdot}67 + 1{\cdot}0 \times 4{\cdot}0) \times 10^3 \times 3^2/8 = 7{\cdot}5 \times 10^3$$

$$\phi_{max} = M/E_cI_e = (7{\cdot}5/21{\cdot}4 \times 117) \times 10^{3-9+6} = 3{\cdot}0 \times 10^{-3}$$

$$y_{max} = (5/48)L^2\phi_{max} \quad \text{(as for example 10.2.2}(a))$$

$$= (5/48) \times 3^2 \times 3{\cdot}0 \times 10^{-3} = 2{\cdot}81 \times 10^{-3}$$

The additional deflection for a permanent load of 2·67 kN/m applied at 14 days for $A_{sc} = 0$ is, from table 10.4.1,

$$2{\cdot}0 \times (2{\cdot}67/6{\cdot}67)y_{max} = 2{\cdot}25 \times 10^{-3}$$

The additional deflection for a permanent load of 0·4 × 4·0 kN/m applied at 90 days is

$$1{\cdot}5 \times (1{\cdot}60/6{\cdot}67)y_{max} = 1{\cdot}01 \times 10^{-3}$$

Hence the maximum deflection is

$$(2{\cdot}81 + 2{\cdot}25 + 1{\cdot}01) \times 10^{-3} = 6{\cdot}07 \text{ mm} \qquad \text{(Answer)}$$

Comments

Notice that in appendix B, the long-term deflection must not be greater than span/250, i.e. 12 mm, and the additional deflection after application of finishes not greater than span/350, which is 8·6 mm. Hence the deflections are satisfactory despite the fact that the above calculations generally overestimate them by between 30 and 100% for the reasons stated in the next section.

Notice also that the additional deflection due to self-weight outstanding after 90 days (a time interval of 90 — 14 = 76 days in table 10.4.2 giving, say, 45%) is

$$0{\cdot}45 \times 2{\cdot}0 \times (2{\cdot}67/6{\cdot}67)y_{max} = 1{\cdot}02 \times 10^{-3}$$

The elastic deflection due to a non-permanent imposed load of 0·6 × 4 kN/m is

$$(2{\cdot}4/6{\cdot}67)y_{max} = 1{\cdot}01 \times 10^{-3}$$

Hence the possible additional deflection after application of finishes is

$$(1{\cdot}02 + 1{\cdot}01 + 1{\cdot}01) \times 10^{-3} = 3{\cdot}04 \times 10^{-3}$$

This is considerably less than the value of 8·6 mm.

*Accurate analysis**

An accurate estimation of the final deflection is sometimes necessary, so an appendix to the Code describes how the support sections of continuous beams, as well as actual experimental values for ε_{sh} and C_t, can be considered. Additional deflections due to shrinkage occur when the free shrinkage of the concrete is resisted by non-symmetrical reinforcement, thus inducing a shrinkage curvature, ϕ_{sh}, into the member. The Unified Code gives values of $\phi_{sh}D/\varepsilon_{sh}$ for various steel ratios. The deflection due to shrinkage, Δ_{sh}, is given by the double integral of ϕ_{sh} with respect to the distance from the support, hence

$$\Delta_{sh} = k'\phi'_{sh}L^2$$

* Omit on first reading.

where $k' = \frac{1}{2}$ for cantilevers, $\frac{1}{8}$ if simply supported and $\frac{1}{16}$ if fixed both ends

ϕ'_{sh} = maximum value of ϕ_{sh} for cantilevers and simple beams, and average ϕ_{sh} for midspan and support sections of continuous members.

The creep deflection, Δ_{cp}, is related to the instantaneous deflection (due to the sustained loading) Δ_i by

$$\Delta_{cp} = k_r C_t \Delta_i$$

where k_r is the reduction factor which allows for the fact that the ratio of creep curvature to instantaneous curvature is less than the ratio of creep strain to instantaneous strain ($k_r = 0{\cdot}61$, $0{\cdot}43$ and $0{\cdot}28$ for $A_{sc} = 0$, $\frac{1}{2}A_{st}$ and A_{st} respectively). Reference should be made to the Unified Code itself if this method is to be used.

10.5 Limiting span/effective-depth ratios

The methods of calculation given in the Unified Code predict the short-term and long-term deflections of laboratory specimens to within $\pm$ 20% but can overestimate the deflection of an actual structure by over 100%. The Code therefore states that the requirements of the limit state of deflection may be deemed to be satisfied if the structure conforms with the limiting span/effective-depth ratios, based on practical experience, given in appendix C. The scope of the clauses giving limiting span-depth ratios are sufficiently extensive to cover almost all structures, thus enabling the calculation procedures to be restricted to use in abnormal circumstances or to justify the use of members of increased slenderness.

Errors in calculating deflections

The assumptions and simplifications for calculating deflections are conservative for the following reasons.

(i) Estimates of the restraint provided by supports are based on oversimplified assumptions and must necessarily be safe estimates.

(ii) The stiffening effect of finishes and partitions is neglected.

(iii) Calculations of rigidity and cracking are based on characteristic strengths which are less than the strengths of the materials averaged over the structure.

(iv) The precise loading, especially that part of the loading which is of long duration, is not known accurately and tends to be overestimated.

Alternative Unified Code approach

The alternative Unified Code approach allows the limit state of deflection to be deemed to be satisfied if the ratio of span to effective depth, L/d_1, of a member does not exceed either 60 or the product of four appropriate factors suitably defined, whichever is the lesser. This alternative approach is outlined in appendix C.

Additional factors

The normal increases in creep coefficient associated with lightweight concrete are accounted for by an additional factor of 0·85 (unless it can be demonstrated experimentally that this is not necessary). Similarly the increased flexibility associated with flat slabs without drops must also be allowed for by additional factor of 0·90. Factors involving the characteristic strength required of the concrete, the actual area of compression reinforcement and the distinction between slabs spanning in one or both directions were considered in drawing up the Code proposals and were shown to have an insignificant influence [10.1].

Support conditions

A member should be considered as "simply supported" in table C7 if the reinforcement at the supports is insufficient to develop the bending moments implied by "partial" or "full" fixity. Although degree of fixity also depends on the relative loading of different spans, full fixity may be assumed in centre spans of continuous members and also in end spans when the combined stiffness of all other members connected to the end of the beam or slab is not less than the stiffness of the beam or slab being considered. The end span of a continuous beam or slab should otherwise be assumed to be partially fixed. A member should only be considered as continuous if its span does not differ by more than 15% from the spans of adjacent members.

Reinforcement

Notice that in table C8 the Code permits the assumption of a steel strength of less than the actual value, if this lower strength would be adequate to satisfy the limit state of collapse, provided that it is never less than 250 MN/m^2. For a flanged beam or a ribbed slab the effective width of the flange should be used in calculating the gross area at midspan. The differences in stiffness which occur between beams and slabs is reflected in the lower percentages of tension steel provided in slabs, so stiffness need not be considered separately.

Long-term effects

Table C9 was based on the following assumptions:

(i) Time-dependent deflection $= k_1 k_2 y_i$

where y_i = short-term deflection
k_1 = ratio of permanent load to total load
k_2 = creep multiplier, obtained from table 10.4.1.

(ii) The ratio l/d_1 is inversely proportional to the total deflection $(1 + k_1 k_2)y_i$.

The values in table C9 are inversely proportional to $(1 + k_1 k_2)$ and their absolute values depend on the assumption that the value of unity applies to the case when the permanent load is 75% of the total load and the amount of compression steel is low. The values of k_2 used in association with the three values of k_1 were 1·6, 1·1 and 0·8 respectively [10.1].

Mix proportions and environment

Table C9 applies to "normal" concrete. Table C10 gives general guidance to help the designer select the appropriate category. Creep and shrinkage coefficients are increased by such factors as high cement content, high water content, high temperature, low humidity and early loading. The properties of the aggregate can also have a pronounced effect. In addition to its influence on creep and shrinkage coefficients, a temperature crossfall through a member can provide an additional increase in deflection.

Deflections must be considered critical when their values must not exceed span/250 (total) and span/350 (after finishes). The 20% increase for non-critical applications is in agreement with increased deflections of span/200 (25% increase) and span/300 (17% increase) respectively. Even lower multipliers should be used in table C10 for "supercritical" conditions (i.e. very brittle finishes). It is expected, however, that the majority of structures will be in category 1 (see table C10) but judgement will inevitably be necessary in many cases.

EXAMPLE 10.5.1

Consider the span-depth ratios for the beam described in example 10.4.1 and hence show that the limit state of deflection is satisfactory.

Solution

Actual span/effective-depth ratio $= L/d_1 = 3/0{\cdot}3 = 10$
Basic l/d_1 ratio from table C7 $= 20$ for deflection $\not> L/250$
X-multiplier from table C8 for $r' < 0{\cdot}5\%(r = 0{\cdot}5\%) = 1{\cdot}55$

Permanent-load/total-load $= (2{\cdot}67 + 0{\cdot}4 \times 4{\cdot}0)/6{\cdot}67 = 0{\cdot}64$
Hence Y-multiplier, from table C9 $= 1{\cdot}15$
Z-multiplier from table C10 for "critical" deflection $= 1{\cdot}0$
Hence $XYZ(l/d_1) = 1{\cdot}55 \times 1{\cdot}15 \times 1{\cdot}0 \times 20 = 35{\cdot}7$

The actual span-depth ratio $\leqslant 35{\cdot}7$ and $\leqslant 60$, hence the limit state of deflection is satisfactory. This result agrees with that for example 10.4.1 since the deflections were much less than the permitted values.

REFERENCE

10.1 BEEBY, A. W., and MILES, J. R. Proposals for the control of deflection in the new Unified Code for structural concrete. *Concrete* vol. 2 no. 12 Dec. 1968, pp 511–15 and vol. 3 no. 3 March 1969, pp 101–10.

Exercise

1. A reinforced concrete beam of 10 in (254 mm) width and 20 in (508 mm) effective depth is singly reinforced with 4 in^2 (2581 mm^2) of steel throughout its length. Its effective span is 20 ft (6·1 m). Find the central deflection due to the self-weight alone according to elastic theory, assuming that the concrete in the tension zone is fully cracked over the whole span.

The beam carries a live load which is uniformly distributed over the central 10 ft (3·04 m) of the beam, thus extending 5 ft (1·52 m) either side of the centre line. If the deflection due to the live load alone is restricted to 3 times that calculated for the self-weight, find the load per foot run that the beam can carry.

Examine the basis of the deflection calculations and explain how they are likely to be in error.

Elastic moduli of steel and concrete are 30×10^6 and 5×10^6 lbf/in^2 (207 and 34·5 GN/m^2). Self-weight of concrete is 144 lbf/ft^3 (22·6 kN/m^3). (UL2 1969, 36 min)

Chapter 11

Limit states of local damage

The limit states of local damage are concerned with damage to the structure which could occur due to the application of normal working loads, that is, loads which are not expected to exceed the characteristic values. The partial safety factors for load are therefore, in general, unity, as for the limit state of deflection. The partial safety factors for material strength are also unity, at least in principle. However, it is necessary to use a value of $\gamma_m = 1{\cdot}3$ for concrete cube strength in order to obtain, approximately, the uniaxial compression strength. Local damage can occur due to either excessive tension or compression, in either the concrete or the steel. Tension in the concrete can result in the limit state of crack formation and in various limit states of crack width. In most structures it is sufficient to ensure that the crack widths do not become excessive but in water-retaining structures it is often necessary to ensure also that the limit state of crack formation is not reached.

11.1 Spalling of concrete and tension in steel

The stresses under working loads in both concrete and steel must be restricted if compression spalling or microcracking of the concrete and excessive tension in the steel are to be avoided. This approach is adopted by conventional elastic theory and provides a suitable basis for analysis for local damage provided that suitable allowance is made for creep. The Unified Code assumes that the long-term modulus for concrete equals half the short-term modulus.

Compression spalling of concrete

Spalling of the concrete must clearly be avoided under working load conditions. The difficulty here, however, is that spalling of the concrete rarely occurs much before final collapse. The Unified Code does not give a definite requirement for reinforced concrete but states that the concrete stress should not exceed $\frac{1}{2}u$ at transfer for prestressed concrete members in flexure (to avoid undesirable microcracking). Although this value is also intended to prevent excessive prestressing losses, it will be assumed for convenience to be suitable for preventing spalling in reinforced concrete.

Excessive tension in steel

The Unified Code stipulates that the steel strain in reinforced concrete beams must not normally exceed either 1·5 millistrain or $\frac{3}{4}f_y/E_s$ (see appendix B). Assuming that E_s is 200 GN/m^2 this means that the steel stress must not exceed either $\frac{3}{4}f_y$ or 300 MN/m^2. The Code considers steel strain rather than stress because it is really concerned with the tensile strain (and hence crack widths) in the concrete. A short-term tensile stress of $\frac{3}{4}f_y$ does not produce visible damage in the steel any more than a short-term compressive stress of $0{\cdot}5u$ produces spalling of the concrete. Even if the full characteristic loads were applied throughout the working life of the structure (which is highly improbable) it is also very unlikely that the concrete would show any spalling at the end of this period.

Design stresses and design strengths

The design stresses given above for the concrete and the steel have been assumed to be $\frac{1}{2}u$ and $\frac{3}{4}f_y$ respectively. These are the maximum stresses considered desirable for working load conditions. They should be regarded as quite distinct from the values for material strength. Hence the coefficients of $\frac{1}{2}$ and $\frac{3}{4}$ should not be confused with the reciprocals of the partial safety factors for material strength. For example, the partial safety factor of 1·3 given in table B1 refers to material strengths only and is irrelevant in this instance. It is therefore not incompatible with the above coefficient of $\frac{1}{2}$ for which the reciprocal is 2 and *not* 1·3.

Comparison with elastic theory

The permissible values used in conventional elastic theory were given by

$$p_{st} = \tfrac{1}{2}f_{yt} \quad \text{and} \quad p_{cb} = u/3$$

which are only 2/3 of the values given above. Hence all stresses at working load are increased by 50% and the moments of resistance given in table 4.1.1 are also increased by 50%. Although the increases for high-yield steels are less, since the design stress cannot exceed 300 MN/m^2, it is clear that this limit state for local damage permits much higher loads than conventional elastic theory. Hence although the limit state of collapse permits higher working loads than conventional elastic theory, they are usually less than those permitted by the limit state of local damage for spalling of concrete and tension in steel for beams.

EXAMPLE 11.1.1

A simply supported beam carries a total uniformly distributed load w per unit length. The midspan section had been designed by conventional elastic theory to be at ESR. If $u = 22{\cdot}5$ MN/m², $f_{yt} = 250$ MN/m² and $m = 18{\cdot}7$ (i.e. long-term E_c) compare the loadings which can be carried for the limit states of collapse and local damage. Assume: (*a*) all dead loading and (*b*) all live loading.

Solution Calculations according to elastic theory provide the values given in table 4.1.1(*b*), i.e.

$$r_e = 1{\cdot}59\% \quad \text{and} \quad M_r = 1{\cdot}63bd_1^2 \quad (\text{MN/m}^2 \text{ units})$$

Hence for limit state of local damage, loading equals $1{\cdot}5w$ and moment of resistance of section equals $2{\cdot}45bd_1^2$.

For collapse,

$$_*w = \text{either } (a)\ 1{\cdot}4w' \quad \text{or} \quad (b)\ 1{\cdot}6w'$$

where w' is the working load permitted by limit state of collapse. Resistance moment at collapse is given by the steel, hence from equation 8.2.3

$$M_{su} = \frac{rf_{yt}}{1{\cdot}15}\left(1 - \frac{1{\cdot}1rf_{yt}}{u}\right) bd_1^2 = 2{\cdot}78bd_1^2$$

after substituting the appropriate values into the equation. Hence in terms of the working load, w', the moment of resistance is $1{\cdot}99bd_1^2$ or $1{\cdot}74bd_1^2$. But $M_r = 1{\cdot}63bd_1^2$ corresponded to a working load of w, hence the working loads corresponding to the limit state of collapse are $1{\cdot}22w$ and $1{\cdot}07w$ respectively. The limit state of collapse therefore permits greater working loads in both of these two cases than conventional elastic theory but neither is as great as that permitted by the limit state of local damage. Thus

$$w < 1{\cdot}07w \quad \text{and} \quad 1{\cdot}22w < 1{\cdot}5w \qquad \text{(Answer)}$$

Assumptions for local damage

The analysis of sections for the limit state of local damage assumes that plane sections remain plane and that stress varies as strain. In addition, the Unified Code assumes that

(i) Concrete modulus is given in MN/m² units by the long-term value of $2250\sqrt{u}$ (i.e. one-half the short-term value).

(ii) Modulus of elasticity of reinforcement = 200 GN/m²

(iii) Tensile strength of concrete is ignored.

11.2 Limit states of crack width for beams

Excessive tension in the reinforcement can cause excessive cracking in the adjacent concrete. Beams subjected to flexural tension due to the action of transverse loads exhibit a series of distributed flexural cracks even at working loads. These cracks are in general harmless unless the width becomes excessive. The Unified Code specifies a maximum crack width of 0·3 mm for any position at the surface. In addition, for members exposed to a particularly aggressive environment the maximum surface crack width at points nearest to main reinforcement shall not exceed 0·1 mm.

Effect of stress gradient

The distribution of flexural cracks in a beam is controlled by the stress gradient rather than by the bond stress between the concrete and the steel. It follows therefore that little, if any, bond slip is likely to occur adjacent to flexural tension cracks in beams. Base *et al.* [11.1] carried out a very extensive series of crack measurements on 133 beams and showed that although some slipping occurs it is convenient to regard the concrete at the surface of the reinforcement as a "point of zero strain". That is, the steel strain and the actual bond slip are relatively small, so the crack width at the surface of the reinforcement can be regarded as zero and the cracks may be assumed to increase in width towards the face of the beams. They also found that for stresses within the elastic range of the steel the crack widths were closely proportional to the steel stress. At very low levels of stress there was a transition from the apparently uncracked beam to the cracked beam, which was to be expected. Within the ranges examined the effect of bar size, steel ratio, concrete strength and method of curing could all be neglected, thus supporting the "no-slip" theory. The width of a crack was therefore determined by the elastic recovery of the concrete due to relief of tension and by the restraining effect of the nearest reinforcing bar.

Crack widths for rectangular beams

The draft Code gives the following general relation for the crack width at a point on the surface of the tension zone of a rectangular beam.

$$\text{Maximum surface crack width} = 2{\cdot}3ce \qquad (11.2.1)$$

where c = distance from nearest "point of zero strain"
e = apparent longitudinal tensile strain in concrete.

The nearest "points of zero strain" are given by either the surface of a longitudinal bar of at least 10 mm diameter or the neutral axis.

Stiffening effect of concrete in tension zone

The position of the neutral axis and the magnitude of the calculated longitudinal strain, ε_c, are obtained using the assumptions given at the end of section 11.1. With reference to figure 11.2.1, if P is distance y from the neutral axis and ε_s is the strain in the tension steel, then the strain in the concrete, ε_c, at P is

$$\varepsilon_c = \varepsilon_s y/(d_1 - d_n)$$

ε_s is calculated in the normal way, neglecting the stiffening effect of the concrete in the tension zone. However this stiffening effect

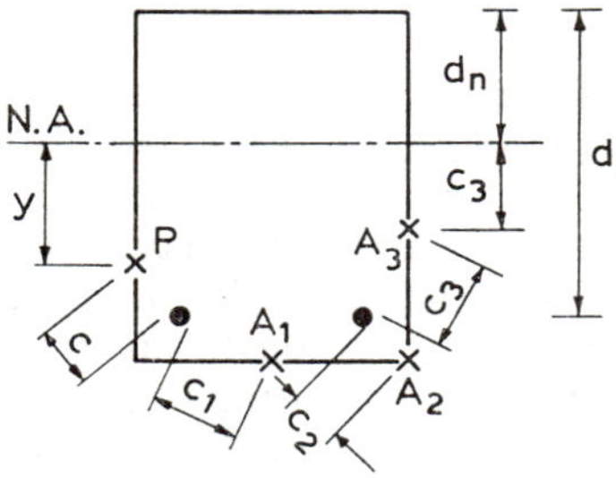

Fig 11.2.1 Determination of crack widths for rectangular beams.

should be allowed for when determining the apparent longitudinal strain in the concrete, e, as follows:

$$e = \varepsilon_c - 4 \times 10^{-6}/r \qquad (11.2.2)$$

where r = steel ratio, $A_{st}/b_r d_1$
b_r = width of section *at centroid of tensile steel.*

Maximum crack width

Although termed "maximum crack width" by the Code, it is in fact the crack width that has an acceptable probability of not being exceeded. Base *et al.* found that only about one crack in a hundred exceeded twice the mean crack width. The Unified Code has increased the probability from 1% to at least 5% in obtaining the coefficient of 2·3 in equation 11.2.1.

Type of bar

Base *et al.* found that the crack widths for plain round bars were approximately 13% greater than for deformed bars. This was no doubt due to the fact that some slight bond slip did occur. However, it is clear that even large differences in bond strength have a relatively small effect. The Unified Code ignores this small difference in assuming that equation 11.2.1 is valid for both types of bar.

Crack widths for T-beams

The "no-slip" theory was verified for beams with steel ratios of 0·93% and over. For T-beams the values of c can become very large and the following rule should be noted [11.2]: "The primary crack pattern which forms is a function of the member geometry. The addition of reinforcement locally modifies this according to equation 11.2.1. If the bars are close enough together this modified pattern dominates."

Protection against corrosion

Stirrup reinforcement, nominal or otherwise, tends to act as a crack inducer (since it is at right angles to the main reinforcement) and such steel is therefore vulnerable to attack. This possibility is covered by the general local damage requirement that the limit state of crack width of 0·3 mm should always be satisfied. For members exposed to a particularly aggressive environment the additional requirement that the maximum crack widths at points nearest to the main steel in regions of maximum tensile strain shall not exceed 0·1 mm is also necessary.

Concrete cover and crack width

It is usually assumed that protection against corrosion is proportional to the depth of the concrete cover. Resistance to corrosion also depends on crack width which, according to equation 11.2.1, increases with depth of cover. However, reference to figure 11.2.1 shows that the most critical positions on the concrete surface are such points as A_1, A_2 or A_3 which depend more on the distribution of the bars than on the concrete cover. Furthermore it should be remembered that once the nominal cover has been selected, as for example when referring to mix design tables (see reference 11.3) the cover to the main steel is then determined. The maximum crack width requirement is therefore considered *in addition* to the concrete cover.

EXAMPLE 11.2.1

Calculate the maximum crack widths for the rectangular beam shown in figure 11.2.2. Assume that $E_s = 200\ \text{GN/m}^2$, $f_y = 420\ \text{MN/m}^2$, $f_{st} = 250\ \text{MN/m}^2$, $r = 0{\cdot}98\%$, $d_n = 175$ mm and bar diameter $= 20$ mm. Are the crack widths and tensile strains satisfactory with respect to protection against corrosion in an aggressive environment for the limit states of local damage?

Solution For soffit of beam (P_1, P_2 and P_3)

$$y = 440 - 175 = 265\ \text{mm}$$

At point P_1

$$c_1 = \sqrt{[(30 + 10)^2 + (40 + 10)^2]} - 10 = 54\ \text{mm}$$

Similarly for P_2

$$c_2 = \sqrt{(40^2 + 75^2)} - 10 = 75 \text{ mm}$$

For P_3

$$c_3 = 30 \text{ mm}$$

For P_4

$$y \simeq \tfrac{1}{2}(400 - 175) = 112 \text{ mm} \quad \text{i.e.} \quad c_4 = 112 \text{ mm}$$

(Exact y given by

$$y = \sqrt{[(225 - y)^2 + 50^2]} - 10 \quad \text{or} \quad y = 113 \text{ mm}).$$

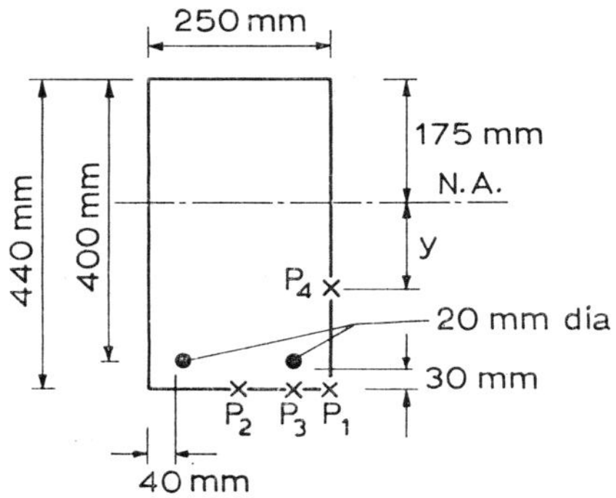

Fig 11.2.2 Example 11.2.1.

Actual ε_s is

$$f_{st}/E_s = (250/200) \times 10^{6-9} = 1{\cdot}25 \times 10^{-3}$$

(Satisfactory since it is $\not> 0{\cdot}0015$ and $\not> \tfrac{3}{4}f_y/E_s$ which equals $1{\cdot}575 \times 10^{-3}$.) $r = 0{\cdot}98\%$, hence from equation 11.2.2 for P_1, P_2 or P_3

$$e = (440 - 175)\varepsilon_s/225 - (4/9{\cdot}8) \times 10^{-6+3}$$

$$= (1{\cdot}47 - 0{\cdot}41) \times 10^{-3}$$

For P_4

$$e_4 = (112/225)\varepsilon_s - 0{\cdot}41 \times 10^{-3} = (0{\cdot}63 - 0{\cdot}41) \times 10^{-3}$$

Hence from equation 11.2.1

$$\Delta_1 < \Delta_2 = 2{\cdot}3 \times 75 \times 1{\cdot}06 \times 10^{-3-3} = 0{\cdot}183 \times 10^{-3}$$

$$\Delta_3 = 2{\cdot}3 \times 30 \times 1{\cdot}06 \times 10^{-3-3} = 73 \times 10^{-6}$$

$$\Delta_4 = 2{\cdot}3 \times 112 \times 0{\cdot}22 \times 10^{-3-3} = 57 \times 10^{-6}$$

Maximum surface crack width (at P_2) = 0·18 mm < 0·3 mm.
Maximum width at minimum cover (at P_3) = 0·07 mm < 0·1 mm.
Hence crack widths are satisfactory for an aggressive environment.
(Answer)

Note: the nominal cover to 10 mm stirrups, for example, would only be 20 mm. Hence from mix design tables [11.3] it is clear that this would be inadequate with any grade of concrete and would have to be increased.

11.3 Thermal and shrinkage cracks in walls

All structures are subject to temperature, creep and shrinkage effects but these are frequently inconsequential compared with the stresses due to gravitational and similar external loads. This is not so, however, when the external loading is small and only nominal reinforcement is considered necessary. The contraction which occurs due to shrinkage, heat of hydration of the cement, and seasonal variations in temperature, can produce excessive tension in lightly reinforced members due to the external restraint which invariably exists. Thin sections such as walls and slabs are particularly susceptible to drying shrinkage, for example, and must therefore contain a certain minimum quantity of steel, in *both* directions, if they are not to be rendered unserviceable by wide cracks. There should therefore be sufficient reinforcement to ensure that a number of distributed cracks can occur such that the limit state of crack width, appropriate to the particular conditions of exposure, is satisfied. The reinforcement must ensure that

(i) All cracking is fully controlled, and
(ii) The width of a fully developed crack is not excessive.

Heat of hydration

All concrete expands after placing due to the heat of hydration of the cement. This expansion is followed by contraction as the concrete cools to the temperature of its surroundings. It is the subsequent thermal contraction which combines with the seasonal temperature variations and the relatively small drying shrinkage to produce the total thermal and shrinkage movement.

Distribution of cracks

The total shrinkage and thermal movement in plain concrete can produce tension which exceeds the tensile strength of the concrete and which produces a wide, single crack. This cracking can be controlled by the introduction of a relatively small amount of reinforcement in the form of small-diameter bars or fabric. The function of the reinforcement is to distribute the cracks since excessive cracking at one section can relieve the direct tension which would otherwise occur at other sections along the wall. This situation for direct tension is therefore very different to that for flexural tension, considered previously, where distributed cracking was assured by the stress gradient across the beam. A further difference, of course, is that the steel ratios for beams are generally much higher than those for walls and similar members.

Unrestrained shrinkage

Consider a reinforced concrete member with perfect bond between concrete and steel as shown in figure 11.3.1. If the free shrinkage

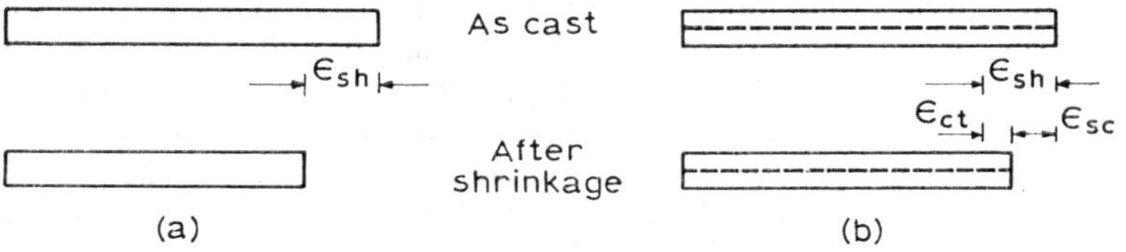

Fig 11.3.1 Strains before and after shrinkage for both plain and reinforced concrete (*a*) Plain concrete. (*b*) Reinforced concrete.

strain for the plain concrete is ε_{sh} then the inclusion of reinforcement produces a tension strain, ε_{ct}, in the concrete and a compression strain, ε_{sc}, in the steel such that

$$\varepsilon_{sh} = \varepsilon_{ct} + \varepsilon_{sc} = (f_{ct}/E_c) + (f_{sc}/E_s) \qquad (11.3.1)$$

Equating internal forces and assuming that there is no external restraint:

$$A_c f_{ct} = r' A_c f_{sc} \qquad (11.3.2)$$

Substituting for f_{ct} in equation 11.3.1, putting $m = E_s/E_c$ and rearranging gives

$$f_{sc} = \varepsilon_{sh} E_s/(1 + mr') \qquad (11.3.3)$$

EXAMPLE 11.3.1

Calculate the theoretical steel and concrete stresses in a reinforced concrete column if $E_s = 200\ \text{GN/m}^2$, $m = 15$, $r' = 1\%$ and $\varepsilon_{sh} = 250$ microstrain. Assume (*a*) no restraint and (*b*) full restraint.

Solution

(*a*) From equation 11.3.3

$f_{sc} = 250 \times 200 \times 10^3/(1 + 0{\cdot}15) = 43{\cdot}5\ \text{MN/m}^2$ (Answer)

From equation 11.3.2

$$f_{ct} = 0{\cdot}01 \times 43{\cdot}5 \times 10^6 = 435\ \text{kN/m}^2 \qquad \text{(Answer)}$$

Notice that these stresses are relatively small even for an extremely high shrinkage strain. Although ε_{sh} and m can vary within wide limits, the stresses in *unrestrained* members, in practice, can be readily sustained by both the concrete and the steel.

(*b*) $f_{sc} = 0$ (Answer)

and

$$f_{ct} = (200/15) \times 250 \times 10^3 = 3{\cdot}33\ \text{MN/m}^2 \qquad \text{(Answer)}$$

The calculated stress for the concrete is far too high and means that the concrete must crack in at least one position. It is now necessary to consider what happens adjacent to this first crack.

Fully restrained shrinkage cracks

When the concrete first cracks, local slipping between the concrete and steel must occur. The amount of slipping and hence the probable spacing between cracks depends upon the properties of the materials and the amount of steel present. If the ends are fully restrained then the steel cannot be in compression until the concrete cracks, since the overall strain is zero. When the concrete fractures, the steel at the crack is in tension and strains locally, resulting in compression strains elsewhere in the steel, since the uncracked lengths of concrete try to contract as shown in figure 11.3.2.

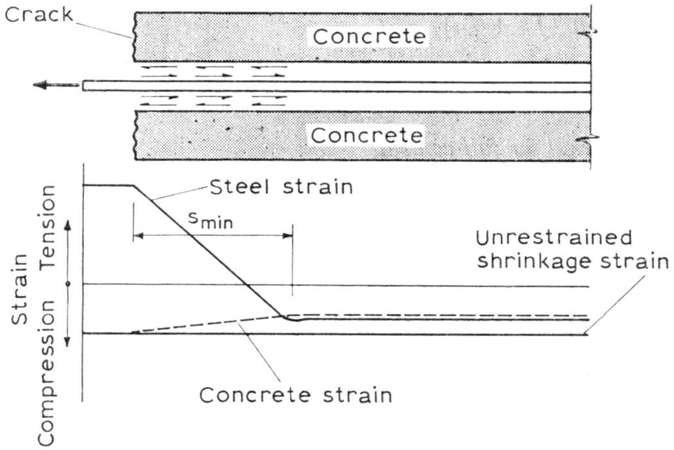

Fig 11.3.2 Probable strain distribution in concrete and steel adjacent to a crack.

Even if the concrete achieved the full shrinkage strain of 250×10^{-6} quoted in example 11.3.1, the steel would attain a stress of about 50 MN/m^2 only. Since the restrained concrete would have cracked at this strain, either a very large number of fine hair cracks form in the concrete or bond slipping (i.e. "bond creep") occurs along the bar.

Relative strengths of steel and concrete

After the first crack has occurred the equilibrium of the portion of the member adjacent to the crack (e.g. between sections 1 and 2 in figure 11.3.3) is given by

$$A_s f_{st} = A_c f_{ct} - A_s f_{sc}$$

or

$$f_{ct} = r(f_{st} + f_{sc})$$

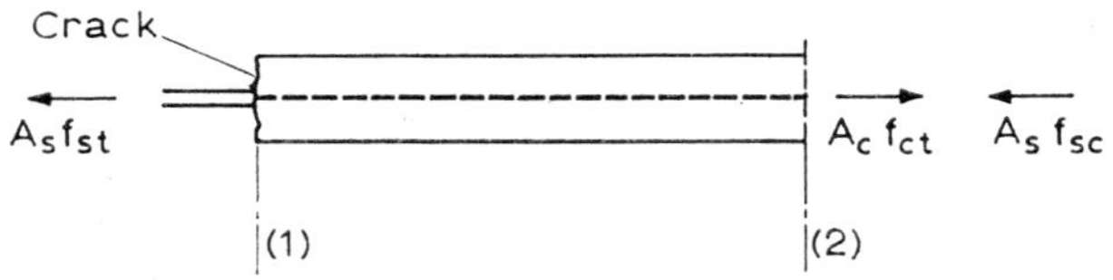

Fig 11.3.3 Forces acting on portion of member adjacent to a crack.

The critical steel ratio, r_{crit}, is given when the tension stresses in the concrete and steel reach their maximum values simultaneously. Hence when $f_{ct} = u_t$ and $f_{sc} = f_{yt}$

$$r_{crit} = u_t/(f_{yt} + f_{sc}) \tag{11.3.4}$$

u_t and f_{yt} are known or can be estimated, while the compressive stress in the steel, f_{sc}, can be determined as shown below.

Effective period for full control

The heat of hydration is largely dissipated after about three days and base restraints can subsequently assist the effective control of cracking very significantly. The period which is most critical and which should be fully controlled by the steel alone may be taken as the first three days only. Hence u_t may be taken as the tensile strength of the concrete at three days and not some larger value.

Compressive stress in steel for no slip

If no bond creep occurs between the concrete and steel in regions of the wall *which are remote from a crack*, equation 11.3.1 can still apply even though the equilibrium condition is given by equation

11.3.4 instead of equation 11.3.2. Substituting $f_{ct} = u_t$ in equation 11.3.1 for the limiting case and transposing gives

$$f_{sc} = E_s\varepsilon_{sh} - mu_t$$

Hence equation 11.3.4 becomes

$$r_{crit} = u_t/(f_{yt} + E_s\varepsilon_{sh} - mu_t) \qquad (11.3.5)$$

Although the final compression strain in the steel can exceed 200 microstrain, the value at about three days appears to be appreciably less. A reasonable simplification which includes a small margin of safety is therefore to assume that $f_{sc} = 0$.

Low steel ratios

If $r' < r_{crit}$ the strength of the concrete exceeds the strength of the steel and the latter continues to yield in tension. It follows that the cracks tend to be widely spaced and unevenly distributed along the wall. A few very wide cracks occurring at quite random positions are clearly undesirable.

Moderate steel ratios

If $r' > r_{crit}$ the strength of the steel exceeds the strength of the concrete so that cracks occur at intermediate sections and the maximum stress in the steel is less than the yield stress. The subsequent cracks in the concrete form when the bond force between steel and concrete equals the tensile strength of the concrete. That is,

$$\rho s\sum o \geqslant u_t A_c \qquad (11.3.6)$$

where $\sum o$ = total perimeter of steel reinforcement
o = perimeter of one bar ($= \pi D$ for round bar)
ρ = average bond strength between concrete and reinforcement

o/A $(= \pi D/\frac{1}{4}\pi D^2) = 4/D$ for round bars so that

$$\sum o = A_s(o/A) = 4r'A_c/D$$

Substituting for $\sum o$ in equation 11.3.6 and rearranging gives

$$s \geqslant u_t D/\rho . 4r'$$

The minimum crack spacing, s, is given by the equality sign and the maximum crack spacing is twice this value (since a further crack can form once the development length on both sides equals the minimum value of s), hence

$$\frac{u_t D}{\rho . 2r'} \geqslant s \geqslant \frac{u_t D}{\rho . 4r'} \qquad (11.3.7)$$

For plain round bars and concrete at early ages, u_t/ρ is approximately unity.

High steel ratios

If $r' \gg r_{crit}$ then s becomes very small. However, high steel ratios are only necessary for structural reasons. If such tensile loads are applied suddenly little "bond creep" can occur and an even larger number of cracks tend to form. Equation 11.3.7 is valid for gradually applied stresses only. It is not valid for impact loads.

EXAMPLE 11.3.2

The horizontal reinforcement in a reinforced concrete wall, 240 mm thick, consists of 10-mm diameter mild steel bars at 100 mm centres near each face. Calculate the probable mean crack spacing for the wall if it is fully restrained at each end. Assume the following values:

$$u_t = \rho = 1{\cdot}5\ \text{MN/m}^2 \quad f_{yt} = 250\ \text{MN/m}^2 \quad E_s = 200\ \text{GN/m}^2$$

$$m = 20 \quad \varepsilon_{sh} = 100\ \text{microstrain.}$$

Solution

$$r' = 2 \times \tfrac{1}{4}\pi \times 10^2/(240 \times 100) = 0{\cdot}655\%$$

and from equation 11.3.5

$$r_{crit} = 1{\cdot}5/(250 + 0{\cdot}2 \times 100 - 20 \times 1{\cdot}5) = 0{\cdot}625\%$$

Hence cracking is controlled by the steel since $r' > r_{crit}$. Substituting in equation 11.3.7 and simplifying gives

$$0{\cdot}764 \geqslant s \geqslant 0{\cdot}382$$

The probable mean spacing for intermediate cracks is

$$0{\cdot}382 \times 3/2 = 0{\cdot}573\ \text{m} \qquad \text{(Answer)}$$

Note: If compressive stress in the steel is ignored in this example then

$$r_{crit} = 1{\cdot}5/250 = 0{\cdot}6\%$$

This value is lower since $(E_s\varepsilon_{sh} - mu_t)$ is negative in this case.

Differential thermal movement

If there is no external restraint it is only the *differential* movement between concrete and steel which produces stress. However, since the coefficient of thermal expansion for the steel is very approximately equal to that for the concrete, any differential movement should be fairly small and cracking is highly unlikely. The combined shrinkage and differential thermal strain is given by

$$\varepsilon_{sh} + \theta(\alpha_c - \alpha_s)$$

where θ is the *fall* in temperature, and this expression should be substituted for ε_{sh} in equation 11.3.3 and 11.3.5 whenever differential thermal strains are significant. Notice however that the actual thermal contraction of the concrete is fully effective in producing the first crack for *restrained* members.

EXAMPLE 11.3.3

The coefficients of thermal expansion of mature concrete and steel in a reinforced concrete wall both equal 10 microstrain per °C. If $u_t = 2$ MN/m² and $E_c = 16$ GN/m², calculate the minimum fall in temperature which would produce the first crack in the concrete, unaided by any shrinkage contraction, if the wall is fully restrained at each end.

Solution The ultimate tensile strain in the concrete is

$$(2/16) \times 10^{-3} = 125 \text{ microstrain}$$

Minimum fall in temperature required to crack concrete is

$$125/10 = 12{\cdot}5°\text{C} \qquad \text{(Answer)}$$

Notice that this temperature drop is relatively small and can frequently occur in practice. Thermal effects can therefore produce very pronounced "first" cracks. If $r' < r_{crit}$, the "first" crack tends to become excessively wide. If $r' > r_{crit}$ the excess overall thermal contraction produces further (intermediate) cracks once the "first" crack becomes fully developed.

Actual contraction strain

The crack spacing formula, equation 11.3.7, assumes that the actual contraction strain due to thermal and shrinkage movement is sufficiently large to produce a crack at all the positions which are theoretically possible. In practice the actual contraction strain is usually very much less, so that many of the cracks predicted by the theory do not in fact occur. Nevertheless it is the width of a fully developed crack which is of interest since it is the cracks of maximum width which cause most concern.

Width of fully developed crack

The total contraction is given by the sum of the thermal strain and the shrinkage strain. The tensile stress in the concrete builds up from zero at a cracked section to a maximum at a section which is a distance of s_{min} (or more) from the crack. Hence if the steel ratio exceeds r_{crit} and the total contraction strain exceeds the ultimate tension strain, the mean tension strain in uncracked concrete of length s_{min} adjacent to crack is $\frac{1}{2}\varepsilon_{ult}$. This is shown in figure 11.3.2.

The concrete strain adjacent to the crack on the left hand side, relative to the unrestrained shrinkage strain, is zero. The strain increases linearly to a maximum value of ε_{ult}, hence the mean tension strain over s_{min} is $\frac{1}{2}\varepsilon_{ult}$

If a second crack occurs at a mean distance of $1{\cdot}5s_{min}$ (rather than $2s_{min}$) from the first crack (s lies between s_{min} and $2s_{min}$), then the mean strain in uncracked concrete of length $\frac{1}{2}s = \frac{3}{4}s_{min}$ adjacent to the crack will reduce to $\frac{3}{8}\varepsilon_{ult}$ (since a bond length of $\frac{3}{4}s_{min}$ can only maintain a maximum strain of $\frac{3}{4}\varepsilon_{ult}$). Hence the final crack width is given by

$$\Delta = s(\varepsilon_{sh} \pm \varepsilon_{th} - \tfrac{3}{8}\varepsilon_{ult}) \qquad (11.3.8(a))$$

However, it is usually preferable to consider the probable maximum crack width given by the maximum crack spacing, i.e.

$$\Delta_{max} = s_{max}(\varepsilon_{sh} \pm \varepsilon_{th} - \tfrac{1}{2}\varepsilon_{ult}) \qquad (11.3.8(b))$$

Notice that $\varepsilon_{sh} \pm \varepsilon_{th} = \varepsilon_{sh} + \theta\alpha_c$ in *all* cases since it is the overall contraction of the *concrete* which determines the actual width of the cracks. θ is the fall in temperature as before. This is in contrast to equations 11.3.3 and 11.3.5 where the *differential* thermal strain affects f_{sc} and therefore r_{crit}.

Permissible crack widths for walls

The maximum surface crack width permitted by the Unified Code is 0·3 mm, generally, or 0·1 mm for a particularly aggressive environment. However, the permissible width of 0·3 mm is somewhat excessive for external walls since the cracks are not merely at the surface. The maximum crack width for very exposed members of 0·2 mm recommended by CEB-FIP [1.7] for the most unfavourable combination of permanent and variable loads, would seem appropriate in this instance.

Construction joints

The first crack occurs in a given bay of concrete due to both shrinkage and thermal contraction. Consider, for example, concrete which is cast between two existing bays. The new bay undergoes a rise in temperature (due to heat of hydration), causing compression and therefore considerable creep in the immature concrete. When the temperature eventually falls the concrete has had time to harden so that its rate of creep is less: hence tension stresses are induced. The bond strength between new and mature concrete is generally less than the tensile strength of the concrete. Hence the first crack is most likely to occur at the construction joint. Such cracks therefore tend to be wide and fully developed. They can become excessive if $r' < r_{crit}$.

Intermediate cracks

The overall thermal contraction of a bay can be satisfied by relatively large cracks at one (or possibly both) construction joints at the end of the bay if extensive bond slipping can occur. However, provided that $r' > r_{crit}$ these first cracks will be restricted in width to the fully developed value and the *surplus* contraction strain transferred to adjacent intermediate cracks. The possible development of intermediate cracks *before the first crack is fully developed* is unlikely. That is, intermediate cracks do not usually appear (if at all) until after the formation of a fully developed crack at the construction joint, since the shrinkage plus *differential* thermal strain only is usually less than the ultimate tension strain of the concrete. Although this conclusion can be modified by local restraints as discussed below it does explain why the worst cracks are usually observed either at the construction joints or at points of stress concentration (e.g. large changes in section).

EXAMPLE 11.3.4

Determine the probable maximum crack spacing for the wall described in example 11.3.2. Is the probable maximum crack width likely to exceed the value of 0·3 mm permitted by the Unified Code if the coefficients of expansion of the steel and the concrete are given by

$$\alpha_c = 12 \quad \text{and} \quad \alpha_s = 10 \text{ microstrain per } ^\circ\text{C?}$$

Assume that the concrete temperature falls by 20°C.

Solution

The probable maximum crack spacing = 0·764 m as before (Answer)

The ends are assumed to be fully restrained. Hence the overall contraction strain of the concrete is given by

$$\varepsilon_{sh} + \varepsilon_{th} = 100 + 12 \times 20 = 340 \text{ microstrain}$$

This is more than sufficient to produce the first crack.* Intermediate cracks can also be formed once the first crack is fully developed since any further contraction can be transferred to adjacent cracks

* The differential thermal contraction strain of concrete in excess of steel is

$$(12 - 10) \times 20 \times 10^{-6} = 40 \text{ microstrain.}$$

Hence the total strain (shrinkage + differential thermal strain) is 140 microstrain. Notice that this is slightly less than the ultimate tension strain in the concrete for example 11.3.2 of $u_t/E_c = 150$ microstrain. Hence take r_{crit} as $1{\cdot}5/250 = 0{\cdot}6\%$ rather than use equation 11.3.5 directly (i.e. put $f_{sc} = 0$, since cracks are initiated by *overall* thermal contraction). r' therefore exceeds r_{crit} and from equation 11.3.7, s_{min} is 0·382 m as before.

because $r' > r_{crit}$. Hence the probable maximum width of each fully developed crack (from equation 11.3.8(*b*)) is given by

$$\Delta_{max} = 0{\cdot}764(340 - 150 \times \tfrac{1}{2}) \times 10^{-6} = 202\ \mu\text{m} \simeq 0{\cdot}2\ \text{mm}$$

The probable maximum crack width is therefore satisfactory for both the Unified Code (0·3 mm) and the CEB-FIP code (0·2 mm). (Answer)

Local restraint

An upper lift of a wall is usually restrained along its base by the lift below. The shear force along the base is an additional term which could be included in equation 11.3.4 as follows:

$$r_{min} = (u_t - R/A_c)(f_{yt} + f_{sc}) \qquad (11.3.9)$$

where r_{min} is the minimum steel ratio required to give controlled cracking in a particular location and R the force due to local restraint. The local restraints can become smaller as the steel ratio approaches r_{crit}. The minimum steel ratio required to give controlled cracking in accordance with condition 11.3.7 therefore varies as R varies along the length of the member. The cracking can only be *completely* controlled by the steel, however, if r' is not less than r_{crit}.

Design charts for local damage

The compressive stress which occurs in the steel in the regions remote from a crack has been shown to be small and can usually be neglected. Furthermore, if it is assumed that the direct tensile strength of concrete at 3 days is 50% of the estimated flexural strength values at 7 days given in the Unified Code, then the grade designation of the concrete in terms of the 28-day characteristic cube strength can be included as shown in figure F1.

In practice, base restraints will usually offer some slight assistance in most cases, hence a constant value of 100×10^{-6} can be assumed for $\frac{1}{2}\varepsilon_{ult}$ in equation 11.3.8(*b*). This has been done in the design chart given in figure F2.

EXAMPLE 11.3.5

Repeat examples 11.3.2 and 4 using the design charts in appendix F. Determine the maximum crack width

(*a*) assuming average contraction strain equals

$$340 - \tfrac{1}{2} \times 150 = 265 \times 10^{-6}$$

(*b*) considering only the combined shrinkage and thermal strain of 340×10^{-6} and allowing a constant value of 100×10^{-6} for base restraint and $\frac{1}{2}\varepsilon_{ult}$.

Solution

$u_t = 1{\cdot}5\ \mathrm{MN/m^2} \quad f_{yt} = 250\ \mathrm{MN/m^2}$

Hence from figure F1, $r_{crit} = 0{\cdot}6\%$ (Answer)

$D = 10\ \mathrm{mm} \quad r' = 0{\cdot}655\%$ using plain round bars

Hence from figure F1

(*a*) if $\varepsilon_{sh} \pm \varepsilon_{th} - \frac{1}{2}\varepsilon_{ult} = 265 \times 10^{-6}$ then

$\Delta_{max} = 0{\cdot}202\ \mathrm{mm}$ (Answer)

(*b*) if $\varepsilon_{sh} \pm \varepsilon_{th} = 340 \times 10^{-6}$ then

$\Delta_{max} = 0{\cdot}185\ \mathrm{mm}$ (Answer)

Variable cracking behaviour

It is clear that variations in local restraints can produce large variations in the cracking behaviour of essentially similar members. If such members contain appreciably less steel than the critical value, as has been common in the past, then the behaviour could vary from completely uncontrolled to *apparently* controlled cracking depending on the local restraint and the temperatures at the time of casting.

Thermal movement and shrinkage

It is common for concrete to set at a temperature of over 30°C, at the end of a hot summer's day, and subsequently fall to 0°C during the following winter. The anticipated thermal strain for such a case could be of the order of 300 microstrain. The autogenous shrinkage of the concrete would be of the order of 75 microstrain only. Results indicate that the apparent shrinkage is small for normal structures exposed to normal climatic conditions and drying from one face only, even for one or two years [11.4,5]. This is probably due to a combination of only partial drying out and to tensile creep greatly reducing the resultant shrinkage strain.

It is evident that thermal movement can be considerably greater than the shrinkage movement. Hence it follows that, while elaborate procedures for curing the concrete for as long as possible can delay the eventual drying out (or partial drying out) of the concrete, relatively simple procedures can be adopted to obtain an *actual* reduction in the probable thermal strain. The emphasis in curing should therefore be on reducing the temperature of the freshly placed concrete, rather than on wet curing the concrete for extended periods *after* it has reached its maximum temperature due to the heat of hydration.

Temperature rise

The rate of hydration of cement increases with temperature. It follows that the heat of hydration is liberated more quickly, and that the subsequent temperature rise is greater, in summer than in winter. The lowering of the temperature of the freshly placed concrete, to below that of the surrounding mature concrete, can prevent the early onset of cracking both by reducing the actual temperature rise and by offsetting some of the rise against the lower initial temperature relative to the surroundings.

Practical precautions

When placing concrete in summer it is desirable to reduce the temperature of the concrete as much as possible, and preferably down to 20°C, if ordinary Portland cement is being used. Rapid-hardening Portland cement should be avoided. Recommended precautions are:

(i) Shield the concreting materials and the concrete from direct sunlight wherever possible by erecting canvas canopies, with open sides, over the aggregate piles etc. The point of concrete placing should also be shielded if possible. Also add ice to the mixing water if necessary.

(ii) *Fully effective* water curing of the concrete should be undertaken for about 30 hours (i.e. till the end of the day after placing). This minimizes both shrinkage and *compressive* creep strain. The heat of hydration must be dissipated quickly if the maximum temperature rise (which can occur after 14 hours or even earlier) is to be reduced. This means that shutters need to be sprayed with cold water before they can be removed and that steel shutters are far preferable to timber shutters in this instance due to their superior thermal conduction. Once the shutters have been removed it is important to avoid thermal shock at the surface of the concrete: i.e. intermittent torrents of ice-cold water (rather than continuous spraying of cool water) must be avoided.

(iii) Adopt "successive bay construction" rather than "alternate bay construction", since the former enables one end to contract before the next bay is cast against it. The infill bays for alternate bay construction receive no relief in restraint at all. Successive bay construction tends, however, to produce a wider crack at the base of the lift since the shear restraint along the base tends to oppose the relief given by the free end.

11.4 Water-retaining structures

Structures which are to contain liquids must not normally crack on the liquid-retaining face under the normal working load. Conventional elastic theory for reinforced concrete assumes that the concrete cracks and therefore retains no tensile strength. If cracking is to be prevented the tensile stress in the concrete must be limited to a suitable permissible value. That is, the requirements of the limit state of crack formation must be satisfied. It is clear that the concrete must also be well compacted and impermeable and that any joints intended to accommodate relative movement between adjoining parts of a structure require special provision for maintaining the water-tightness of the joint.

Tensile strength of concrete

Consider an axially loaded reinforced concrete tension member. Provided that the concrete does not crack, the same approach that has been used for axially-loaded columns can also be applied here. If F is the applied load and f_{ct}, f_{st} are the tensile stresses in the concrete and steel respectively, then

$$F = f_{ct}A_c + f_{st}A_s$$

Putting $f_{st} = mf_{ct}$, $A_s = r'A_c$ and re-arranging gives

$$f_{ct} = \frac{F}{(1 + mr')A_c} \tag{11.4.1}$$

Cracking of concrete

The maximum permissible stress in the steel is severely restricted if the concrete is to remain uncracked. For example, if $m = 15$ and $f_{ct} = 2$ MN/m^2 then f_{st} cannot exceed 30 MN/m^2: an extremely low proportion of the steel strength. If the concrete cracks, the steel can readily take the additional load, but the size of these cracks must be controlled in order to prevent rapid corrosion of the reinforcement and to contain the liquid effectively. Cracking of the concrete in liquid-retaining structures can allow constant leaching out of lime, rendering the steel in relatively narrow cracks liable to corrosion. To ensure satisfactory serviceability for some structures it is considered necessary to satisfy the limit state of crack formation.

Uncracked concrete beams

Consider the equivalent concrete section for an uncracked beam section and express all stresses in terms of the extreme concrete fibre stress in tension, f_{ct}, as shown in figure 11.4.1. Let the depth of the neutral axis of the uncracked section be Nd. Taking first moments of area about the neutral axis and assuming that $m - 1$ approximately equals m gives

$$bd^2(N - \tfrac{1}{2}) = mr'bd^2(k - N)$$

where $k = d_1/d$ and $r' = A_{st}/bd$. Hence

$$N = (\tfrac{1}{2} + mr'k)/(1 + mr') \tag{11.4.2}$$

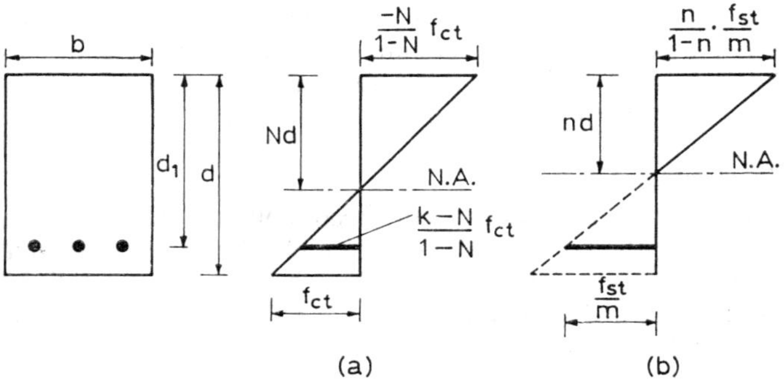

Fig 11.4.1 Concrete stress distributions in uncracked and cracked beam sections. (*a*) Uncracked. (*b*) Cracked.

Now

$$I_e = bd^3[1/12 + (N - \tfrac{1}{2})^2 + mr'(k - N)^2]$$

Hence the moment of resistance of concrete in tension is given by

$$M_{rct} = \frac{f_{ct}I_e}{(1 - N)d} = \frac{f_{ct}bd^2}{1 - N}[\tfrac{1}{3} - N + N^2 + mr'(k - N)^2] \tag{11.4.3}$$

Cracked concrete beams

Elastic theory can be used to obtain the depth to the neutral axis of a cracked section in terms of the overall depth, nd, as follows:

$$n = [\surd(1 + 2k/mr') - 1]mr' \tag{11.4.4}$$

This equation corresponds to equation 3.2.4, except that there n_1 was in terms of the effective depth, d_1.

If all stresses are expressed in terms of the maximum tensile steel stress f_{st}, then the moment of resistance of the steel for the cracked section is given by

$$M_{rt} = f_{st}r'bd^2(k - n/3) \tag{11.4.5}$$

Permissible stresses for no cracking

The permissible concrete stresses, p_{ct}, for calculations relating to the resistance to cracking as given in the former code, CP2007:1960 [11.5] are approximately as follows (in MN/m² units):

Direct tension: $0{\cdot}69 + u/40$

Flexural tension: $1{\cdot}4(0{\cdot}69 + u/40)$

The ratio u_t/p_{ct} for flexural tension varies from 1·4 to 1·6, the value being based on the flexural strength values for u_t given in the Unified Code for cube strengths of 22·5 and 45·0 MN/m² respectively. The ratio is therefore similar to that for the limit state of local damage for concrete in compression, that is $u/p_{cb} = 2$ (i.e. f_c'/p_{cb} is about 1·5). The probability of a crack occurring is therefore no higher than that of an excessive crack occurring in a normal beam. Nevertheless it is desirable to take additional precautions in order to ensure that the consequences of failure—i.e. cracks, if they should occur—satisfy the appropriate limit state of crack width.

Additional precautions against cracking

CP2007 limited the permissible stress in the steel to 12000 lbf/in² (8·3 MN/m²) for "strength" (i.e. calculations for the cracked section using equation 11.4.5). This would imply that γ_m for steel, for the limit state of collapse, should be increased by about 50%. However, the logical limit state theory approach is to consider crack widths for the cracked section in addition to crack formation for the uncracked section. Leakage from water-retaining structures is likely to be considerable if crack widths through the thickness of the wall exceed about 0·1 mm. Furthermore, such cracks would tend not to heal autogenously, due to the leaching action of the water, and would subsequently suffer from corrosion of the reinforcement. The crack width should therefore not exceed about 0·1 mm and the calculations should be carried out in accordance with section 11.2 (flexure), or 11.3 (direct tension), as appropriate. This ensures that if cracks should occur they will be sufficiently fine as not to endanger, seriously, the serviceability of the structure. Autogenous healing of fine cracks in water-retaining structures can occur if the pressure difference across the section can be made relatively small for a sufficiently long period—for example, if a reservoir can be filled gradually over a period of weeks or even months.

EXAMPLE 11.4.1

If the permissible concrete stress to prevent cracking, p_{ct}, for flexural tension is given by $1{\cdot}4\ (0{\cdot}69 + u/40)$, in MN/m² units, determine the moment for the limit state of local damage for a water-retaining structure for the uncracked midspan section considered in example 11.1.1(*a*). Assume that $d_1/d = 0{\cdot}8$. What load can be carried by the beam before the section cracks, expressed as a proportion of the elastic theory working load w referred to in example 11.1.1?

Solution

From figure 11.4.2

$$p_{ct} = 1{\cdot}4(0{\cdot}69 + u/40) = 1{\cdot}75 \quad (\text{MN/m}^2 \text{ units})$$

$$r = 1{\cdot}59\% \quad \text{therefore} \quad r' = 1{\cdot}59 \times 0{\cdot}8\% = 1{\cdot}27\%$$

$$mr' = 18{\cdot}7 \times 1{\cdot}27\% = 0{\cdot}2375 \quad \text{and}$$

$$mr'k = 0{\cdot}8 \times 0{\cdot}2375 = 0{\cdot}190$$

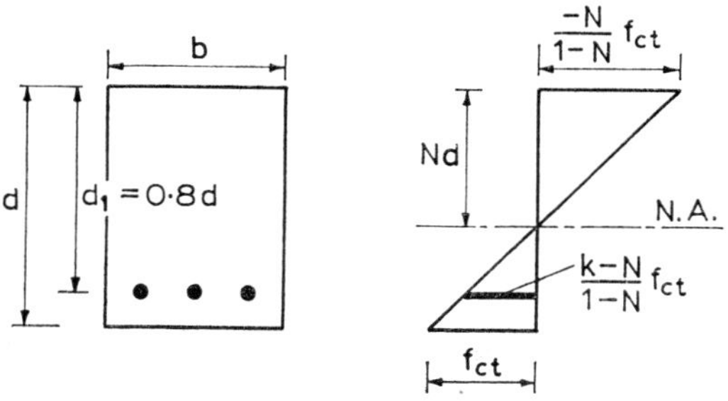

Fig 11.4.2 Example 11.4.1.

Hence from equation 11.4.2

$$N = (0{\cdot}5 + 0{\cdot}190)/1{\cdot}2375 = 0{\cdot}558$$

Now

$$I_e = bd^3[1/12 + (N - \tfrac{1}{2})^2 + mr'(k - N)^2] = 0{\cdot}101bd^3$$

Therefore

$$M_{rct} = p_{ct}I_e/(1 - N)d = 1{\cdot}75 \times 0{\cdot}101bd^3/0{\cdot}442d$$

$$= 0{\cdot}400bd^2 \quad (\text{MN/m}^2 \text{ units})$$

Proportion of elastic working load w is given by

$$(0{\cdot}400bd^2/1{\cdot}63bd_1^{\,2})w = (0{\cdot}400/1{\cdot}63 \times 0{\cdot}8^2)w = 0{\cdot}383w$$

(Answer)

Expansion and contraction joints

The provision of properly constructed expansion and contraction joints permits the use of larger-diameter bars and a possible reduction in the steel ratio below the critical value. In no case, however, should the steel ratio be less than about $\frac{2}{3}r_{crit}$ since it is possible for a crack to occur midway between contraction joints if the latter are at a spacing of s_{max} or more. Such a crack would be uncontrolled for low steel ratios but the elimination of end restraint at the expansion joints appears to justify a reduction from r_{crit} to $\frac{2}{3}r_{crit}$ provided that not more than one contraction joint is used between adjacent expansion joints. An expansion joint permits repeated movements of expansion and contraction. A contraction joint need only be a deep groove which acts as a crack inducer and which permits convenient sealing of the crack subsequently.

EXAMPLE 11.4.2

Determine the maximum distance between properly constructed expansion joints for the retaining walls of a water-retaining structure for which $r' = \frac{2}{3}r_{crit} = 0{\cdot}35\%$ and the bar diameter must not exceed 20 mm, if:

(*a*) no cracks are permitted;

(*b*) the crack width must not exceed 0·1 mm and the effective total contraction $\varepsilon_{sh} \pm \varepsilon_{th} - \frac{1}{2}\varepsilon_{ult}$ equals 200×10^{-6} after allowance for base restraint.

Solution

(*a*) From figure F4 for $D = 20$ mm, $r' = 0{\cdot}35\%$ and plain mild steel bars: $s_{max} = 2{\cdot}84$ m. If joint spacing is slightly less than s_{max} no intermediate crack can form, hence let joint spacing = 2·80 m (Answer)

(*b*) If joint spacing does not exceed $2s_{max}$ then width of intermediate cracks is effectively halved by the expansion joints. Hence from figure F4 if

$$\Delta_{max} = 0{\cdot}2 \text{ mm} \quad \text{and} \quad \varepsilon_{sh} \pm \varepsilon_{th} - \tfrac{1}{2}\varepsilon_{ult} = 200 \times 10^{-6}$$

then $s_{max} = 1{\cdot}0$ m.

Maximum joint spacing $= 2s_{max} = 2{\cdot}0$ m (Answer)

Notice that the spacing for (*b*) is closer than that for (*a*). Hence if expansion joints are to be introduced it is economical in most cases to avoid cracks completely and use the largest permitted bar size in order to increase the joint spacing as much as possible. If a simple contraction joint is introduced midway between the more elaborate expansion joints, then the maximum spacing between expansion joints for case (*a*) becomes 5·6 m.

Structural members uncracked in direct tension

Calculations based on elastic theory for the uncracked equivalent (or gross) section, using permissible stresses for the concrete in direct tension equal to two-thirds of the indirect tension strengths given in table 3.3 of the Unified Code, can ensure that concrete sections remain uncracked. Where structural loading calculations for direct tension are necessary and the required steel ratios are significantly greater than r_{crit}, no further calculations are generally necessary for thermal and shrinkage effects in the same direction.

Structural members cracked in direct tension

The calculation of the crack widths for members loaded in tension cannot generally be determined as accurately as for cracks, due either to flexural loading or to thermal and shrinkage effects. The formula which applies for flexural loading assumes that bond slipping can be ignored and provides a satisfactory value if there is a strain gradient across the section. This ensures that additional cracks form as the strains increase. For a member loaded in direct tension, however, there is a greater tendency for bond slipping to occur since additional cracks at adjacent sections are not induced by a strain gradient as in flexural members. Equally, however, the degree of bond slipping which can occur at later ages is much less than that at very early ages due to restrained thermal and shrinkage contraction. Hence neither theory is entirely satisfactory for mature members loaded in tension and the existing requirements in CP2007:1960 of a maximum steel stress of 12000 lbf/in^2 (82.8 MN/m^2) is therefore appropriate for direct tension. The minimum steel ratio should exceed the r_{crit} values in all cases, as noted above for uncracked sections.

REFERENCES

11.1 BASE, G. D., REED, J. D., BEEBY, A. W. and TAYLOR, H. P. J. *An investigation of the crack control characteristics of various types of bar in reinforced concrete beams.* (Research Report 18 parts 1 and 2, CACA, London Dec. 1966).

11.2 BEEBY, A. W. Private communication.

11.3 HUGHES, B. P. Contribution to discussion on *Specification of concrete* by D. C. Teychenné. *Struct. Eng.* vol. 46 no. 10 Oct. 1968 pp 323–4.

11.4 EVANS, E. P. and HUGHES, B. P. Shrinkage and thermal cracking in a reinforced concrete retaining wall. *Proc. Inst. Civil Engineers* vol. 39 Jan. 1968 pp 111–25.

11.5 HUGHES, B. P. and MILLER, M. M. Thermal cracking and movement in reinforced concrete walls. *Proc. Inst. Civil Engineers* (supplement paper 7254S) Vol. 45, Feb. 1970.

11.6 BRITISH STANDARDS INSTITUTION (London) CP2007:1960 *Design and construction of reinforced and prestressed concrete structure for the storage of water and other aqueous liquids*. 50 pp.

Exercises

1. (i) Show that the crack spacing, s, for a continuous length of reinforced concrete wall, fully restrained at the ends and without any contraction joints, is given by

$$\frac{u_t}{\rho}\cdot\frac{D}{2r} \geqslant s \geqslant \frac{u_t}{\rho}\cdot\frac{D}{4r}$$

when

$$r' > u_t/(f_{yt} + f_{sc})$$

D = diameter of steel bars
ρ = average bond strength between concrete and steel
u_t = ultimate tensile strength of concrete
f_{sc} = stress for steel in compression
f_{yt} = yield strength of steel in tension
r' = steel ratio

(ii) A wall contains 0·6% mild steel (f_{yt} = 250 MN/m²) in the longitudinal direction. The concrete sets at a temperature of 30°C and then falls to 10°C with an effective coefficient of thermal expansion of half its normal value (5×10^{-6} strain/°C). Calculate the mean crack width at 0°C a few months later if ρ and u_t both equal 1·3 MN/m², the shrinkage strain of concrete is 120×10^{-6}, ultimate tensile strain of concrete (elastic + creep) is 200×10^{-6}, coefficient of thermal expansion (steel or concrete) is 10×10^{-6} strain/°C and D = 8 mm. (Modified from UB2 1966, 36 min)

2. (i) Calculate the shrinkage strain of a concrete mix having specific volume ratios (volume fractions) given by $a = 0{\cdot}55$, $b = 0{\cdot}20$, $c = 0{\cdot}10$, $w = 0{\cdot}15$ if the shrinkage strain for the corresponding neat cement paste is $3{\cdot}0 \times 10^{-3}$. You may assume that

$$S_c = S_p(c + w)^n$$

where the symbols have the usual meaning and $n = 1{\cdot}6$.

(ii) The longitudinal steel ratio for a reinforced concrete wall is 0·45%, yield strength of steel equals 40000 lb/in² (276 MN/m²), ultimate tensile strain of concrete (elastic + creep) equals 200 $\times 10^{-6}$, ultimate tensile strength of concrete equals 200 lb/in² (1·38 MN/m²) and bond strength between concrete and steel equals 180 lb/in² (1·24 MN/m²).

If the shrinkage strain for the concrete in the wall is 50% of that calculated for the small specimens considered in (i), calculate the compressive stress in the steel if the wall is completely unrestrained, modular ratio equals 15 and modulus of elasticity of steel equals 30×10^6 lb/in^2 (207 GN/m^2).

What would be the mean crack spacing for the wall if it were fully restrained at the ends and the diameter of the reinforcing bars $= \frac{3}{8}$ in (9·5 mm)? (UB2 1967, 36 min)

3. A typical section through the retaining wall of a rectangular, reinforced concrete reservoir is as shown in figure A. The maximum

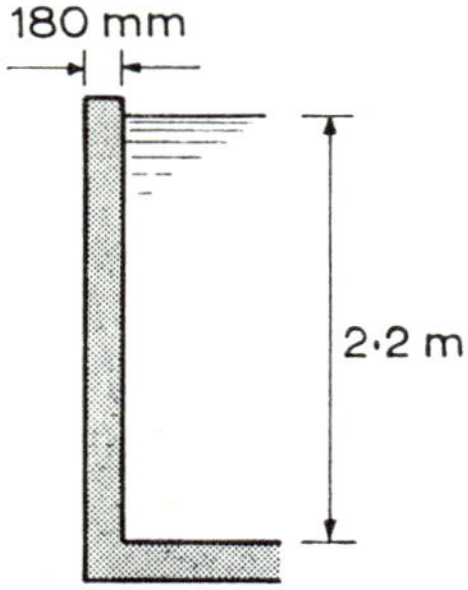

Ch. 11 Fig A Question 3.

depth of fresh water (weighing 9·81 kN/m^3) is 2·2 m and the wall thickness is 180 mm. The water-retaining face of the wall near the base is reinforced vertically with 25-mm diameter mild steel bars at 200-mm centres. Calculate the stresses in the concrete and the steel at the base of the wall assuming that the section is (*a*) uncracked and (*b*) cracked. Assume that

Effective depth $d_1 = 125$ mm

$E_{steel} = 200$ GN/m^2

$E_{concrete} = 12{\cdot}5$ GN/m^2 ($u = 30$ MN/m^2)

Also calculate the maximum surface crack width and comment on the suitability of the wall as a water-retaining structure according to the following criteria: (*a*) Flexural tensile stress in concrete $\not>$ 2·07 MN/m^2, and (*b*) crack widths $\not>$ 0·1 mm. You may assume, without proof, the following formula for surface crack widths provided that you clearly define the symbols used:

$$\Delta = 2{\cdot}3c[\varepsilon_c - 4 \times 10^{-6}/r]$$

(UB3/H 1970 amended, 45 min)

Chapter 12

Torsion

Torsion is not, in general, of major importance in structures of conventional beam and slab construction and can often be ignored if sufficient flexural, axial and shear strength has been provided. Torsion can be a primary effect, however, wherever unbalanced loading occurs across beams. For example, edge beams which are loaded on one side only; secondary beams which are not continuous over main beams; beams with horizontal projections; and beams cranked or curved in plan. The Code considers torsion, like shear and bond, in terms of the limit state of collapse. The structures should be analysed on the basis of elastic theory and the flexural rigidity $E_c I_o$ should be based on the gross section. The torsional rigidity GJ may be calculated taking G as $0{\cdot}4E_c$ and the torsion constant, J, as $\frac{1}{2}J_o$ where J_o is the polar second moment of area based on the gross concrete section.

12.1 Rectangular beams without torsion reinforcement

A rectangular section subjected to pure torsion within the elastic range exhibits the greatest shear stress at the middle of the longer sides. The diagonal tension stress on one side of the web, due to torsion, therefore augments that due to transverse shear. It is therefore essential to consider the combination of transverse and torsional shear.

Elastic theory

The elastic axial torque for rectangular sections has been obtained by St Venant (e.g. see reference 12.1) in terms of the angle of twist per unit length, θ, as follows:

$$T = \beta x^3 y G\theta \qquad (12.1.1)$$

where G = shear modulus of the material
x = shorter side of rectangle
y = longer side of rectangle
β = a function of x/y

The maximum elastic shear stress, q_t, occurs at the centre of the longer side, y, and is given by

$$q_t = \gamma x G\theta = T/\alpha x^2 y \qquad (12.1.2)$$

where γ is another function of x/y and $\alpha = \beta/\gamma$. Typical values of the functions α, β and γ are shown in table 12.1.1.

Table 12.1.1 Elastic torsion constants for rectangular sections

y/x	α	β	γ
1·0	0·208	0·141	0·675
1·4	0·227	0·187	0·822
2·0	0·246	0·229	0·930
5·0	0·292	0·291	0·999
∞	0·333	0·333	1·000

Plastic theory

The axial torque, for rectangular sections with a uniform shear stress over the whole section, is given by [12.3]:

$$q_t = T/\tfrac{1}{2}x^2(y - x/3) \qquad (12.1.3)$$

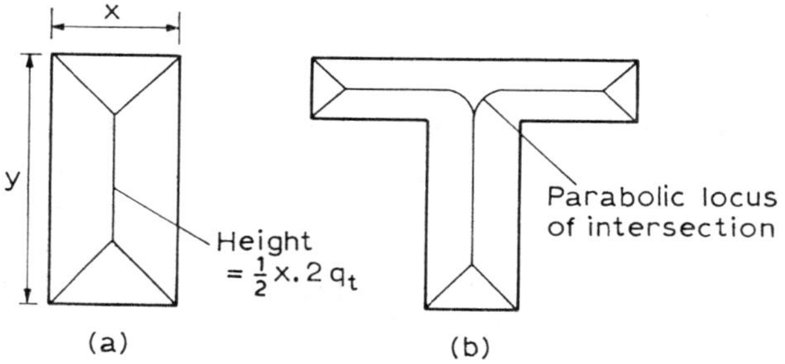

Fig 12.1.1 "Sand heap" analogy for sections in torsion. (*a*) Rectangular beam. (*b*) T-beam.

This result can be obtained by use of the "sand heap analogy", i.e. the fact that the torsional resistance of the cross-section is proportional to the volume of a sand heap of constant maximum slope on the section, the slope being equal to twice the constant shear stress. Reference to figure 12.1.1(*a*) indicates that the volume of the sand heap is given by

$$[\tfrac{1}{2}x(y - x) + \tfrac{1}{3}x^2]q_t x$$

Hence

$$T = \tfrac{1}{2}x^2(y - x/3)q_t$$

as given in equation 12.1.3.

Elasto-plastic behaviour

At collapse, a uniform shear stress probably exists over a considerable proportion of a concrete section since tests indicate that some stress redistribution occurs as the ultimate torque is approached. The extent to which plastic deformation occurs is somewhat uncertain. It has therefore been usual for codes of practice to adopt one of two extreme assumptions. Either:

(i) No plastic yield (i.e. elastic theory for which maximum stress nowhere exceeds a certain value), or

(ii) Extensive plastic yield (i.e. average stress for entire section does not exceed a particular value).

Accuracy of torsion calculations

The calculation of the torsional moments according to elastic theory depends upon the relative stiffnesses of the members of the structure. Such calculations are inevitably approximate. Design calculations for the torsional strengths of sections are also approximate and are based largely on experimental test results. A suitable choice of values for stress therefore enables the essentially different elastic and plastic theories to predict collapse loads which are reasonably consistent for a given set of experimental results.

Combined torsion and bending

Beams are rarely subjected to torsion only. The presence of transverse shear implies that bending stresses must also be present since shear equals rate of change of bending moment. Cowan has suggested that

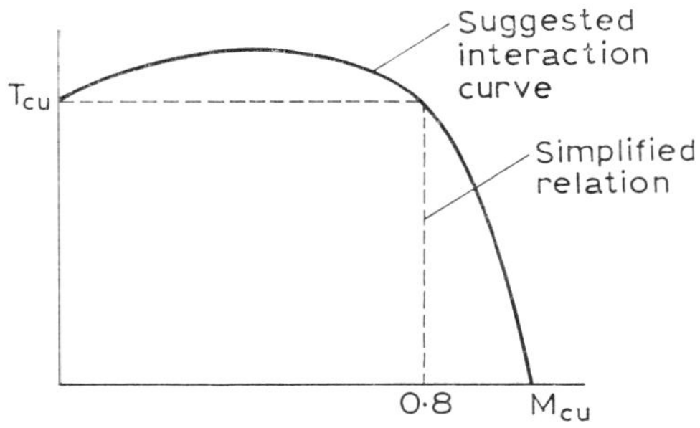

Fig 12.1.2 Relation between torque and bending moment at collapse (reference 12.4).

the variation of the ultimate torsional moment with the ultimate flexural moment is as indicated in figure 12.1.2 (see reference 12.4 or 12.5). The simplified, dotted, line therefore gives a safe value of the ultimate torsional moment provided that the bending moment is less than about 80% of the ultimate flexural value.

Visco-elastic limit

Cowan's suggested interaction curve is not strictly based on the collapse load, since failure in his tests was classified as the commencement of large-scale cracking, which was called the visco-elastic limit. This value can coincide with collapse although it is usually rather less and is therefore a safe estimate of the true ultimate torsional moment. This visco-elastic limit for a member subjected to pure torsion is defined as being the point at which the maximum tensile stress, which has the same value as the maximum shear stress, reaches the ultimate tensile strength of the concrete.

Torsional resistance of rectangular sections

The Unified Code assumes a fully plastic section at collapse and therefore gives the following equation:

$$_{*}T_c = \tfrac{1}{2}q_t x^2(y - x/3) \qquad (12.1.4)$$

where $_{*}T_c$ = design torsional moment for *collapse*
q_t = torsional shear stress
x, y are as before.

Provided that the flexural moment does not exceed 80% of the collapse moment then equation 12.1.4 can be assumed to give a safe estimate of the torsional moment at collapse.

Torsional resistance of concrete

No special provision of torsion reinforcement need be made if only "nominal torsion" is applied to the section, since nominal torsion reinforcement, in the form of minimum shear reinforcement and "hanger bars", assists the torsional strength of the concrete. Where torsion is a major effect then special torsional reinforcement must be provided. In all cases, however, the values of q_t and $(q + q_t)$ must not exceed the appropriate maximum values for *design* strength given in table C4. The maximum values of $(q + q_t)$ in table C4 correspond approximately with values of $4q_c$ for high steel ratios in table C3 or with $u_k/5\gamma_m$ where this is more critical.

12.2 Beams with torsion reinforcement

There is an analogy between shear and torsion for sections without shear or torsion reinforcement, since the Unified Code considers the principal tensile stress in the concrete in both cases. This similarity ceases, however, for sections with torsion reinforcement. This is because considerable rotation of the sections must occur before the torsion reinforcement becomes fully effective. Under these circumstances the torsional resistance of the cracked concrete must be ignored and the total torsion taken by the torsion reinforcement. Torsion reinforcement is therefore more analogous to flexural reinforcement in this respect.

Hoop reinforcement

If the required torsional resistance exceeds that for the concrete section, torsional reinforcement is necessary and is required by the Unified Code to take the total torsion. This reinforcement is normally a combination of vertical hoops and additional longitudinal steel, although it can be in the form of spirals in the appropriate direction. The hoops are vertical *closed* stirrups with the ends suitably anchored. The area of the closed hoops, which must be additional to any stirrups required for shear due to normal bending, is obtained from

$$A_{sh} = {}_*T_s s/(0{\cdot}60 x_1 y_1 \times 0{\cdot}87 f_{yh}) \qquad (12.2.1)$$

where ${}_*T_s$ = torsion taken by torsion steel

A_{sh} = cross-sectional area of both legs of hoop

s = spacing of hoops in longitudinal direction of member ($\not> x_1$ and $\not> \frac{1}{2}y_1$)

x_1 = smaller centre-line dimension of hoop

y_1 = larger centre-line dimension of hoop

f_{yh} = characteristic strength of hoop steel, which should not be assumed greater than 425 MN/m²

The Australian Code of Practice corresponding to CP114 has a similar expression [12.6]. Equation 12.2.1 effectively replaces 0·80/2 in the Australian Code with $0{\cdot}60/\gamma_l\gamma_m$ which is approximately 0·35. The Unified Code is therefore very conservative compared with its Australian predecessor with respect to torsional reinforcement, especially since the latter only requires the hoop steel to take the *excess* torsion. A justification for the Unified Code's approach is that design for torsion is less certain than for other loadings so that a partial safety factor for "service" has been effectively included as well.

Additional longitudinal steel

The area of longitudinal reinforcement, which is additional to that required for flexure, is given by the volume of steel which equals that for the hoops, suitably adjusted for any differences in yield strength. That is

$$A_{sa} f_{ya} s = A_{sh} f_{yh}(x_1 + y_1) \qquad (12.2.2)$$

where A_{sa} and f_{ya} refer to the additional longitudinal steel required to resist the longitudinal component of the diagonal tension. The longitudinal reinforcement should be distributed evenly round the inside perimeter of the hoops. At least four bars should be provided, one in each corner of the hoops. This is best effected by increasing the sizes of longitudinal bars at the levels of the tension and compression reinforcement above that which would be required for flexure alone.

If $y_1 > 0{\cdot}3$ m, bars should be provided at points along the larger face such that the centre-to-centre distance between bars does not exceed 0·3 m. The torsional steel should extend a distance $x_1 + y_1$ along the beam beyond points where such steel is theoretically necessary according to tables C3 and C4.

EXAMPLE 12.2.1

An RC beam, 0·30 m × 0·50 m overall and 0·45 m in effective depth, is subjected to a torsional moment of 20 kNm due to the characteristic *imposed* load.

(*a*) Calculate the design torsional moment for the concrete section without torsional reinforcement. Assume a design shear force of 27 kN, a longitudinal steel ratio of 1·0% and a characteristic

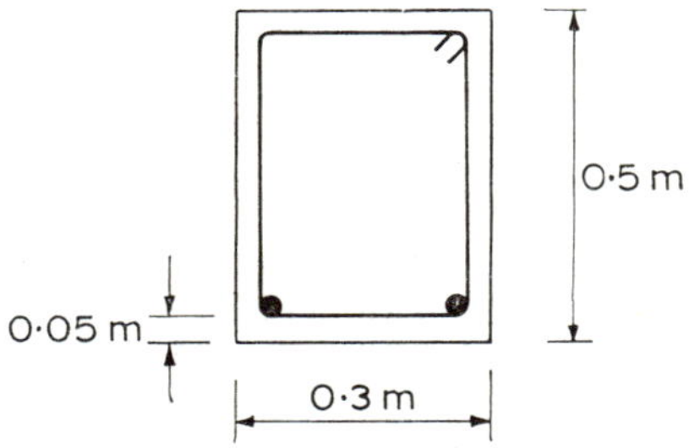

Fig 12.2.1 Example 12.2.1.

cube strength of 30 MN/m² for the section concerned and assume that the torsional moment of resistance of the section with only nominal shear stirrups and hanger bars is given by the total diagonal tension in the concrete due to both shear and torsion.

(*b*) Calculate the required spacing of 16 mm diameter hoops (if any) and additional area of longitudinal steel to resist the entire torsional moment as required by the Unified Code. Assume minimum cover of 25 mm, bar diameter of 16 mm for hoops, and characteristic yield strength for both hoops and main steel of 250 MN/m².

Solution

$$T_k = 20 \text{ kNm} \quad \text{and} \quad \gamma_l = 1{\cdot}6$$

Therefore

$$_*T = 20 \times 1{\cdot}6 \times 10^3 = 32 \times 10^3$$

Substituting in equation 12.1.4 and transposing gives

$$q_t = 2 \times 32 \times 10^3/0{\cdot}3^2(0{\cdot}5 - 0{\cdot}1) = 1{\cdot}78 \times 10^6$$

From tables C3 and C4 for $u = 30 \times 10^6$:

$$max\ q_t = 3{\cdot}00 \quad max\ (q + q_t) = 4{\cdot}10 \quad \text{and}$$
$$q_c = 0{\cdot}70 \quad \text{(all MN/m}^2\text{ units)}$$

Hence $q_t < max\ q_t$. Also

$$q = 27 \times 10^3/0{\cdot}30 \times 0{\cdot}45 = 0{\cdot}20 \times 10^6$$

Hence

$$q + q_t = (0{\cdot}20 + 1{\cdot}78) \times 10^6 = 1{\cdot}98 \times 10^6 < max\ (q + q_t)$$

(*a*) Torsional resistance of concrete section is given by

$$q_c - q = 0{\cdot}5 \times 10^6$$

or

$$_*T_c = \tfrac{1}{2} \times 0{\cdot}5 \times 0{\cdot}3^2 \times 0{\cdot}4 \times 10^6$$
$$= 9{\cdot}0 \text{ kNm} \qquad \text{(Answer)}$$

(Hence notice the justification for only considering torsion when it is a major effect).

(*b*) $10^3x_1 = 300 - 2(25 + 8) = 234$

$10^3y_1 = 500 - 2(25 + 8) = 434$

and

$$s \not> \tfrac{1}{2}y_1 = 0{\cdot}217 \text{ and } \not> 0{\cdot}2$$

Also

$$A_{sh} = 2 \times \tfrac{1}{4}\pi \times 16^2 \times 10^{-6} = 402 \times 10^{-6}$$

Therefore

$$s = 0{\cdot}6x_1y_1 \times 0{\cdot}87f_{yq} \times A_{sh}/{}_*T_c \quad (k = 0{\cdot}60)$$
$$= 0{\cdot}60 \times 0{\cdot}234 \times 0{\cdot}434 \times 0{\cdot}87 \times 250 \times 402 \times 10^{6-6-3}/32$$
$$= 166 \times 10^{-3} \quad \text{say, 165 mm}$$
$$s = 165 \text{ mm} \qquad \text{(Answer)}$$
$$A_{sa} = 402(0{\cdot}234 + 0{\cdot}434) \times 10^{-6}/0{\cdot}165$$
$$= 1\,630 \text{ mm}^2 \qquad \text{(Answer)}$$

T-, L- and I-beams

Bach's approximation for elastic analysis of complex sections (e.g. see reference 12.2) divides the section into its component rectangles, such that equation 12.1.1 becomes

$$T = G\theta\sum\beta x^3 y \qquad (12.2.3)$$

A similar approach could be used for plastic theory such that equation 12.1.3 becomes

$$T = \tfrac{1}{2}q_t\sum x^2(y - x/3) \qquad (12.2.4)$$

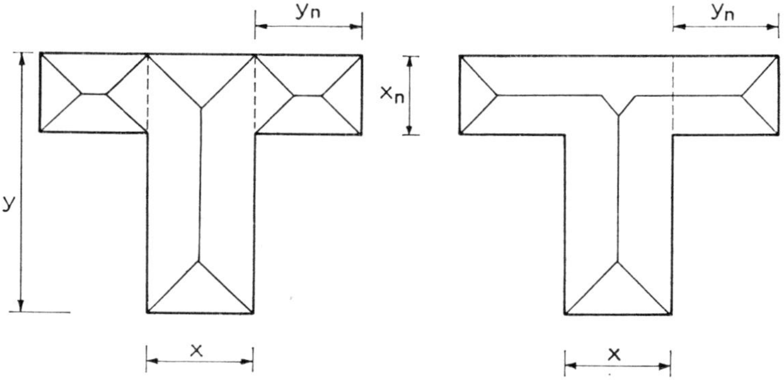

Fig 12.2.2 Approximations to "sand heap" analogy. (*a*) Component rectangles. (*b*) Linear approximation.

Reference to the "sand heap" analogy in figure 12.1.1 and to figure 12.2.2(*a*) shows, however, that this would be conservative. Figure 12.2.2(*b*) is a closer approximation in which the volume for the rectangle y_n by x_n is increased from

$$\tfrac{1}{2}q_t x^2(y - \tfrac{1}{3}x) \quad \text{to} \quad \tfrac{1}{2}q_t x_n^2(y_n - x_n/6)$$

due to the elimination of the end slope. There is a corresponding increase in volume for the central rectangle, hence the expression for all the additional rectangles should be $\tfrac{1}{2}q_t x_n^2 y_n$. The expression for the central rectangle x_1 by y_1 should remain as $\tfrac{1}{2}q_t x^2(y - \tfrac{1}{3}x)$, hence a close approximation is given by

$$T = \tfrac{1}{2}q_t x^2(y - \tfrac{1}{3}x) + \sum_{n=2}^{n=N} \tfrac{1}{2}q_t x_n^2 y_n \qquad (12.2.5)$$

where

$x_1 > x_2$, x_3, etc

The draft Code gives a simplified expression as follows

$$T = 0{\cdot}4q_t\sum x^2 y \qquad (12.2.6)$$

Reinforcement for component rectangles

Where torsional reinforcement is provided each component rectangle of the section should contain all, or part of, a reinforcing cage consisting of rectangular closed hoops and longitudinal steel as for rectangular beams. The cage provided in the rib of the beam is extended into the flange (T-beam) or flanges (I-beam). The torsional resistance of each cage is given by equation 12.2.1. The proportion of the total torsion taken by each cage should be made approximately equal to

$$x_n^2 y_n T / \sum x_n^2 y_n \qquad (12.2.7)$$

where x_n = smaller dimension of component rectangle being considered
y_n = larger centre-line dimension of hoops in cage being considered

For example, for a T-beam there will be two component rectangles: the rib and the flange, as shown in figure 12.2.3.

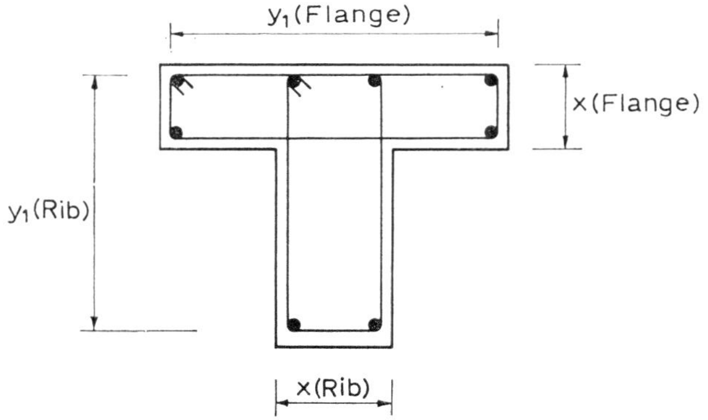

Fig 12.2.3 Reinforcement for component rectangles.

Closed box sections

Closed hollow sections of circular or square shape are the most efficient for resisting torsion and should be used if the torsional moments are very large. The average shear stress due to torsion for a thin-walled hollow section where the wall thickness is small compared with the overall dimension is given by

$$_*T_c = 2Abq_t \qquad (12.2.8)$$

where A = area enclosed by median line of walls of box section
b = minimum wall thickness

The reinforcement can be designed as for rectangular sections.

Other sections

The shear stresses due to torsion in multi-cellular sections may be calculated using shear flow analysis. Each box should be reinforced to resist the torsion arising from its shear flow. Torsional stress for complex sections should be calculated assuming plastic distribution of shear stresses. Where torsional reinforcement in the form of non-rectangular reinforcing cages is required, the formulae for rectangular beams may be used except that x_1y_1 shall be replaced by the cage area, and $x_1 + y_1$ by half the cage perimeter.

REFERENCES

12.1 TIMOSHENKO, S. *Strength of Materials* vol. 1 (Macmillan, 1938).
12.2 TIMOSHENKO, S. and GOODIER, J. N. *Theory of Elasticity* (McGraw-Hill, 1961).
12.3 NADAI, A. *Theory of flow and fracture of solids* (McGraw-Hill, 1950).
12.4 COWAN, H. J. Elastic theory for torsional strength of rectangular reinforced beams. *Magazine of Concrete Research* vol. 2 1950 pp 3–8.
12.5 COWAN, H. J. and ARMSTRONG, S. Experiments on the strength of reinforced and prestressed concrete beams and of concrete-encased steel joints in combined bending and torsion. *Magazine of Concrete Research* vol. 7 1955 pp 3–20.
12.6 CRAWLEY, D. B. Commentary on the code for the use of concrete in buildings CA2:1963. *Constructional Review Technical Supplement* Australia, Oct. 1968.

Chapter 13

Reinforced concrete slabs

A structural slab is essentially a thin, plane-sided member which carries transverse loads. Reinforced concrete slabs are eminently suitable for structural floors, bridge decks, retaining walls, etc., since they can offer a combination of strength, ductility and resistance to fire, corrosion, abrasion, or sound transmission, which is superior to that of other materials of similar cost. For solid slabs the strength in shear of the concrete is usually more than adequate, even for concentrated loads. In addition, since the minimum depth of the slab is usually governed by the limit states of deflection, the flexural compressive strength of the concrete is also more than adequate so that the section is appreciably under-reinforced. It can therefore be assumed that the rotational capacity of slabs is very high.

Many proprietary floors reduce self-weight by forming voids in the lower surface of the slab with hollow clay pots, lightweight permanent shuttering, removable shuttering with suitable inserts, waffle construction and similar techniques which are suitable for larger spans. Such floors can be designed by a combination of the methods used for beams with those for the uniform solid slabs considered in this chapter. Practical rules for the design of ribbed and hollow block slabs are given in appendix C.

13.1 Equilibrium of thin plates in bending

A two-way spanning slab is a redundant structure. The distribution of moments throughout the slab can therefore be determined most conveniently from either an elastic or a plastic analysis. In both cases, however, an exact solution must satisfy the equilibrium equations. Comprehensive discussions of plate theory are given elsewhere, e.g. references 13.1 and 2, so a brief summary is adequate for the present introduction to the subject. Plate theory for slabs considers thin uniform slabs which are subjected to loads and support reactions which are normal to the plane of the undeformed slab.

Notation and sign convention

Axes x and y are taken in the plane of the slab and the load acts at right angles along the positive direction of the z-axis. That is, the z-axis is vertically downwards for a horizontal slab. The notation which will be used is only applicable to plates and slabs and is shown in figure 13.1.1. It is widely used although it is a little confusing at first for those who are familiar with the general mathematician's notation (in which the vectorial direction of the moment also controls the notation, i.e. M_x would be a moment which acts clockwise about the positive x-axis direction and *not* as given below).

M_x = normal bending moment *acting upon* x-plane
M_y = normal bending moment *acting upon* y-plane
M_{xy} = torsional moment *acting upon* x-plane
M_{yx} = torsional moment *acting upon* y-plane

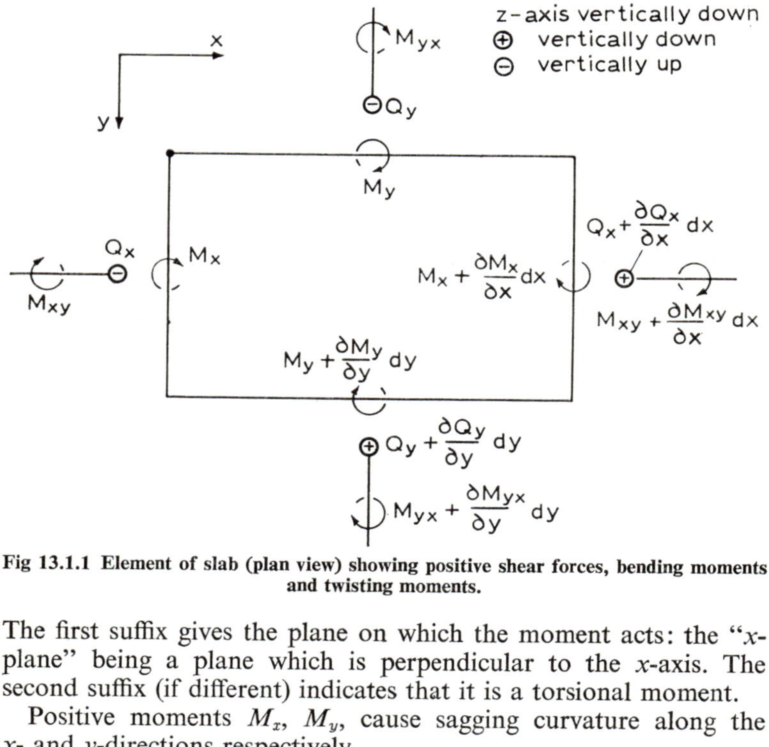

Fig 13.1.1 Element of slab (plan view) showing positive shear forces, bending moments and twisting moments.

The first suffix gives the plane on which the moment acts: the "x-plane" being a plane which is perpendicular to the x-axis. The second suffix (if different) indicates that it is a torsional moment.

Positive moments M_x, M_y, cause sagging curvature along the x- and y-directions respectively.

Positive shear forces Q_x, Q_y, act in the positive z-direction on planes whose outward normals are in the positive x- and y-directions respectively.

Positive torsional moments M_{xy}, M_{yx} act in a clockwise direction about the outward pointing normals.

Note: The x-plane is parallel to both the y- and z-axes and could therefore also be referred to as a plane parallel to the yz-plane.

Equilibrium equations

Consider the equilibrium of a small element, δx by δy, subjected to a uniformly distributed load per unit area, w. Let Q and M be the shear force per unit length and moment per unit length respectively. For example, Q_x is the shear force per unit length of plate along the y-direction and M_x the bending moment per unit length along the y-direction.

Resolving vertically,

$$\frac{\partial Q_x}{\partial x}\,dx\delta y + \frac{\partial Q_y}{\partial y}\,dy\delta x + w\delta x\delta y = 0$$

Taking moments about y-axis (and ignoring 2nd-order terms):

$$\frac{\partial M_x}{\partial x}\,dx\delta y + \frac{\partial M_{yx}}{\partial y}\,dy\delta x - Q_x\delta y\,.\,\delta x = 0$$

Taking moments about x-axis (and ignoring 2nd-order terms):

$$-\frac{\partial M_y}{\partial y}\,dy\delta x + \frac{\partial M_{xy}}{\partial x}\,dx\delta y + Q_y\delta x\,.\,\delta y = 0$$

Hence

$$\frac{\partial Q_x}{\partial x} + \frac{\partial Q_y}{\partial y} = -w \tag{13.1.1}$$

$$\frac{\partial M_x}{\partial x} + \frac{\partial M_{yx}}{\partial y} = Q_x \tag{13.1.2}$$

$$\frac{\partial M_y}{\partial y} - \frac{\partial M_{xy}}{\partial x} = Q_y \tag{13.1.3}$$

In addition, it is shown later that

$$M_{xy} = -M_{yx} \tag{13.1.4}$$

Hence by differentiating equations 13.1.2 and 13.1.3, w.r.t. x and y respectively, and substituting in equation 13.1.1, the basic equilibrium equation for a thin plate or slab is obtained as follows:

$$\frac{\partial^2 M_x}{\partial x^2} + \frac{\partial^2 M_y}{\partial y^2} - 2\,\frac{\partial^2 M_{xy}}{\partial x \partial y} = -w \tag{13.1.5}$$

Equilibrium of triangular element

Let a triangular element ABC be cut out of a thin plate, of thickness h, which is subjected to pure bending moments M_x and M_y in two perpendicular directions. Consider a thin triangular lamina abc which is in the z-plane as shown in figure 13.1.2. The stresses acting on the face ac can be resolved into a component σ_n along the normal direction and a shearing component τ_{nt} along the tangential direction. Summing the normal stresses, σ_n, for all laminas such as abc over the thickness of the plate gives the bending moment acting, per unit length, on the section AC of the plate; that is,

$$M_n = \int_{-\frac{1}{2}h}^{\frac{1}{2}h} \sigma_n z dz \tag{13.1.6}$$

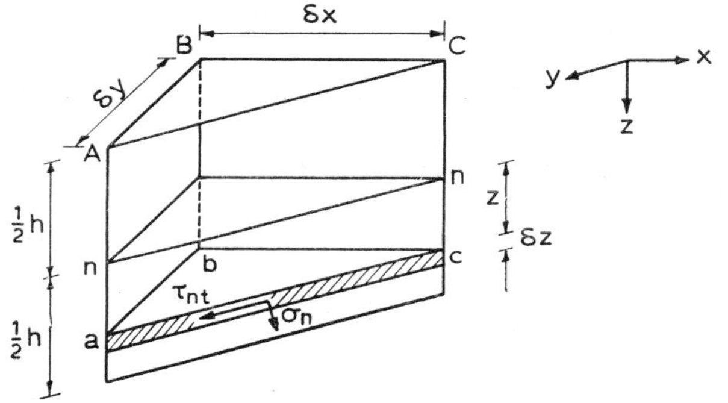

Fig 13.1.2 Triangular element in plate subjected to moments in two perpendicular directions.

The shearing stresses τ_{nt} give the twisting moment acting, per unit length, on the section AC of the plate, that is

$$M_{nt} = -\int_{-\frac{1}{2}h}^{\frac{1}{2}h} \tau_{nt} z dz \tag{13.1.7(a)}$$

Thus it is evident that the twisting moment per unit length along a section is due to the horizontal shearing stresses continuously distributed along the section.

For the n-plane shown in figure 13.1.2, τ_{nt} acts in a direction at 90° clockwise from the n-axis. For the t-plane, however, τ_{tn} acts in a direction at 90° anticlockwise from the t-axis. Hence

$$M_{tn} = +\int_{-\frac{1}{2}h}^{+\frac{1}{2}h} \tau_{nt} z dz \tag{13.1.7(b)}$$

Hence since the complementary horizontal shear stresses are equal, i.e. $\tau_{tn} = \tau_{nt}$, it follows that $M_{tn} = -M_{nt}$. Similarly for the x- and y-axes, $M_{xy} = -M_{yx}$.

Note: Reference 13.1 has a different sign convention for shear stresses, such that $\tau_{nt} = -\tau_{tn}$. However, equation 13.1.5 is still obtained in the same form.

Moment vectors

Bending moments are vectorial quantities and can therefore be resolved in the same way as other vectors. The direction of the vector is given by the clockwise rotation of the moment about its axis, as shown in figure 13.1.3. It is essential to remember this,

Fig 13.1.3 Vectorial representation of a moment.

even though it is convenient, as shown above, to adopt a *notation* which relates moments to the planes on which they act, rather than to the direction of the axis. When drawing diagrams for which moment vectors are considered it is desirable to show arrows at

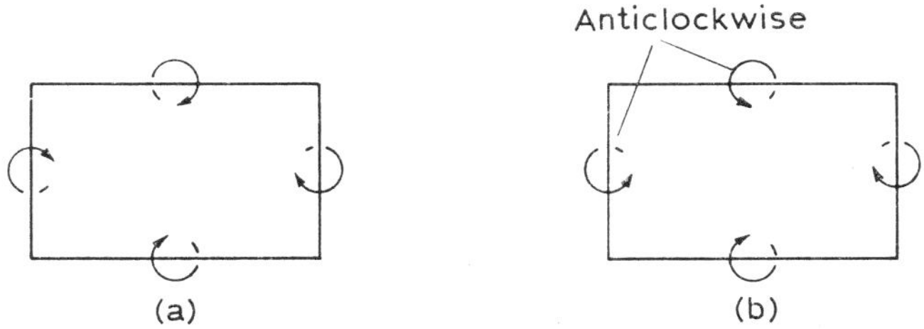

Fig 13.1.4 Methods of showing moments prior to vectorial representation. (*a*) Preferred. (*b*) Not recommended.

the clockwise extremity in all cases as in figure 13.1.4(*a*). Although sagging moments can be clearly indicated with arrows at either end, as in (*b*), errors in vectorial representation are more likely.

Resolution of moment vectors

The moments acting on a plane whose *external* normal makes some angle ϕ with the x-axis can be determined by considering the equilibrium of small triangular elements as shown in figure 13.1.5. Notice that a similar notation can be used for rectangular axes n, t, making angles of ϕ with the x-, y-axes respectively, as was done above. Clockwise rotation about the *outward* normals gives the positive directions of the torsion moments in all cases.

Resolving along t-axis for element (i):

$$M_n c = M_x b \cos\phi + M_y a \sin\phi - M_{xy} b \sin\phi + M_{yx} a \cos\phi$$

$$b/c = \cos\phi \quad a/c = \sin\phi \quad \text{and} \quad M_{yx} = -M_{xy}$$

Hence

$$M_n = M_x \cos^2\phi + M_y \sin^2\phi - M_{xy} \sin 2\phi \qquad (13.1.8)$$

Resolving along n-axis for element (ii), (or putting $\phi = 90 + \phi$ and writing M_t for M_n in above) gives

$$M_t = M_x \sin^2\phi + M_y \cos^2\phi + M_{xy} \sin 2\phi \qquad (13.1.9)$$

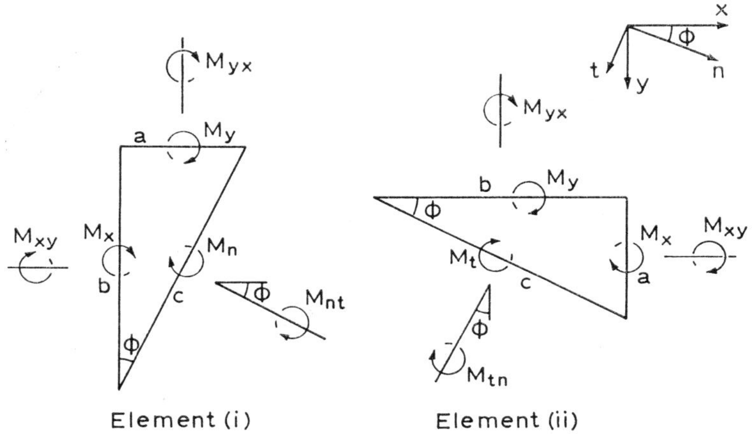

Fig 13.1.5 Triangular elements with faces having external normals along *n* and *t* axes.

Resolving along n-axis for element (i):

$$M_{nt}c = M_x b \sin\phi - M_y a \cos\phi + M_{xy} b \cos\phi + M_{yx} a \sin\phi$$

Hence

$$M_{nt} = \tfrac{1}{2}(M_x - M_y) \sin 2\phi + M_{xy} \cos 2\phi \qquad (13.1.10)$$

Resolving along t-axis for element (ii), (or putting $\phi = 90 + \phi$ in above) gives

$$M_{tn} = -\tfrac{1}{2}(M_x - M_y) \sin 2\phi - M_{xy} \cos 2\phi \qquad (13.1.11)$$

Principal moments

The maximum and minimum bending moments act on planes at right angles to each other and can be found by differentiating equation 13.1.8 (or 9):

$$dM/d\phi = -M_x \sin 2\phi + M_y \sin 2\phi - M_{xy}\, 2 \cos 2\phi$$
$$= 0$$

or

$$\tan 2\phi = -2M_{xy}/(M_x - M_y) \qquad (13.1.12)$$

Substituting for ϕ in equations 13.1.10 and 11 shows that $M_{nt} = 0$ and $M_{tn} = 0$, which confirms that no torsional moments act on the principal planes.

Substituting

$$\cos^2 \phi = \tfrac{1}{2}(1 + \cos 2\phi), \text{ etc.,}$$

in equation 13.1.8 and then substituting

$$\sin 2\phi = -2M_{xy}/[(M_x - M_y)^2 + 4M_{xy}]^{\frac{1}{2}}, \text{ etc.,}$$

gives

$$M_n = \tfrac{1}{2}(M_x + M_y) + \tfrac{1}{2}[(M_x - M_y)^2 + 4M_{xy}^2]^{\frac{1}{2}} \qquad (13.1.13)$$

Similarly, M_t is given by the negative square root.

Alternatively, if x and y are assumed to be the principal axes, then $M_{xy} = 0$, and putting $\phi = -\phi$ (since ϕ is measured in a clockwise direction) then

$$M_x, M_y = \tfrac{1}{2}(M_n + M_t) \pm \tfrac{1}{2}[(M_n - M_t)^2 + 4M_{nt}^2]^{\frac{1}{2}} \qquad (13.1.14)$$

and

$$\tan 2\phi = 2M_{nt}/(M_n - M_t) \qquad (13.1.15)$$

Mohr's circle

It will be evident that equations 13.1.6 to 13 are similar to those for plane stress given in standard text books for structures and strength of materials. Mohr's circle construction is therefore equally applicable, as shown in figure 13.1.6.

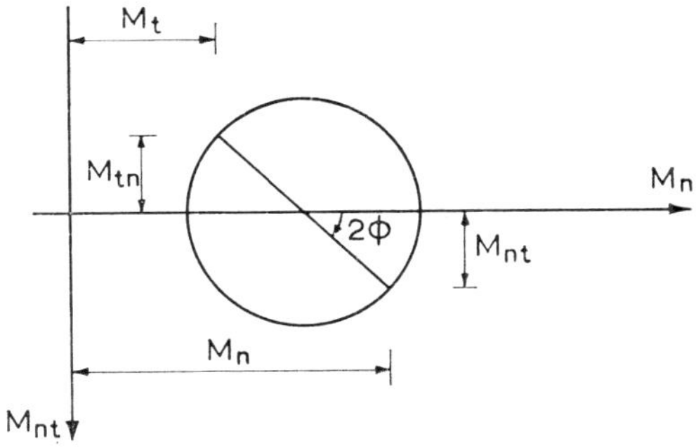

Fig 13.1.6 Mohr's circle construction for moments.

Terminal forces

If a torsional moment M_{nt} per unit length acts on the diagonal of a slab ABCD as shown in figure 13.1.7(a) it can be effectively replaced by equal and opposite vertical shear forces of M_{nt} acting unit distance apart as shown in (*b*). This replacement of the distributed horizontal force by a statically equivalent system of two vertical forces cannot cause any appreciable disturbance in the plate, from Saint Venant's principle, except within a distance comparable with the plate thickness (which is assumed small). Similarly a torsional moment of $M_{nt} + \partial M_{nt}/\partial t$ per unit length can also be replaced as shown.

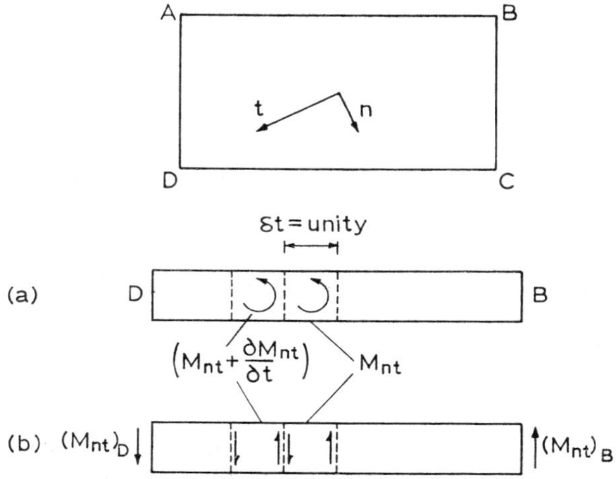

Fig 13.1.7 "Terminal" forces for one edge (or section) of a plate.

The torsional moment along the diagonal section BD is therefore statically equivalent to a distributed vertical shear force $-\partial M_{nt}/\partial t$ per unit length together with concentrated "terminal" forces M_{nt} at the ends B and D. Notice that if the sides AD and DC of the slab were subjected to torsional moments of M_{xy} and M_{yx} then since $M_{xy} = -M_{yx}$ there are unbalanced corner forces of magnitude $2M_{nt}$ at each corner. The corners of a loaded slab therefore tend to lift if the supports cannot provide adequate downwards restraint. If the corners are restrained then they should be provided with suitable torsion reinforcement. This is theoretically required along or perpendicular to the diagonal, according to whether it is placed in the bottom or the top of the slab.

Boundary conditions at a free edge

Let the external reaction at a supported edge be V_n per unit length and the internal shear in the slab be Q_n per unit length. Hence from figure 13.1.7.

$$V_n - Q_n = -\partial M_{nt}/\partial t \qquad (13.1.16)$$

However, $V_n = 0$ for a free edge, hence $Q_n = \partial M_{nt}/\partial t$. This implies that neither Q_n, $\partial M_{nt}/\partial t$ nor M_{nt} need be zero. This is because simple plate theory does not distinguish between stresses due to rate of change of torsional moment and those due to vertical shear. Although the edge conditions $M_{nt} = Q_n = 0$ cannot be specified completely in simple plate theory, it is adequate for practical purposes. In fact, it is shown later that some design methods in practical use do not even satisfy other requirements for the torsional moments.

13.2 Elastic design of slabs

The present section title refers to the determination of the distribution of moments throughout the slab, and *not* to the design of the cross-section of the slab itself. Hence elastic theory solutions can be used for the limit states of collapse, as for any other redundant structure. The Unified Code requirements for local damage are outlined first to emphasize the fact that it is collapse which is of primary concern.

Local damage for slabs

The draft Code states that the requirements of the limit states of local damage may be deemed to be complied with for solid slabs provided that the following conditions are satisfied:

(i) Compliance with requirements for limit states of collapse (which requires that assumed distribution of moments should be very approximately equal to that given by elastic theory).

(ii) Characteristic strength of tensile reinforcement used in calculations for limit states of collapse does not exceed 460 MN/m².

(iii) If overall depth of slab exceeds 250 mm then clear distance between longitudinal bars should be between 150 and 300 mm. It should also not exceed $1{\cdot}5/r'$ mm if r' exceeds 1% of the gross concrete section.

The elastic theory solutions given in the Code are therefore expressed in terms of the limit states of collapse.

One-way slabs

Elastic theory for plates indicates that the effective flexural rigidity of a strip of unit width is $EI/(1 - \nu^2)$ and not simply EI as for a beam. This increase in stiffness is due to the transverse strain, which is associated with the longitudinal stress, being restrained by the adjacent slab. Poisson's ratio for concrete is only about 0·15 so that ν^2 is negligible and this "Poisson's ratio effect" may be neglected. The effect may also be ignored for the steel reinforcement since this in the form of discrete bars. One-way spanning reinforced concrete slabs can therefore be considered as beams, as shown in earlier chapters, provided that suitable provision is made for distribution reinforcement. Such reinforcement is required to distribute the effects of point loads over an adequate breadth and to control the cracking due to thermal and shrinkage effects.

Theoretical solutions

Theoretical elastic analyses for thin plates, based on the equilibrium and compatibility equations and the elastic stress-strain relations, require the solution of a fourth-order partial differential equation. Available solutions are therefore restricted to simple shapes, such as rectangles, circles, triangles, etc., and are usually obtained by Fourier series or finite difference techniques. Such a solution, even when available, takes no account of variations in the flexural rigidity of the slab due to cracking of the concrete or curtailment of the steel, or of the effects of thermal movement, shrinkage and creep.

Semi-empirical methods

A more practical approach is to use a theoretical analysis which is adjusted according to experimental data. This approach is much simpler to use and could even take into account such effects as membrane action if sufficient experimental data were available. The Unified Code gives two semi-empirical methods for rectangular slabs, which are described below.

Rectangular slabs simply supported on four sides

The Grashof-Rankine method (see, e.g., reference 13.3) assumes that a uniformly distributed load on a rectangular slab is shared between strips running in the two directions parallel to the sides of the slab. If all torsional moments in equation 13.1.5 are neglected, as well as any interaction between the x-strips and the y-strips, the strips can be designed as beams for which

$$d^2M_x/dx^2 = -w_x \quad d^2M_y/dy^2 = -w_y \quad \text{and} \quad w_x + w_y = w$$

where w is the total design load per unit area for limit state of collapse.

The ratio of w_x to w_y is determined by a compatibility requirement that the midspan deflections of the central strips in the two directions should be equal. Notice that this method neglects the effects of torsional moments and therefore gives relatively high bending moments for the strips. The Code states that for simply supported slabs, where adequate provision to resist torsion at the corners and to prevent the corners from lifting is not made, the bending moments at midspan should be taken as follows:

$$M_x = \alpha_x w l_x^2 \quad \text{and} \quad M_y = \alpha_y w l_x^2$$

where l_x = length of shorter side
l_y = length of longer side (see table 13.2.1)
α_x, α_y = coefficients as shown in table 13.2.1.

Table 13.2.1 Bending moment coefficients for slabs spanning in two directions at right angles, simply supported on four sides

l_y/l_x	1·0	1·1	1·2	1·3	1·4	1·5	1·75	2·0	2·5	3·0
$10^3\alpha_x$	62	74	84	93	99	104	113	118	122	124
$10^3\alpha_y$	62	61	59	55	51	46	37	29	20	14

Notice that M_y is given in terms of l_x. Notice also that $\alpha_x = 0{\cdot}124$ for $l_y/l_x = 3{\cdot}0$ is virtually the same as for a one-way spanning slab (for which $\alpha_x = 1/8$). At least 50% of the tension reinforcement provided at midspan should extend to the supports and the remaining 50% should extend to within $0{\cdot}1l_x$ or $0{\cdot}1l_y$ of the support, as appropriate.

Rectangular slabs restrained on four sides

Westergaard's method (see, e.g., reference 13.3) is based on theoretical analyses supplemented by experimental data, and allows for some moment redistribution. It is applicable to rectangular slabs with any combination of continuous or simply supported edges provided that all four corners are prevented from lifting and suitable provision is made for torsion as specified in the Unified Code.

The slabs are considered as being divided in each direction into middle strips and edge strips, as shown in figure 13.2.1, where

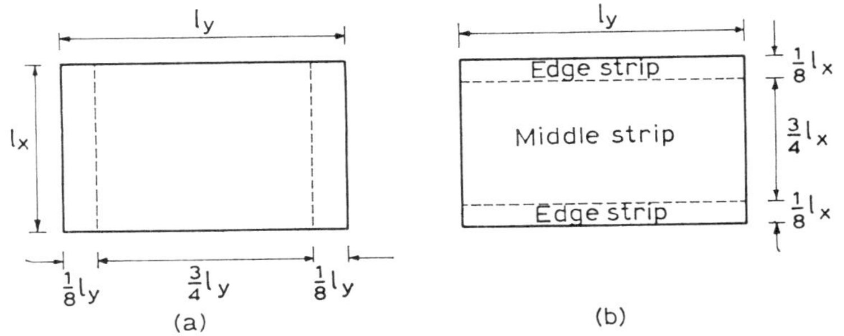

Fig 13.2.1 Division of slab into "middle" and "edge" strips. (*a*) For span l_x. (*b*) for span l_y.

width of middle strip usually equals 3/4 of lateral span. However, if l_y/l_x exceeds 4·0 then the width of the middle strip spanning the short direction is $l_y - l_x$ and the width of each edge strip is $\frac{1}{2}l_x$. The maximum bending moments per unit width in the middle strip of the slab are given by:

$$M_x = \beta_x w l_x^2 \quad \text{and} \quad M_y = \beta_y w l_x^2$$

where M_x, M_y are the maximum bending moments on strips of unit width spanning l_x and l_y respectively, and β_x, β_y are coefficients (of l_x^2), given in table 13.2.2.

Table 13.2.2 Bending moment coefficients for rectangular panels supported on four sides with provision for torsion at corners

Type of panel and position of negative or positive moment	$10^3 \times$ short-span coefficient β_x							
	Values of l_y/l_x							
	1·0	1·1	1·2	1·3	1·4	1·5	1·75	2·0+
Case 1. Interior panels								
Negative moment at continuous edge	33*	40	45	50	54	59	71	83
Positive moment at midspan	25*	30	34	38	41	45	53	62
Case 2. One short or long edge discontinuous								
Negative moment at continuous edge	41*	47	53	57	61	65	75	85
Positive moment at midspan	31*	35	40	43	46	49	56	64
Case 3. Two adjacent edges discontinuous								
Negative moment at continuous edge	49*	56	62	66	70	73	82	90
Positive moment at midspan	37*	42	47	50	53	55	62	68
Case 4. Two short edges discontinuous								
Negative moment at continuous edge	56	61	65	69	71	73	77	80
Positive moment at midspan	44*	46	49	51	53	55	58	60
Case 5. Two long edges discontinuous								
Negative moment at continuous edge	(56)*	—	—	—	—	—	—	—
Positive moment at midspan	44*	53	60	65	68	71	77	80
Case 6. Three edges discontinuous (one short or long edge continuous)								
Negative moment at continuous edge	58*	65	71	77	81	85	92	98
Positive moment at midspan	44*	49	54	58	61	64	69	74
Case 7. Four edges discontinuous								
Positive moment at midspan	50*	57	62	67	71	75	81	83

* Long-span coefficient, β_y, for all values of l_y/l_x is equal to the value of β_x for $_y/l_x$ = 1·0.

Tension reinforcement for restrained slabs

All the tension reinforcement provided in the middle strip at midspan should extend in the lower part of the slab to within $0{\cdot}25l$ of continuous edges and $0{\cdot}15l$ of discontinuous edges. At least 50% of this reinforcement should extend to discontinuous edges or to within $0{\cdot}15l$ of continuous edges.

The tension reinforcement provided over the continuous edges of the middle strips should extend in the upper part of the slab a distance of $0{\cdot}15l$ from the supports and at least 50% should extend a distance of $0{\cdot}30l$.

The negative moment to be assumed to occur at discontinuous edges depends upon the degree of rotational restraint at the nominally free edge. Nominal reinforcement equal to 50% of the midspan reinforcement and extending $0{\cdot}1l$ into the slab should be adequate. No reinforcement parallel to the adjacent edges of the slab need be inserted in the edge strips other than for satisfying the requirements for shrinkage, etc. (never less than 0·12%) and for torsion (see below). If the ends of the slab are fully restrained and edge beams are not present to take the longitudinal tension then r' in the edge strips should be not less than the value required for the limit state of crack width (i.e. r_{crit} or even higher). This requirement occurs, for example, in cantilever slabs and especially canopies where temperature differences between the outside and the inside of the building are also present. Very unsightly blemishes can occur if design for local damage is neglected, as shown in the frontispiece.

Torsion reinforcement for restrained slabs

Torsion reinforcement must be provided at corners contained by edges over neither of which the slab is continuous. It shall consist

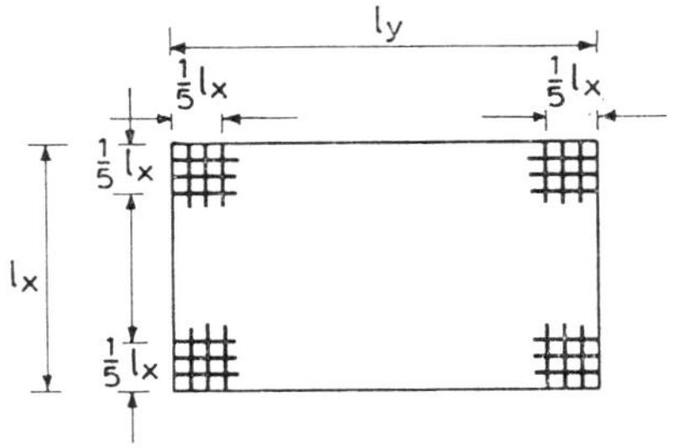

Fig 13.2.2 Torsion reinforcement for "restrained" simply supported slab.

of top and bottom reinforcement, each level containing bars placed parallel to the sides of the slab and extending in these directions for a distance of one-fifth of the shorter span, as shown in figure 13.2.2.

The area of the bars in each of the two layers at each level, per unit width of slab, should be 3/4 of the area required for the maximum midspan moment in the slab. Torsion steel equal to half the preceding amount should be provided at corners contained by edges over only one of which the slab is discontinuous. No torsion steel need be provided at corners contained by edges over both of which the slab is continuous.

Concentrated loads on slabs

The bending moments due to concentrated loads may be calculated from any methods based on elastic theory, such as Pigeaud or Westergaard (see, e.g., reference 13.3). Alternatively, the Unified Code states that they may be calculated by the use of Johansen's yield-line method or Hillerborg's strip method, which are discussed later. In the case of precast floors constructed of narrow precast units such that loads can be transferred between the units the same methods may be employed. The width of floor assumed by the code to be effective in withstanding the loads must not exceed the width of the loaded area by more than:

(i) width of three precast units and joints,

(ii) width of four precast units and joints where there is a topping at least 30 mm thick.

Flat slabs

Normal beam and slab construction requires supporting the slab on all four sides on either beams or walls. Beams can seriously reduce the headroom and so impose restrictions on the use of the space beneath. This disadvantage can be avoided by supporting the slab directly on the columns. The term "flat slab" means a reinforced concrete slab, with or without "drops", supported by columns with or without flared column heads. As for a normal slab, a flat slab may be solid or may have recesses formed in the soffit to provide a series of ribs in two directions. The recesses may be formed by removable or permanent filler blocks.

It is clear that severe shear stresses must occur in a flat slab at the perimeter of each column, due to the possibility of the column punching through the slab. The requirements for concentrated loads on slabs are therefore directly applicable (see appendix C). Such considerations usually control the required thickness of the slab. Self-weight is usually a high proportion of the total load and it is often desirable to thicken the slab locally around columns, to form a "drop" as shown in figure 13.2.3, rather than use a uniform thickness throughout. A flared column head increases the shear perimeter and can therefore also reduce the shear stress. A theoretical elastic analysis

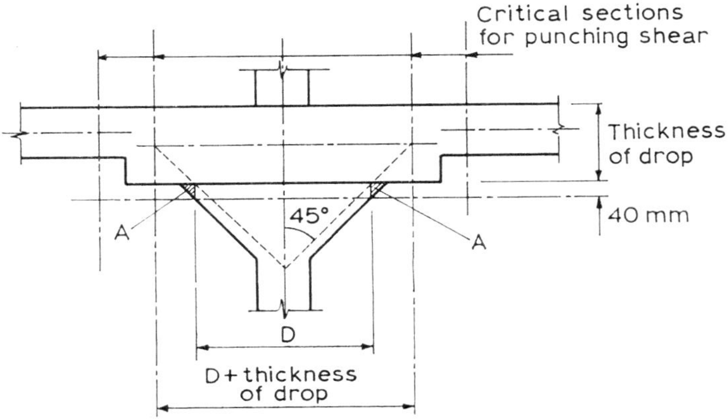

Fig 13.2.3 Slab with drop and column with column head showing critical sections for punching shear.

The presence of any concrete in the shaded areas A does not affect the calculation. Critical sections for shear (diagonal tension) are further out than those for punching shear.

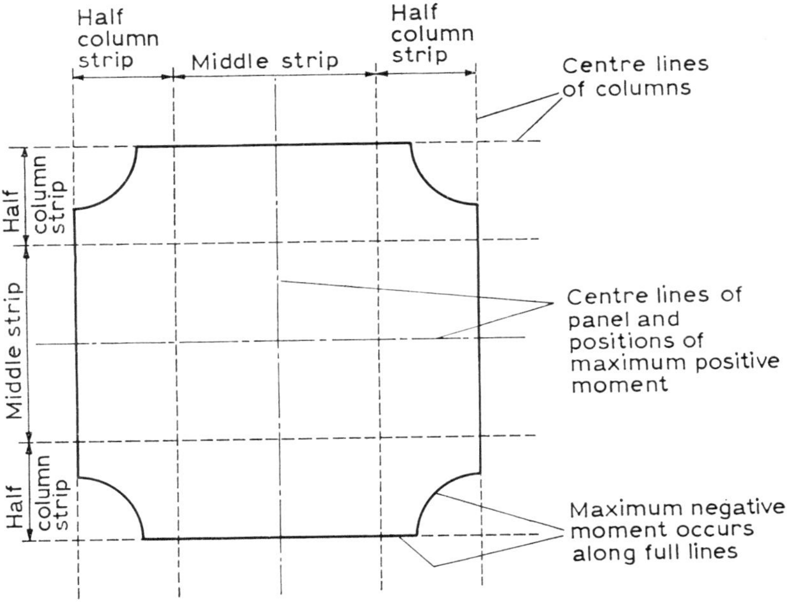

Fig 13.2.4 Division of flat slab panels into column strips and middle strips.

for flat slabs is complicated and so design methods as given in the Unified Code divide the panels into "column strips" and "middle strips" as shown in figure 13.2.4. However, yield-line theory, which is discussed in the next section, is equally suitable for *either* flat slabs *or* beam-and-column construction. Practical design of flat slabs using the continuous-frame method of analysis is outlined in appendix C. The Unified Code itself also contains an alternative empirical design method.

13.3 Introduction to collapse theories for slabs

Elastic theory solutions for slabs suffer from two distinct disadvantages. Theoretical solutions are complicated and often require the use of computers, electrical analogues or similar techniques, while semi-empirical methods are strictly limited in their applicability. The Unified Code still prefers elastic theory solutions for simple rectangular shapes but recommends that bending moments may be determined using Johansen's yield-line method or Hillerborg's strip method for more complicated shapes. These two methods are applicable to slabs of any shape, loading and edge conditions so that their use is far preferable for all but the simplest configurations. The Code requires, however, that the ratio between support moment M_i and midspan moment M_v in continuous or fixed slabs designed by these methods should be approximately equal to that obtained from elastic theory; the recommended values for M_i/M_v being as follows:

(i) 1·5 to 2·0 for slabs effectively fixed about the support
(ii) 1·0 to 1·5 if major restraint to rotation is provided by an adjacent slab of comparable thickness.

The rules for detailing of the reinforcement should also still follow the principles inherent in the elastic analysis of slabs in order to ensure that the limit state of local damage is satisfactory. However it is recognized that, as for other structural members, very extensive redistribution of bending moments can occur before the collapse load is reached. The collapse load of a slab in flexure may be determined by either an upper-bound method, a lower-bound method, or by an exact solution.

Historical background

Yield-line theory was initiated by Ingerslev [13.4] and substantially pioneered by Johansen who produced his thesis, in Danish, in 1943. Extensive design experience had been gained in Scandinavian countries by the time Hognestad published a summary of yield-line theory, in English, in 1953. It had a very controversial introduction into English-speaking countries, due very largely to the difficult theoretical justification for Johansen's so-called "equilibrium method". Anomalies were apparent and extensive research continued following the publications, from 1961–2, by Wood [13.2], Jones [13.5] and the English translation of Johansen's thesis [13.6]. A special publication appeared in 1965 [13.7] and a book by Jones and Wood [13.8] in 1967 has resolved the earlier lack of agreement.

Limit theory for slabs

The basic assumptions for limit theory, discussed in chapter 6, are that failures due to shear, bond and sudden flexural compression are prevented. That is, it is assumed implicitly that the structure has a long horizontal portion to its moment-rotation curve, otherwise a section might fail and lose its moment-carrying capacity before all the plastic hinges had formed. In the case of a slab this means that all yield lines must be capable of reaching the boundaries of the slab before any portion loses its moment-carrying capacity due to excessive rotation. RC slabs are ideally suited to limit theory because the need to limit deflections leads to thicknesses which require relatively low steel ratios for flexure and also low shear stresses. Hence slab sections are usually very under-reinforced so that the steel ratio is often considerably less than $0{\cdot}227u/f_{yt}$ (i.e. $d_n < \frac{1}{2}d_1$). It is now convenient to reconsider upper- and lower-bound solutions with special reference to slabs.

Upper-bound solutions

Upper-bound methods postulate a collapse mechanism and ensure that:

(i) The moments at the plastic hinges (which have been postulated) are not greater than the ultimate moments of resistance of the sections.

(ii) The collapse mechanism is compatible with the boundary conditions.

The elements of the slab between the plastic hinges are *not* examined to ensure that the moments there do not exceed the ultimate moments of resistance of the sections. This method therefore gives a collapse load which may be too high, and is therefore an upper limit or bound to the correct collapse load. If all the possible collapse mechanisms for the slab are examined, the one giving the lowest collapse load is the most likely and is therefore the correct mechanism. The ultimate moments of resistance between the plastic hinges are exceeded only if an incorrect collapse mechanism is postulated.

Lower-bound solutions

Lower-bound methods choose a particular distribution of moments at collapse and ensure that:

(i) The ultimate moments of resistance are not exceeded *anywhere* in the slab.

(ii) The equilibrium equations are satisfied at all positions (which includes compliance with the boundary conditions).

Lower bound methods give a collapse load which may be too low.

Yield-line theory

Yield-line theory gives an upper-bound solution so that it is essential to ensure that the most critical collapse mechanism has been selected if the load-carrying capacity of the slab is not to be over-estimated. The correct collapse mechanisms for most cases are now well known, however, so that the experienced designer is rarely faced with the possibility that a further (more critical) mechanism may exist. Yield-line theory using the virtual work method will be considered in the most detail in the present chapter.

Yield-line theory applies to almost any shape of slab, any load and any edge condition. However, it is restricted as a *design* method to slabs of constant thickness which are uniformly reinforced in each of two mutually perpendicular directions. Thus the steel ratio for the longitudinal direction is constant across the width of the slab and the steel ratio for the transverse direction is constant along the length of the slab. However, the steel ratios may be different in the two directions at right angles and also in the top and bottom. For such slabs the collapse moment of resistance per unit width has a constant value along any straight line in the plane of the slab. Yield-line theory gives no information on the best distribution of the steel within the slab, but can be used to *analyse* a slab where the steel has been distributed according to some other method (e.g. Hillerborg's method).

Hillerborg's method

Lower-bound solutions are much more difficult to obtain, as has been shown by Wood [13.2]. However, a method which has been developed by Hillerborg obtains the moment distribution for the slab by assuming strip action. It has been shown by Wood and Armer [13.9] that Hillerborg's method can provide an exact solution with an unlimited number of simultaneous modes. Hillerborg, however, originally intended it as a lower-bound rather than an exact solution. It is a powerful alternative to yield-line theory since it is essentially a design technique which provides information on the best distribution of the steel within the slab.

13.4 Yield lines and yield-line mechanisms

The progressive development of yield lines is described in order to indicate how yield-line mechanisms are formed. Satisfactory prediction of the possible yield-line pattern is essential if the correct collapse load for the slab is to be calculated. The conditions which assist in determining yield-line mechanisms are outlined and typical examples are illustrated.

Inelastic deformation

A simply supported, two-way spanning slab, subjected to a uniformly distributed load of increasing intensity, initially deforms elastically. The concrete on the tension side of the slab cracks as the load increases, causing the stiffness of the cracked section to be reduced and the distribution of moments in the slab to change. This redistribution produces, for equal increments of load, greater increases in moment at an uncracked section than before cracking occurred.

Yield-line propagation

The initial stages of the inelastic deformation produces yielding of the reinforcement in the central area of the slab, where the moment is greatest. Once the steel in an under-reinforced section has yielded, then although the section continues to deform its resistance moment does not increase by any appreciable amount. Consequently an even

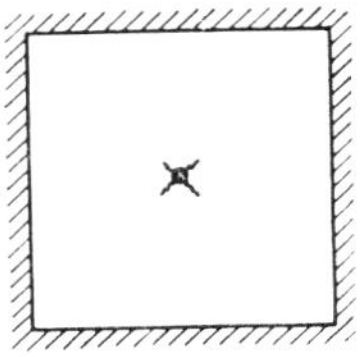

Fig 13.4.1 Initial propagation of yield lines.

greater redistribution of moments must take place after yielding of the steel commences. As further loading is applied the steel in the sections adjacent to the central area also yields, eventually, since an increasing proportion of the moment has to be carried by these sections. Lines along which the steel has yielded are therefore propagated from the point at which yielding initially occurred. The yield lines at this stage could be as shown in figure 13.4.1.

Yield-line pattern

Even further increase in load causes the reinforcement in more and more sections to yield and so extend the yield lines. The yield lines eventually reach the boundaries of the slab, at which stage the resistance moments along the yield lines are almost at their ultimate values. Since the yield lines cannot extend any further, the slab is

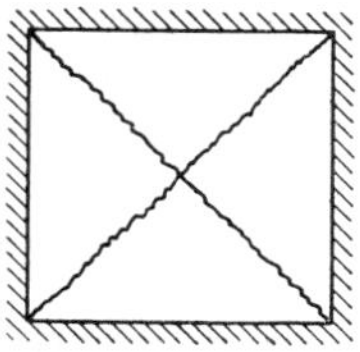

Fig 13.4.2 Final extent of yield lines.

virtually carrying the maximum possible load. The system of yield lines (or fracture lines), such as those shown in figure 13.4.2, is called a yield-line pattern.

Collapse load for slab

A slight increase in load now causes a state of unstable equilibrium; the slab continues to deflect under this load until the curvature at some section along the yield lines is so great that the concrete crushes. This section then loses its ability to carry any moment. This increases the state of unstable equilibrium even more and causes crushing to occur along the whole length of the yield lines. Hence the condition when the yield lines just reach the boundary can be regarded as the collapse criterion of the slab.

Mode of deformation

The curvature of the slab along the yield lines where the steel is yielding is much larger than that for the areas in between, which are still behaving elastically. Consequently, it is reasonable to assume that the elastic deformations are negligible in comparison with the plastic deformations. This means that the slab elements between the yield lines can be assumed to remain plane and that all deformations take place in the yield lines.

Straight yield lines

Consider the slab elements in the deflected state. These elements are inclined planes and therefore intersect in straight lines. It follows that the yield lines are also straight since these are, in fact, the intersections between the plane elements.

Axes of rotation

The plane elements must rotate about certain axes if the slab is to deflect. In figure 13.4.3, element a rotates about AB, element b about BC, c about CD and d about DA. It is also a necessary condition of deformation that the yield line (being a line of intersection between two adjacent elements) passes through the intersection of the axes

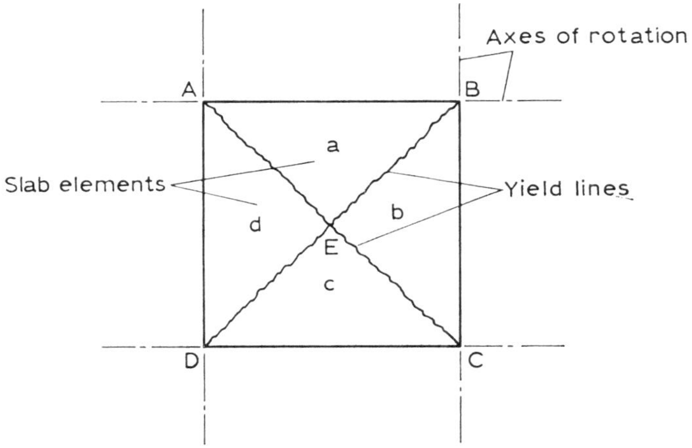

Fig 13.4.3 Fully developed yield-line pattern.

of rotation of these plane slab elements: i.e. yield line EB must pass through point B, the intersection of AB and BC, EC through point C, etc.

Prediction of yield-line pattern

The foregoing can be summarized by the following five conditions which assist the prediction of the yield line pattern for any slab:

(i) Yield lines terminate at a slab boundary.
(ii) Yield lines are straight.*
(iii) Yield lines (produced if necessary) pass through the intersection of the axes of rotation of adjacent slab elements.
(iv) Axes of rotation generally lie along lines of supports and pass over columns.
(v) Any symmetry of the slab is maintained in the yield-line pattern.

* It should be noted however that for some systems of point loads or circular supports a multiple system of yield lines can approximate to a single curved yield line, e.g. see figure 13.4.6(*b*).

Notation

The notation used is shown in figure 13.4.4.

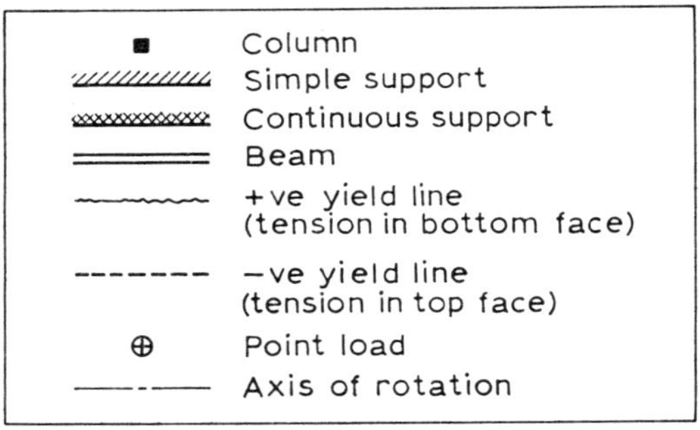

Fig 13.4.4 Standard notation for yield-line diagrams.

Typical failure mechanisms

Typical yield-line patterns of various slabs subjected to uniformly distributed loads are shown in figure 13.4.5 while two examples of a point load are shown in figure 13.4.6. Notice that, in figure 13.4.5, the yield line EF between segments a and c, for a rectangular slab, passes through the intersection of the axes of rotation AB and DC at infinity. Similarly the other yield lines pass through the corners.

The negative yield lines for the continuous rectangular slab must also form along the lines of support before the latter can act as axes of rotation and transform the slab into a mechanism. The more irregularly shaped slabs are seen to follow the same general principles. Notice that only the general patterns have been indicated in some cases. For example, the exact positions of E or F must be found in the process of analysis. Fortunately, however, the analysis can be commenced with only a qualitative assessment of the general yield-line pattern. In other cases it is necessary to consider more than one yield-line pattern. For example, consider the last three patterns in figure 13.4.5, which are all for the same slab layout.

13.5 Normal bending moment criterion for yield-line theory

It is shown in this section that the criteria for failure on which yield line theory is based are effectively considerations of the normal bending moments in the slab. This is consistent with an upper-bound approach. "Classical" yield-line theory, however, ignores any partial kinking of the bars, which can increase the resistance moment by up to 14% for an isotropically reinforced slab. The criterion

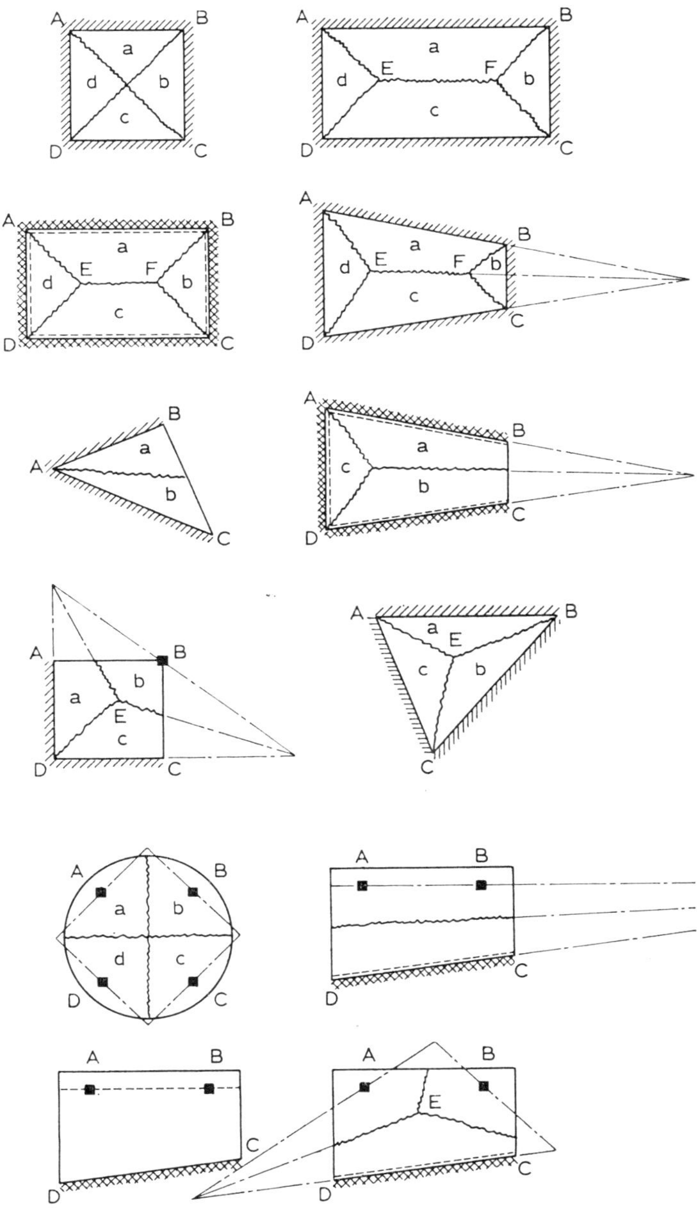

Fig 13.4.5 Various yield line patterns for uniformly distributed loads.

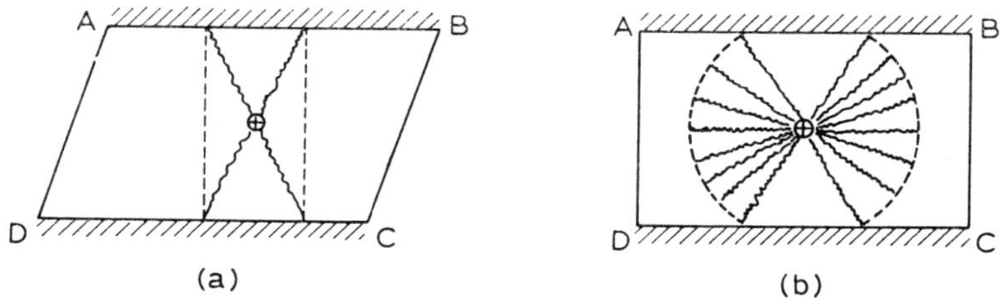

Fig 13.4.6 Typical yield line patterns for point loads.

normally used is therefore seen to be a "lower bound" for an upper-bound method!

Yield principal moments

It is convenient to express the ultimate bending strength in terms of moments per unit length of slab normal to the direction of the reinforcement and to let the x- and y-axes be parallel to the directions of the reinforcement as shown in figure 13.5.1. Let m_x be the yield

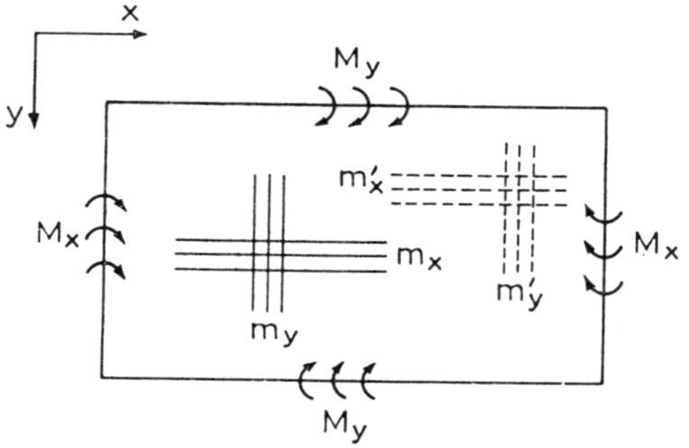

Fig 13.5.1 Bending moments and resistance moments per unit length.

principal moment for the x-direction steel, per unit length of the slab parallel to the y-direction. Similarly, let m_y be the yield principal moment for the y-direction steel, per unit length of slab parallel to the x-direction.

Figure 13.5.1 indicates that the ultimate moment per unit length about a positive yield line parallel to the y-axis is m_x. If the steel is in the bottom face of the slab, giving positive (sagging) bending strength, the reinforcement lines are shown full and they provide resistance along a positive yield line. If the steel is in the top of the slab to give negative (hogging) strength, it is shown by broken lines and it provides resistance along a negative yield line.

Moments along a yield line

Consider a small element of slab subjected to principal moments M_x and M_y. If M_n and M_{nt} are the normal and twisting moments respectively, both per unit length, applied along a yield line which makes an angle ϕ with the x-axis, then their values can be obtained by resolving vectors and putting $M_{xy} = 0$. Hence from equations 13.1.8 and 10

$$M_n = M_x \cos^2 \phi + M_y \sin^2 \phi \tag{13.5.1}$$

and

$$M_{nt} = \tfrac{1}{2}(M_x - M_y) \sin 2\phi \tag{13.5.2}$$

Yield-line criterion

If it is assumed that all the reinforcement which crosses a yield line is yielding, then yield criteria can be obtained from equations 13.5.1 and 2 by putting

$$M_x = m_x = m \quad \text{and} \quad M_y = m_y = \mu m$$

The criterion for classical yield-line theory, however, considers only the normal bending moment as given in equation 13.5.1. That is,

$$m_n = m_x \cos^2 \phi + m_y \sin^2 \phi \tag{13.5.3(a)}$$

or

$$m_n = m \cos^2 \phi + \mu m \sin^2 \phi \tag{13.5.3(b)}$$

Lower-bound solutions

Lower-bound solutions would require that the twisting moment is also satisfied, that is

$$m_{nt} = m(1 - \mu) \sin \phi \cos \phi \tag{13.5.4}$$

Wood has shown, however, that the assumption that all steel is yielding leads to yet a third condition for lower-bound solutions [13.2]. It is extremely difficult to find, in general, a statically admissible yield moment which gives a collapse load at all close to the upper-bound solutions. In addition, Kemp has shown that even for the maximum normal bending moment criterion only, all directions should be considered and not just the yield-line direction [13.10]. It is clear that true lower-bound solutions are very difficult to obtain. However, although simple yield-line theory satisfies one basic condition only (and then in general not exactly), it has been found to give surprisingly satisfactory predictions of the collapse load. This is largely due to the fact that twisting moments at a yield line do little, if any, virtual work.

Square yield criterion

For an isotropically reinforced slab, where $\mu = 1$, equations 13.5.3 and 13.5.4 reduce to

$$m_n = m \tag{13.5.5}$$

and

$$m_{nt} = 0 \tag{13.5.6}$$

so that none of the theoretical complications just referred to arise. Plotting M_n against ϕ according to equation 13.5.1 provides the curve shown in figure 13.5.2. The slab will yield wherever the maximum principal moment equals the yield moment of resistance of the slab. For an isotropic slab the yield moment is constant for any direction, that is, for any value of ϕ. The lines for M_n and m touch

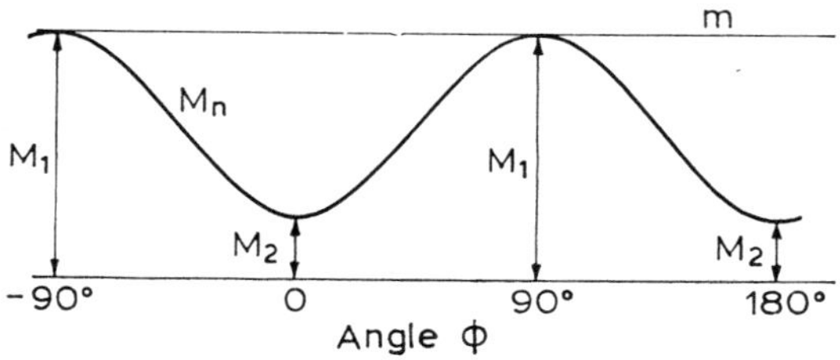

Fig 13.5.2 Graphical representation of normal moment.

at $\phi = \pm 90°$ in figure 13.5.2, hence the slab yields in that direction. If there is reinforcement in the bottom face only the yield criterion is given by:

$$0 \leqslant M_i \leqslant m \quad (i = 1, 2)$$

Negative reinforcement

If isotropic reinforcement is also provided in the top face to produce a negative yield moment of resistance m' per unit length, then the yield criterion becomes

$$m' \leqslant M_i \leqslant m \quad (i = 1, 2)$$

To summarize for an isotropic slab: the normal moment along a positive yield line is the maximum principal moment, i.e. ($M_1 = m$) and the twisting moment is zero. The normal moment along a negative yield line is the minimum principal moment (i.e. $M_2 = m'$) and the twisting moment is zero.

Orthotropic slabs

Kemp has shown that the yield criterion for orthotropically reinforced slabs is essentially an envelope of permissible normal bending moments [13.10]. If the plane of action of the applied principal moment, M_1, makes some clockwise angle, δ, with the x-direction reinforcement, it should be noticed that yield occurs where the two curves just touch, which corresponds to a yield line with a normal

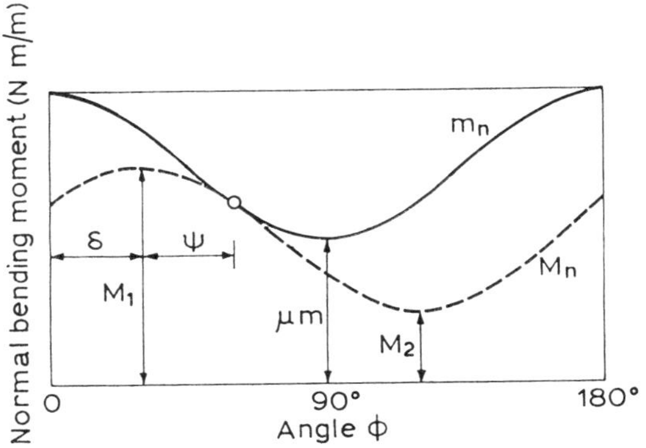

Fig 13.5.3 Positive yield in an orthotropically reinforced slab.

at $\phi = \delta + \psi$ measured clockwise from the x-direction, as shown in figure 13.5.3. In other words, the conditions satisfied by classical yield-line theory are:

$$M_n = m_n \quad \text{and} \quad dM_n/d\phi = dm_n/d\phi$$

Kinking of bars

Equation 13.5.3(a) assumes that the reinforcing bars crossing the crack at a yield line remain straight as shown in figure 13.5.4(a).

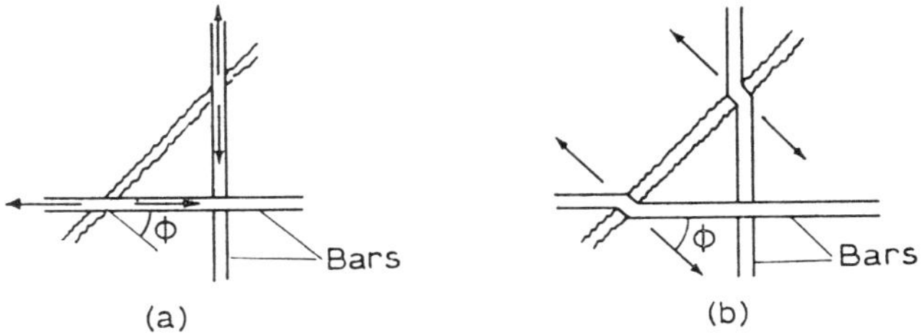

Fig 13.5.4 Bars with (a) no kinking and (b) full kinking.
Note: the crack widths in this diagram are exaggerated.

However if the bars are kinked at the crack, as shown in figure 13.5.4(b), the yield moment is increased. Maximum possible kinking would occur if the bars were forced to cross the crack at right angles.

If this is assumed it is evident that resolving moment vectors along the yield line, for unit length of yield line, now gives

$$m_n = m_x \cos \phi + m_y \sin \phi \tag{13.5.7}$$

Notice that the complication due to kinking is present even for an isotropically reinforced slab. The maximum effect due to kinking occurs if the yield line is at 45° to the bars. Hence putting $m_x = m_y$ and $\phi = 45°$ in equations 13.5.5 and 7 shows that:

$$m_n \text{ for full kinking} = (\sqrt{2}) \times m_n \text{ for no kinking}$$

This is an increase of over 40%. However, full kinking cannot occur, because the high bearing stresses against the concrete tend to allow the bars to straighten and also because the crack width is in fact much smaller than the bar diameter. Although some increase in moment due to kinking is possible this has usually been ignored since it is on the safe side. However, a recent paper by Kemp and Prince enables accurate predictions to be made according to a strain compatibility theory [13.11]. This theory indicates a maximum increase of 14% at $\phi = 45°$.

13.6 Methods of analysis for yield-line theory

There are two well-known methods for yield-line theory:

(i) The virtual work method

(ii) Johansen's "equilibrium" method.

Both methods are based on a postulated yield-line pattern. Johansen's "equilibrium" method is therefore a misnomer in that it is an upper-bound solution and not, as its name implies, a lower-bound solution. It is necessary, for either method, to analyse alternative yield-line patterns in certain cases in order to determine the most critical collapse mechanism.

Virtual work method

When the yield-line pattern has formed completely only an infinitesimal increase in load is necessary to cause the structure to deflect. By the principle of virtual work, the work done on the slab by the loads equals the work done against the yield lines as they rotate. Hence if a point on the slab is given a virtual deflection, δ, displacements in the form of rotations compatible with this deflection must take place along the yield lines. The internal work done on the slab is the sum of the rotations in the yield lines multiplied by the resisting collapse moments. The external work done by the loads is the sum of the loads multiplied by their respective deflections. Equating the internal and external work gives the relation between the collapse resistance moments in the slab and the collapse load.

"Equilibrium" method

A yield-line pattern is postulated, as before, and then the equilibrium of each slab element is studied. The equilibrium equations obtained for each element, by taking moments and resolving vertically, are solved simultaneously to obtain the solution. The equilibrium method cannot be used, however, without further specialized knowledge. The reason for this is that moments cannot be taken about appropriate axes for each individual slab element until the magnitude and distribution of the shear forces (or reactions) between adjacent elements are known. Certain theorems due to K. W. Johansen enable the statical equivalents of these forces to be calculated, so the "equilibrium" method, once understood, usually gives solutions more readily than the virtual work method.

Although the equilibrium method enables column reactions to be calculated it is still restricted, as a design method, to isotropically or orthotropically reinforced slabs. The reader is therefore advised not to consider this method until after studying both the virtual work method and Hillerborg's strip method. It is therefore not discussed further in this book, and reference can be made subsequently to more advanced texts (e.g. reference 13.8).

Uniform steel distribution

Yield-line solutions are greatly simplified by using a mesh type of reinforcement which is constant throughout the slab. In other words, the steel ratio along a given line is constant. The steel ratio for different directions, however, may still vary. Both of the following commonly used types of slab have a constant steel ratio along any given line:

(i) Isotropically reinforced slabs. If the ultimate flexural moment about any two perpendicular lines in the plane of the slab is the same, then the slab is isotropically reinforced.

(ii) Orthotropically reinforced slabs. If the ultimate flexural moments about two perpendicular lines in the plane of the slab are different, then the slab is orthotropically reinforced.

13.7 Yield-line analysis by virtual work

Consider a rigid body which is in equilibrium under the action of a system of forces. If the body is given a small arbitrary displacement the sum of the work done by the forces (i.e. force times its corresponding displacement) is zero since the resultant force is zero. Hence the principle of virtual work can be stated as follows:

"If a rigid body which is in statical equilibrium under a system of forces is given a virtual displacement the sum of the virtual work done by the forces is zero." The virtual displacement is a small arbitrary displacement and the virtual work is the work resulting from the displacement.

External work

Once a yield-line pattern at failure has been postulated a convenient point in the slab is selected and given a virtual displacement δ. The deflections of all parts of the slab can then be expressed in terms of δ and the total work done by the external loads calculated from the relation

$$\text{external work} = \int\int w\delta dxdy$$

which may be written as $\sum(W\delta)$ for convenience. Notice that the reactions at the supports do not contribute to the virtual work since their displacement is zero.

Internal work

The virtual work done by the internal actions at the yield lines is due only to the yield moments. Notice that the twisting moments and the shear forces on each side of a yield line are equal and opposite, hence the virtual work done by the twisting moments and the shear forces when summed over the whole slab is zero. The work done at a yield line is therefore the product of the total moment along the yield line and the rotation at the yield line. Hence the total work done at all the yield lines is given by

$$\text{internal work} = \sum m_n l\theta = \sum(M\theta)$$

where m_n = resistance moment normal to yield line, per unit length along yield line

l = length of yield line

θ = rotation at the yield line.

Virtual work equation

The solution for the slab is obtained by equating the external work done by the loads to the internal work done against the yield line moments. That is:

$$\sum(W\delta) = \sum(M\theta) \qquad (13.7.1)$$

Notice that the virtual work equation (i.e. equation 13.7.1) is applied once the yield-line pattern has been postulated, hence the loads and moments are at their maximum values, e.g. the expression for the external work done is $W\delta$ and not $\frac{1}{2}W\delta$ as for real (i.e. total) elastic deflections.

EXAMPLE 13.7.1

A rectangular slab is simply supported along two opposite edges and subjected to a uniformly distributed load. Determine the collapse load per unit area, w, in terms of the resistance moment per unit width, m. Let L be the span.

Solution Assume a yield-line pattern as shown in figure 13.7.1 and let λL equal the length of each support. Then the average deflection of

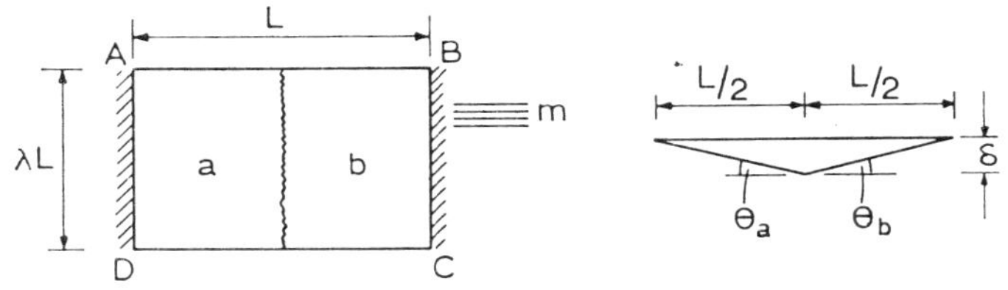

Fig 13.7.1 Example 13.7.1.

the slab elements is $\frac{1}{2}\delta$ if the maximum deflection is δ. The total rotation of the yield line is $(\theta_a + \theta_b) = 2(\delta/\frac{1}{2}L)$, hence from equation 13.7.1

$$w.\lambda L.L.\tfrac{1}{2}\delta = m.\lambda L.2(\delta/\tfrac{1}{2}L)$$

Simplifying and re-arranging gives

$$m = wL^2/8, \qquad \text{(Answer)}$$

a well-known result.

EXAMPLE 13.7.2

A simply supported, isotropically reinforced, square slab is subjected to a uniformly distributed load of w per unit area, as shown in figure 13.7.2. Determine the resistance moment per unit width, m, in terms of w.

Solution Using equation 13.7.1 as before:

$$w.L^2.\delta/3 = 2(2mL\sqrt{2})(2\delta/L\sqrt{2})$$

Simplifying and re-arranging gives

$$m = wL^2/24. \qquad \text{(Answer)}$$

Notice that since δ always cancels out, a unit deflection can be considered at the selected point if so desired. However, the advantage of considering δ is that it provides a partial check and the working is easier to follow.

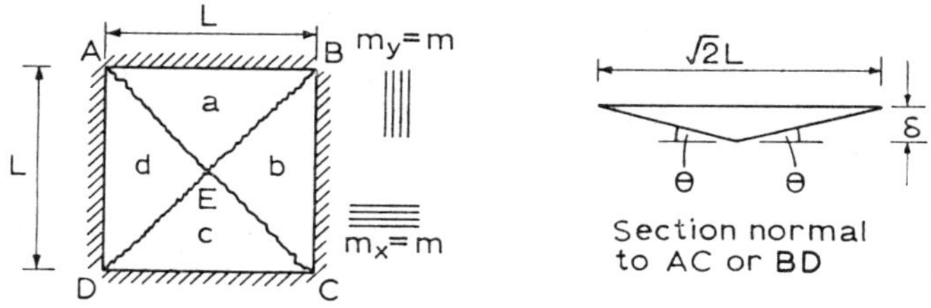

Fig 13.7.2 Example 13.7.2.

Components of work

The work done at a yield line is given by the product of the ultimate normal moment along the yield line and the true rotation of the yield line. However, it is generally much easier to calculate the components of the work done. Considering figure 13.7.3:

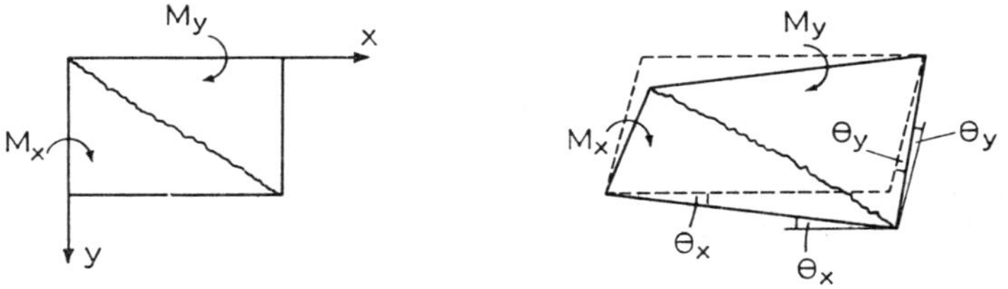

Fig 13.7.3 Components of the work done at a yield line.

$$\sum(M\theta) = \sum(M_x\theta_x + M_y\theta_y)$$

$$= \sum(m_xL_y\theta_x + m_yL_x\theta_y) \qquad (13.7.2)$$

where m_x = collapse moment of x-direction steel per unit length in y-direction

L_x = projected length of the yield line on the x-axis

θ_x = rotation of x-direction steel at yield line.

Similarly for m_y, L_y and θ_y.

Slab-element rotations

The rotation of a yield line is the sum of the rotations of the adjacent slab elements. When resolving along the axes it is usually much more convenient to consider the rotation of the elements instead of the yield lines. For example, reconsider example 13.7.2.

EXAMPLE 13.7.3 (same as example 13.7.2)

Alternative solution Element a rotates about the x-axis (see figure 13.7.2). The portions of the yield lines which surround this element only are AE and EB. Hence by considering only the rotation of these yield lines which is due to the rotation of element a from the horizontal, it is evident that the work component about the x-axis is given by

$$m_y L_x \theta_y \quad \text{where} \quad m_y = m \quad \text{and} \quad \theta_y = \delta/\tfrac{1}{2}L$$

as shown in figure 13.7.4.

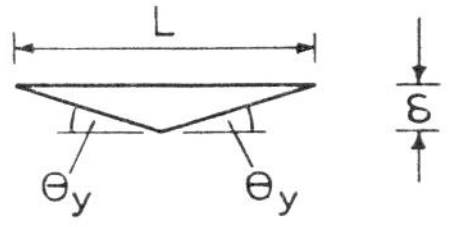

Fig 13.7.4 Example 13.7.3.
Section is normal to AB of figure 13.7.2.

Work component about y-axis for element a is zero since $\theta_x = 0$. Hence work done by element a is

$$mL.2\delta/L = m.2\delta$$

Work done by element b is $m_x L_y \theta_x$ which in this case also equals $m.2\delta$. Hence applying the principle of virtual work to the whole slab:

$$4(m.2\delta) = w.L^2.\delta/3 \quad \text{or}$$

$$m = wL^2/24 \quad \text{as before.} \qquad \text{(Answer)}$$

Minimum load principle

While a general pattern of failure can often be drawn, the exact pattern cannot always be determined by inspection. Consider figure 13.7.5 for example. Length EF is unknown for (*a*), the co-ordinates of E are unknown for (*b*), while α, β and ϕ are unknown

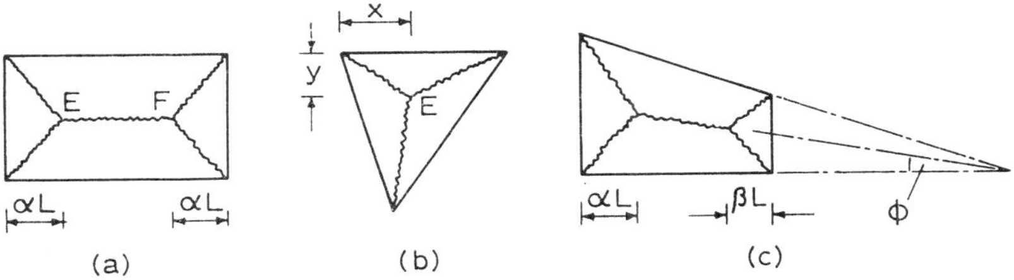

Fig 13.7.5 Yield patterns with different numbers of unknowns. (*a*) One unknown. (*b*) Two unknowns. (*c*) Three unknowns.

for (*c*). Hence the general pattern of yield lines is often defined by unknown parameters, such as ratios or angles, so that the virtual work equation is obtained in terms of these to give an equation of the form

$$m = wf(\alpha,\beta,\gamma \ldots \phi)$$

Only one equation relating m and w is obtained from virtual work hence n more equations are required for n parameters in order to solve for the collapse load. These are given by the turning values relative to each parameter, since m must be a maximum (or w a minimum) for the most probable collapse mechanism. That is,

$$\frac{\partial m}{\partial \alpha} = \frac{\partial m}{\partial \beta} = \ldots = \frac{\partial m}{\partial \phi} = 0 \qquad (13.7.3)$$

It is usually convenient to express m in the form

$$m = wf_1(\alpha,\beta, \ldots)/f_2(\alpha,\beta, \ldots)$$

A turning value is given by putting $\partial m/\partial \alpha = 0$, that is

$$\frac{f_1(\alpha,\beta, \ldots)}{f_2(\alpha,\beta, \ldots)} = \frac{\partial f_1/\partial \alpha}{\partial f_2/\partial \alpha} = \frac{\partial f_1/\partial \beta}{\partial f_2/\partial \beta} = \ldots \text{etc.} \qquad (13.7.4.)$$

EXAMPLE 13.7.4

State clearly the conditions which help to predict the yield-line pattern for a reinforced concrete slab. An orthotropically reinforced rectangular slab is simply supported along all four edges and measures 4·8 m × 6 m. If the effective depth is 0·2 m, A_{st} is 381 and 296 mm² per metre width for the y- and x-direction reinforcement respectively, $u = 22{\cdot}5$ MN/m² and $f_{yt} = 420$ MN/m², determine the ultimate value of w.

Solution Equating the steel and concrete forces for a metre width of slab gives

$$(420/1{\cdot}15) \times 296 \times 10^{+6-6} = 0{\cdot}4 \times 22{\cdot}5 \times d_n \times 10^6$$

or

$$d_n = 12 \times 10^{-3}$$

Therefore

$$m_x = (420/1{\cdot}15) \times 296(200 - \tfrac{1}{2} \times 12) \times 10^{6-6-3} = 21000$$

Similarly

$$m_y = (420/1{\cdot}15) \times 381(200 - 20{\cdot}3 \times 0{\cdot}381) \times 10^{-3} = 26700$$

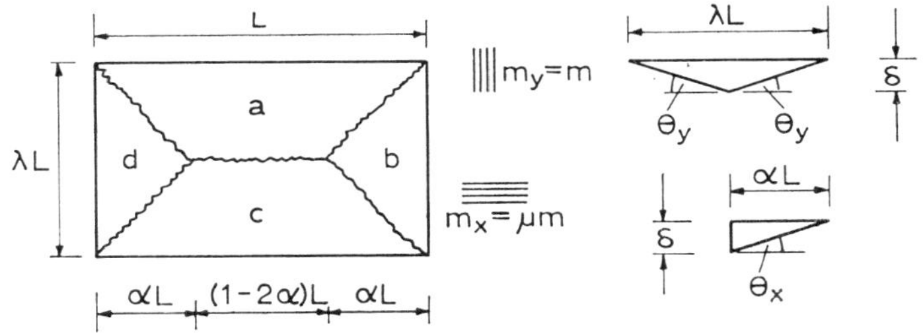

Fig 13.7.6 Example 13.7.4.

Using $\sum W\delta = \sum M\theta$,

$$w \times \tfrac{1}{2}\delta[\lambda L(1 - 2\alpha)L] + w \times \tfrac{1}{3}\delta[\lambda L.2\alpha L]$$
$$= 2(m_x\lambda L)(\delta/\alpha L) + 2(m_y L)(2\delta/\lambda L)$$

Let $m_y = m$ and $m_x = \mu m$ where $\mu = 21/26{\cdot}7 = 0{\cdot}787$. Then

$$w\lambda^2L^2/12 = (\mu\lambda^2 + 2\alpha)m/(3\alpha - 2\alpha^2)$$

$$dw/d\alpha = 0 \quad \text{when } (\mu\lambda^2 + 2\alpha)/(3\alpha - 2\alpha^2) = 2/(3 - 4\alpha)$$

or

$$-\alpha^2 - \mu\lambda^2\alpha + \tfrac{3}{4}\mu\lambda^2 = 0$$

$$\alpha = -\tfrac{1}{2}[\mu\lambda^2 \pm \sqrt{(\mu^2\lambda^4 + 3\mu\lambda^2)}]$$

If $\lambda = 0{\cdot}8$ and $\mu = 0{\cdot}787$ then $\alpha = 0{\cdot}413$. Hence

$$w = \frac{12}{(4{\cdot}8)^2} \times \frac{0{\cdot}787 \times 0{\cdot}64 + 0{\cdot}826}{(3 - 0{\cdot}826) \times 0{\cdot}413} \times (26\,700) = 20\,550$$

$$= 20{\cdot}55 \text{ kN/m}^2 \qquad \text{(Answer)}$$

EXAMPLE 13.7.5

Sketch two possible modes of failure for the uniformly loaded, isotropically reinforced slab shown in figure 13.7.7(*a*). Using either of your assumed modes of failure calculate m in terms of the ultimate uniformly distributed load w if $\mu = \eta = 1$ and $m' = m$.

Solution The possible modes are shown in figures 21.7.7(*b*) and (*c*).

General case for mode A

$$w.\lambda L(1 - \alpha)L(\delta/3) + w\lambda L.\alpha L(\delta/2)$$
$$= (m + m')\lambda L\delta/(1 - \alpha)L + 2(\mu m + \eta m')L(2\delta/\lambda L)$$

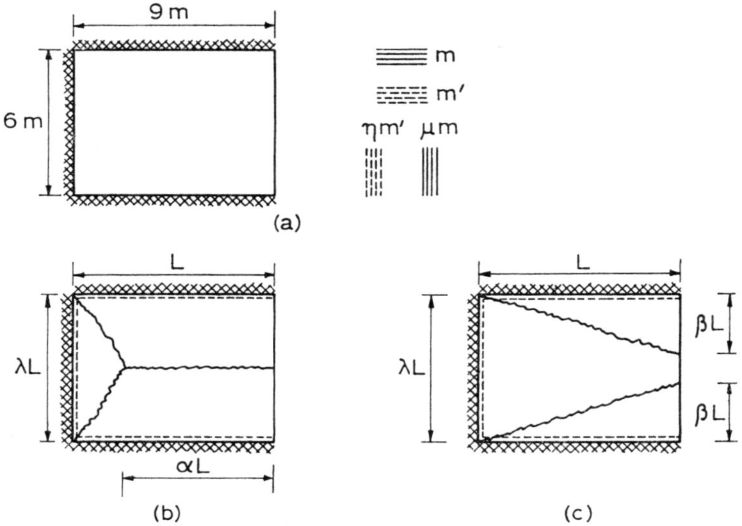

Fig 13.7.7 Example 13.7.5. (*a*) Slab layout. (*b*) Mode A. (*c*) Mode B.

or

$$(w\lambda^2L^2/6)(2 + \alpha)(1 - \alpha) = \lambda^2(m + m') + 4(\mu m + \eta m')(1 - \alpha)$$

$dw/d\alpha = 0$ when

$$(2 + \alpha)(1 - \alpha)[-4(\mu m + \eta m')] = [\lambda^2(m + m') + 4(\mu m + \eta m')(1 - \alpha)](-2\alpha - 1)$$

which simplifies to

$$\tfrac{1}{2}\alpha^2 - (1 + A)\alpha + \tfrac{1}{2}(1 - A) = 0 \quad \text{where}$$

$$A = \lambda^2(m + m')/4(\mu m + \eta m')$$

Hence

$$\alpha = (1 + A) \pm \sqrt{[(1 + A)^2 - (1 - A)]}$$

For this particular slab $\lambda = 2/3$ and $A = 1/9$, hence α equals either 0·522 or a value greater than one. $\alpha = 0{\cdot}522$ is a valid solution. Hence, substituting

$$\alpha = 0{\cdot}522 \quad \lambda L = 6 \quad \mu = 1 \quad \eta = 1 \quad m' = m$$

in the above equation for m and transposing gives

$$m = 1{\cdot}54w \qquad \text{(Answer)}$$

General case for mode B

$$w(\delta/3)(2\beta L^2) + w(\delta/2)L^2(\lambda - 2\beta) = (2m\beta L + m'\lambda L)(\delta/L) + 2(\mu m + \eta m')L(\delta/\beta L)$$

or

$$\frac{wL^2}{6} = \frac{2m\beta^2 + \lambda m'\beta + 2(\mu m + \eta m')}{\beta(3\lambda - 2\beta)}$$

$$dw/d\beta = 0 \quad \text{when}$$

$$[2m\beta^2 + \lambda m'\beta + 2(\mu m + \eta m')](3\lambda - 4\beta) = \beta(3\lambda - 2\beta)(4m\beta + \lambda m')$$

or

$$B\beta^2 + 4\beta - 3\lambda = 0 \quad \text{where}$$

$$B = \lambda(3m + m')/(\mu m + \eta m')$$

Hence

$$\beta = \frac{-4 \pm \sqrt{(16 + 12\lambda B)}}{2B}$$

For this particular slab $\lambda = 2/3$ and $B = 4/3$, hence

$$\beta = 0{\cdot}436$$

But $\beta = 0{\cdot}436$ is not valid since $0{\cdot}436 > \frac{1}{2}\lambda$. Thus mode A is more likely than mode B.

EXAMPLE 13.7.6

An isotropically reinforced triangular slab is supported on two sides at an angle θ as shown in figure 13.7.8. If the lengths of the supported sides are a and b determine the uniformly distributed collapse load w in terms of the resistance moment m.

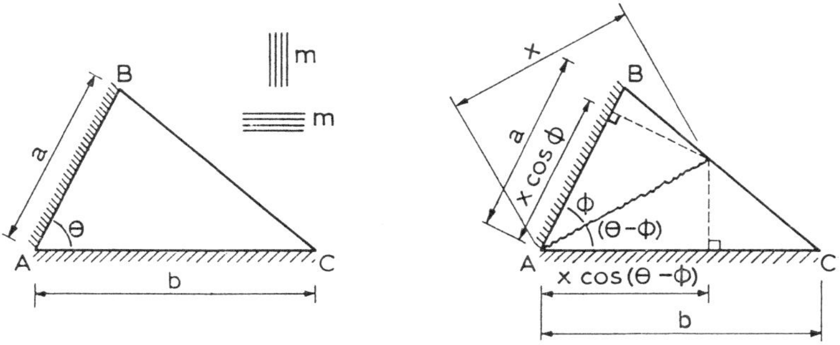

Fig 13.7.8 Example 13.7.6.

Solution Let the yield line form at angle ϕ to the side of length a as shown.

$$\sum W\delta = w(2\delta/3) \times \tfrac{1}{2}ab \sin\theta = \tfrac{1}{3}w\delta ab \sin\theta$$

Now the resistance moment per unit length along AB or AC is m, hence

$$\sum M\theta = mx.\cos\phi.\delta/(x\sin\phi) + mx\cos(\theta - \phi).\delta/(x\sin(\theta - \phi))$$
$$= m(\cot\phi + \cot(\theta - \phi))\delta$$

Putting $\sum W\delta = \sum M\theta$ gives

$$wab\sin\theta = 3m(\cot\phi + \cot(\theta - \phi))$$
$$= 3m(\cot^2\phi + 1)/(\cot\phi - \cot\theta)$$

$\partial w/\partial\phi = 0$ when

$$\frac{\cot^2\phi + 1}{\cot\phi - \cot\theta} = \frac{-2\cot\phi(1 + \cot^2\phi)}{-(1 + \cot^2\phi)} = 2\cot\phi$$

or

$$\cot^2\phi - 2\cot\theta\cot\phi - 1 = 0$$

Hence

$$\cot\phi = \cot\theta \pm 1/\sin\theta = \cot\tfrac{1}{2}\theta \quad \text{or} \quad -\tan\tfrac{1}{2}\theta$$

Thus $\phi = \tfrac{1}{2}\theta$ and

$$w = 3m/ab\sin^2\tfrac{1}{2}\theta \qquad \text{(Answer)}$$

13.8 Special aspects of yield-line analysis

Since yield-line theory gives an upper-bound solution it is essential to select the most probable yield-line pattern if the lowest collapse load is to be determined. Alternative modes of failure which are possible due to point loads, curtailment of reinforcement, composite action of beams and slabs, corner levers, etc., therefore require special consideration.

Point loads

Consider a simply supported square slab as shown in figure 13.8.1(*a*).

Work done by load $= W\delta$

Internal work $= 4mL.2\delta/L = 8m\delta$

Hence $W = 8m$

This is not the correct collapse load, however, since a lower value of W can be obtained if a yield-line mechanism in the form of a circular fan is considered, as shown in figure 13.8.1(*b*). The moment along OB or OA equals mR, since the slab is isotropically reinforced.

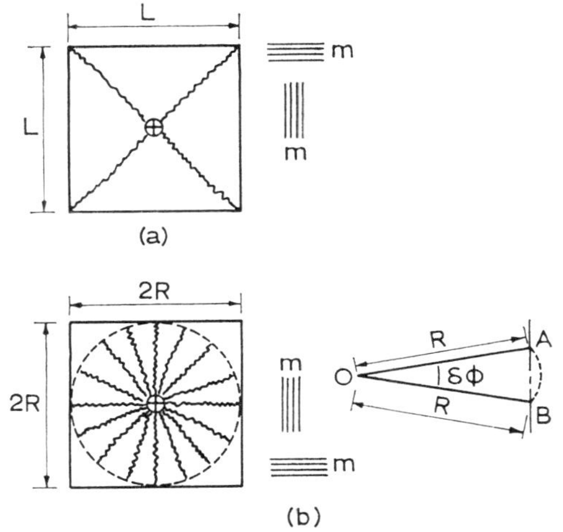

Fig 13.8.1 Central point load for square slab. (*a*) Diagonal yield lines. (*b*) Radial yield lines.

Hence

moment along AB $= mR.\delta\phi$

Rotation about AB $= \delta/R$

Rotation about normal to AB $= 0$

Therefore the internal work is

$$\sum mR\delta\phi(\delta/R) = \delta\int_0^{2\pi} m d\phi = 2\pi m\delta$$

Work done by load $= W\delta$

Hence $W = 2\pi m$

If there is isotropic negative reinforcement with yield moment m':

Moment along AB $= m'(R.\delta\phi)$

Total internal work $= \delta\int_0^{2\pi}(m + m')d\phi = 2\pi(m + m')\delta$

Hence $W = 2\pi(m + m')$

Curtailment of reinforcement

Consider a rectangular slab continuously reinforced with top and bottom steel as shown in figure 13.8.2(*a*). If the top steel is curtailed as shown in figure 13.8.2(*b*) it is necessary to consider the alternative yield-line mechanism. This is illustrated by example 13.8.1.

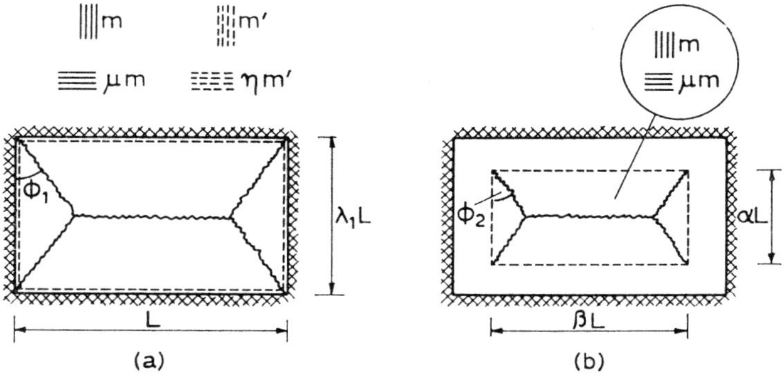

Fig 13.8.2 Example 13.8.1.

EXAMPLE 13.8.1

Determine the uniformly distributed load, w, at collapse of the rectangular slab, $\lambda_1 L \times L$, in terms of the angle ϕ_1 defining the yield-line mechanism shown in figure 13.8.2(*a*). Hence show that the most likely value for ϕ_1 is given by

$$\tan\phi_1 = \sqrt{(\lambda_1{}^2 B^2 + 3B)} - \lambda_1 B$$

where $B = (\mu m + \eta m')/(m + m')$. Hence show also that

$$w = 24(\mu m + \eta m')/\lambda_1{}^2 L^2 \tan^2 \phi_1$$

It is decided to curtail the top reinforcement in the central area, $\alpha L \times \beta L$, of the slab so that the alternative yield-line mechanism shown in figure 13.8.2(*b*) becomes possible. By comparison with the above result the alternative collapse load is given by

$$w = 24\mu m/(\alpha/\beta)^2(\beta L)^2 \tan^2 \phi_2$$

Hence determine the maximum value of α if the curtailment of the top reinforcement is not to reduce the collapse load for the case when $\alpha = \lambda_1 \beta$ and $\mu = \eta = 1$. (UB3/H 1969, 45 min)

Solution

$$\sum M\theta = 2(m + m')L(2\delta/\lambda_1 L) + 2(\mu m + \eta m')\lambda_1 L(2\delta/\lambda_1 L \tan \phi_1)$$

$$\sum W\delta = w\lambda_1 L[\tfrac{1}{3}\delta(\lambda_1 L \tan \phi_1) + \tfrac{1}{2}\delta L(1 - \lambda_1 \tan \phi_1)]$$

Equating $\sum M\theta$ and $\sum W\delta$:

$$\frac{4(m+m')}{\lambda_1} + \frac{4(\mu m + \eta m')}{\tan \phi_1} = \frac{w\lambda_1 L^2}{6}(3 - \lambda_1 \tan \phi_1)$$

Putting $B = (\mu m + \eta m')/(m + m')$ and re-arranging:

$$\frac{w\lambda_1{}^2 L^2}{24(m+m')} = \frac{\tan \phi_1 + \lambda_1 B}{3 \tan \phi_1 - \lambda_1 \tan^2 \phi_1} \qquad (1)$$

A turning value for w is given by $dw/d\phi = 0$, i.e.

$$\frac{(\tan \phi_1 + \lambda_1 B)}{(3 \tan \phi_1 - \lambda_1 \tan^2 \phi_1)} = \frac{\sec^2 \phi_1}{(3 \sec^2 \phi_1 - 2\lambda_1 \tan \phi_1 \sec^2 \phi_1)}$$

Hence

$$\tan^2 \phi_1 + 2\lambda_1 B \tan \phi_1 - 3B = 0$$

$$\tan \phi_1 = \sqrt{(\lambda_1{}^2 B^2 + 3B)} - \lambda_1 B$$

Partial substitution for $\tan \phi_1$ in RHS of equation (1) gives

$$\frac{\sqrt{(\lambda_1{}^2 B^2 + 3B)}}{\tan \phi_1 [3 + \lambda_1{}^2 B - \lambda_1 \sqrt{(\lambda_1{}^2 B^2 + 3B)}]}$$

which simplifies to $B/\tan^2 \phi_1$. Hence

$$w = 24(\mu m + \eta m')/\lambda_1{}^2 L^2 \tan^2 \phi_1 \qquad \text{(QED)}$$

Alternative collapse loads are equal if

$$w = \frac{24(\mu m + \eta m')}{\lambda_1{}^2 L^2 \tan^2 \phi_1} = \frac{24 \mu m}{\alpha^2 L^2 \tan {}^2 \phi_2}$$

or

$$\alpha^2 = [\mu m \lambda_1{}^2/(\mu m + \eta m')] \times (\tan^2 \phi_1/\tan^2 \phi_2)$$

If $\mu = \eta = 1$, then $B = 1$ and $\tan \phi_1 = \tan \phi_2$. Hence

$$\alpha^2 = m\lambda_1{}^2/(m + m_1) \qquad \text{(Answer)}$$

Composite action of beams and slabs

Consider an isotropic square slab, supported on flexible beams, and carrying a uniformly distributed load as shown in figure 13.8.3.

Collapse of slab alone (mode 1) occurs at $wL^2 = 24m$

Collapse of slab and beams together as given by mode 2 is calculated as follows:

Let M_B be the normal collapse moment for each beam at midspan. Resolving along the x- and y-axes in the normal way then, by addition,

$$\sum M\theta = 4(2M_B + mL)(2\delta/L)$$

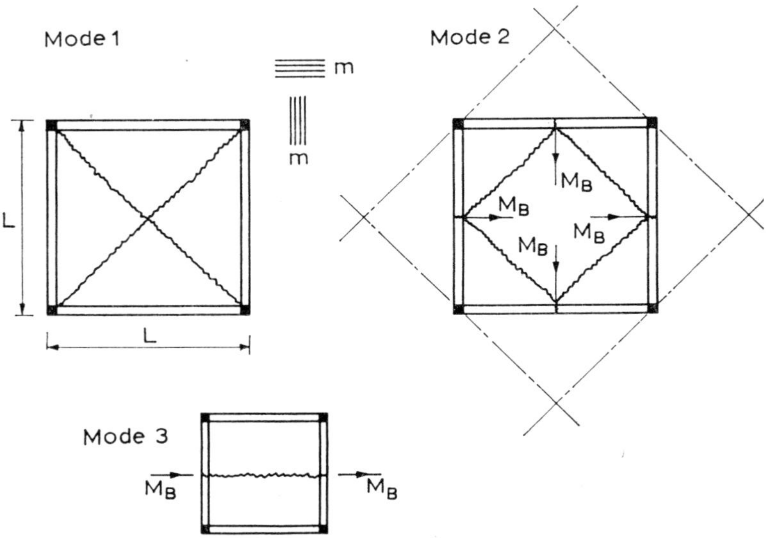

Fig 13.8.3 Alternative modes of failure for beam and slab construction.

Notice that *each* beam has a total rotation of $2 \times (2\delta/L)$ at midspan. If the beams have a collapse moment of resistance at the supports due to their connection to the columns of M_C this should be included in $\sum M\theta$.

$$\sum W\delta = \tfrac{1}{2}wL^2(2\delta/3) + \tfrac{1}{2}wL^2(\delta) = wL^2(5\delta/6)$$

Hence

$$wL^2(5/6) = 8(m + 2M_B/L)$$

$$wL^2 = (48/5)(m + 2M_B/L)$$

A third mode is given by the normal mode of failure at midspan for either pair of beams. In this case there is simply a single yield line at midspan perpendicular to the pair of beams concerned. Hence, putting

$$W\delta = M\theta$$

for mode 3 gives

$$wL^2(\delta/2) = 2(mL + 2M_B)(2\delta/L)$$

$$wL^2 = 8(m + 2M_B/L)$$

If $M_B = mL$ the values of wL^2 for the three modes are as follows:

mode 1 $24m$ as before
mode 2 $28{\cdot}8m$
mode 3 $24m$

Hence it is clear that for this square slab, mode 2 is always less likely than mode 3. Provided that the plastic moment of resistance of each beam at midspan equals that for the full width of the square slab, mode 3 is no more likely than mode 1.

Corner levers and fans

Consider a simply supported, isotropic slab subjected to a uniformly distributed load of w. Alternative modes of failure are shown in figure 13.8.4. Mode (a) is the simple failure pattern usually assumed,

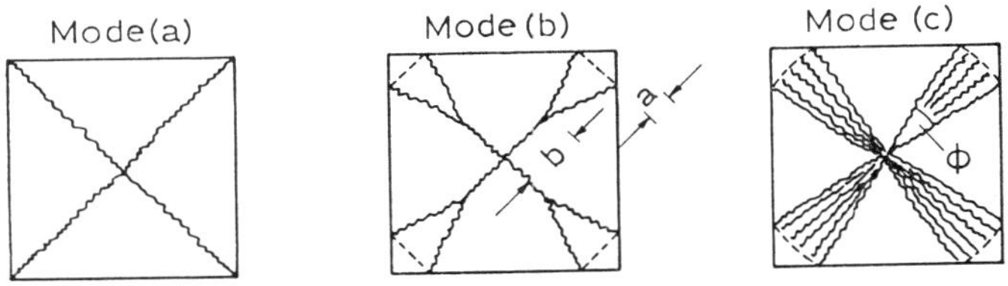

Fig 13.8.4 Alternative modes of failure for square slab.

for which $wL^2 = 24m$. However, even if the corners are held down it is usually possible to obtain lower failure loads by assuming the presence of corner levers, as in mode (*b*), or corner fans, as in mode (c). An indication of the reduction in load is given in table 13.8.1.

Table 13.8.1 Reductions for more critical modes of failure compared with the normally assumed mode

Mode	m'/m	wL^2/m	Reduction
(b) or (c)	1·0	24·0	—
(b)	0·5	23·0	4%
(b)	0·25	23·0	4%
(b)	0	22·0	8%
(c)	0	21·7	10%

Notice that the corner levers in mode (*b*) are eliminated (i.e. $a = 0$) if negative steel equal in magnitude to the positive steel is introduced at the corners. Good detailing procedure should therefore allow for this. Mode (c) gives the lowest value of all when $m' = 0$ and $\phi = 30°$.

Acute angles with free edges

Corner levers can produce greater reductions in ultimate load for acute angles as shown in figure 13.8.5.

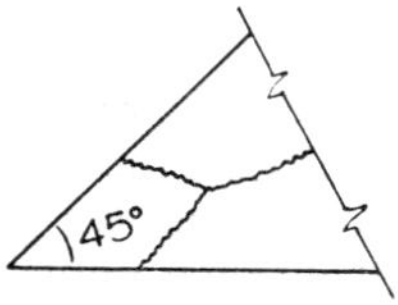

Fig 13.8.5 Corner lever at an acute angle.

Table 13.8.2 Reductions due to corner levers for acute angles with free edges

Angle	m'/m	Reduction
45°	0	26%
45°	1	14%

Biaxial bending

The "square yield criterion" which has been used for yield-line theory is shown by the broken line in figure 13.8.6. Yield in uniaxial bending is given by the four plotted points for which the moment

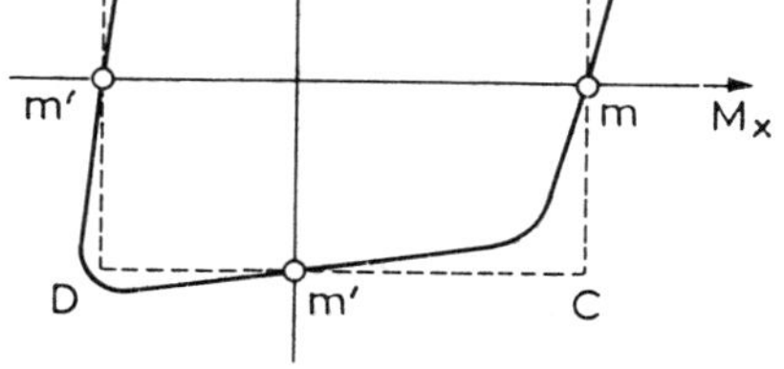

Fig 13.8.6 Yield criteria for a horizontally cast reinforced concrete slab.

equals m or m'. The behaviour of concrete under biaxial stress is dependent upon the magnitude of the lateral stress (e.g. see reference 13.12). Hence although the reinforcement is virtually unaffected by the lateral stress the behaviour of the plain concrete is reflected (to a lesser and limited extent) in the yield behaviour of the reinforced slab as shown by the full line in figure 13.8.6. Only the compressive strength characteristics of the plain concrete are important in figure 13.8.6, hence the "D" quadrant is affected by the compression/compression and not the tension/tension biaxial stress plot for plain concrete. Notice that there is a reserve of strength, using the square yield criterion, for biaxial moments of the same sign, but a deficiency of strength in the tension/compression quadrants (i.e. two moments of opposite sign). This deficiency of strength is made good by not curtailing the steel too rapidly. The use of the square yield criterion, which neglects interaction altogether, for yield-line analysis is satisfactory if slightly generous bond lengths in accordance with the Unified Code are always provided (see chapter 5).

Compressive membrane action due to lateral restraint

Membrane action is the development of forces acting in the plane of the member. Such action can produce increases in the collapse load for slabs of up to ten times the value predicted by yield-line theory [13.2]. Arching action in *deep* beams has been discussed in chapter 5 (see figure 5.7.3). Although this simple form of arching must be negligible in simply supported slabs (due to the high span/depth ratio), similar arching action can be very pronounced if the edges of the slabs are restrained against lateral displacement. This is the case, for example, with interior panels of continuous slabs, where the horizontal thrust can be taken by the adjacent panels (instead of by the steel reinforcement) as shown in figure 13.8.7.

Fig 13.8.7 Compressive membrane (arching) action in laterally restrained slab.

This action is made possible by the fact that slabs are under-reinforced so that the neutral axis depth is small. The centre of compression in the concrete at midspan can therefore be appreciably above that at the support.

Tensile membrane action

Tensile membrane action can also occur in restrained slabs at very large deflections as shown in figure 13.8.8, which shows a typical load-deflection curve for a uniformly loaded two-way rectangular slab with laterally restrained edges. Peak A represents the collapse load due to the fully developed yield-line pattern, enhanced by compressive membrane forces. Trough B occurs due to the reduction, with increasing deflections, in the compressive membrane action in the central region of the slab, and its replacement by tensile membrane forces. Beyond trough B the load is carried almost

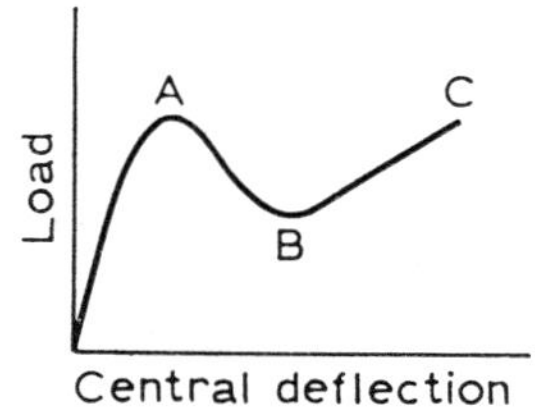

Fig 13.8.8 Typical load-deflection curve for a uniformly loaded, laterally restrained slab.

entirely by the reinforcement acting as a plastic tensile membrane with full depth cracking of the concrete over the central region of the slab due to the extensive stretching of the slab surface. Fracture of the steel finally occurs at point C.

Membrane action in simply supported slabs

Tests *on simply supported slabs* have shown that arching action must be completely absent in this case since the cracks are seen to extend throughout the full depth of the slab as the deflections increase. Nevertheless moderate increases in the collapse load can occur at large deflections due to a modified form of membrane action. Consider a diagonal section AA across the slab shown in figure 13.8.9.

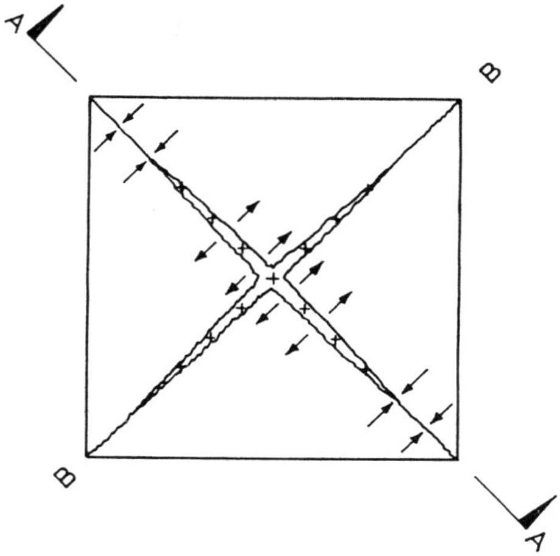

Fig 13.8.9 Membrane action in simply supported slab.

The purely tensile forces across the centre at midspan must be balanced by additional compressive forces in the concrete (which is not separated in the flexural compression zone) near the edges of the slab. The same argument applies to the diagonal BB. The bending moment across AA exceeds the assumed yield value near the supports since the concrete can resist the superimposed compression, and the tension in the steel is relieved. This is illustrated by reference to the design curves for columns, which show increasing collapse moments for increasing axial loads when tension failures occur. The bending moment near midspan is only slightly reduced since the overstressed steel can continue to yield and provide a resistance moment due to the increasing inclination of the member. Tests indicate that the net result is a moderate increase in the collapse load.

Membrane action and design

These membrane effects are difficult to consider in design, either because of their interdependence with adjacent parts of the structure, due to lateral stiffness requirements or introduction of alternative modes of failure, or because of the extremely large deflections which are necessary. Nevertheless a knowledge of their existence increases the confidence which can be placed on upper-bound yield-line analysis.

13.9 Hillerborg's strip method

Yield-line theory is essentially an upper-bound solution which is best suited for analysis. Hillerborg's strip method as modified by Wood & Armer [13.9], however, can be an exact solution, which is ideally suited for design. It gives, in effect, a great simplification of elastic design of slabs and the solution can then be checked by the virtual work method.

Simple strip method

The simple strip method assumes that the load at any point on a slab is carried to the nearest support by bending in the plane perpendicular to that support. For a rectangular slab the load would be allocated to the supporting beams by means of the lines of load discontinuity as shown in figure 13.9.1. This assumption in fact was

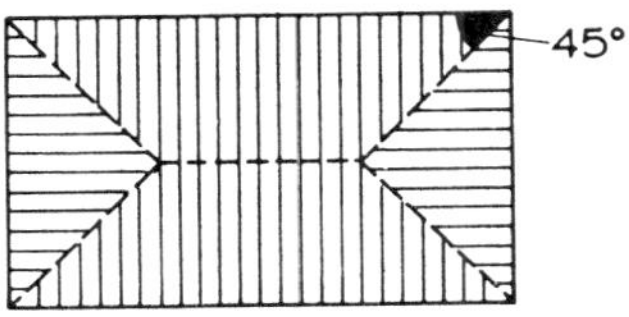

Fig 13.9.1 Lines of load discontinuity.

used for determining the loads on beams supporting two-way slabs long before Hillerborg applied it to the slab itself. Although the elastic deflections at a given point would be dissimilar it can be assumed that plasticity enables the deflections for the two perpendicular spans to become compatible, before collapse occurs, and hence the design method is valid.

Equilibrium moments

The effect of distributing the load according to figure 13.9.1 on the strips XX and YY shown in figure 13.9.2 is to subject them to loading near their ends only. The bending moment diagrams are therefore

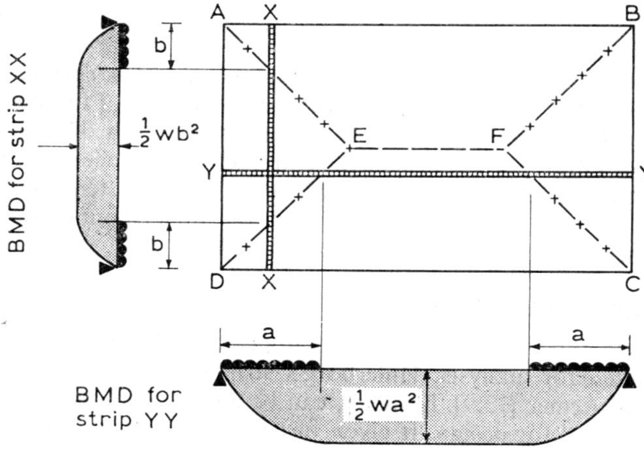

Fig 13.9.2 Bending moment diagrams for rectangular slab (after Hillerborg).

parabolic near the ends and are constant over the central portions of the strips. A set of moments in equilibrium with the applied loading has therefore been formed. Reinforcement can now be provided to ensure that the yield condition is not violated anywhere in the slab. This requires reinforcement in the central portions of a strip even though it may not carry load locally. It is immediately apparent that the strip method attempts to *design* a slab with variable reinforcement, and not *analyse* a slab with given reinforcement.

Equilibrium equation

Hillerborg intended his theory, in its simplest form, to provide a lower-bound solution. This would be valid only if the equilibrium equation was satisfied at all points, i.e.:

$$\partial^2 M_x/\partial x^2 + \partial^2 M_y/\partial y^2 - 2\partial^2 M_{xy}/\partial x\partial y = -w \qquad (13.9.1)$$

where moments M_x, M_y and twist M_{xy} follow the usual notation as before.

It is clear from the above that the strip method attempts to make the stress field coincide with the field of resistance moments due to the reinforcement. This is a radical change in approach from yield-line theory methods and means that for the first time M_{xy} can be made zero, deliberately, so that the directions of the reinforcement (assumed to be along the x- and y-axes) coincide with the principal moment direction. Equation 13.9.1 can therefore be replaced by equations representing twistless beam-strip action:

$$\partial^2 M_x/\partial x^2 = -\alpha w \quad \text{and} \quad \partial^2 M_y/\partial y^2 = -(1-\alpha)w \qquad (13.9.2)$$

where α and $(1 - \alpha)$ are the proportions of load taken in the strips in the x- and y-directions respectively. Usually α would be taken as 1 or 0, i.e. when $\alpha = 1$ all the load is dispersed in the x-direction and when $\alpha = 0$ all the load is dispersed in the y-direction. w could change throughout the slab without affecting the validity of the method, and point loads can be treated as local concentrations of w.

Lines of stress discontinuity

Most of the lines of load discontinuity are also lines of stress discontinuity but notice that line EF in figure 13.9.2 is not a line of stress discontinuity but rather a line of zero shear. The lines of stress discontinuity are quite arbitrary but the continuity of shear must be observed for each strip. That is, the accumulated shear force in each imaginary strip must be taken care of when the load dispersion changes. Notice that a very useful feature of the strip method is that the beam loads are known in intensity and in distribution. By contrast they are not known from true upper-bound solutions for yield-line theory.

EXAMPLE 13.9.1

Use Hillerborg's strip method to determine the required values of m_x and m_y at any point of the rectangular slab shown in example 13.7.4. Check the result by analysing the yield-line pattern which crosses every strip at a point of maximum bending moment (i.e. assume $\alpha = \lambda/2$).

Solution Refer to figure 13.9.3

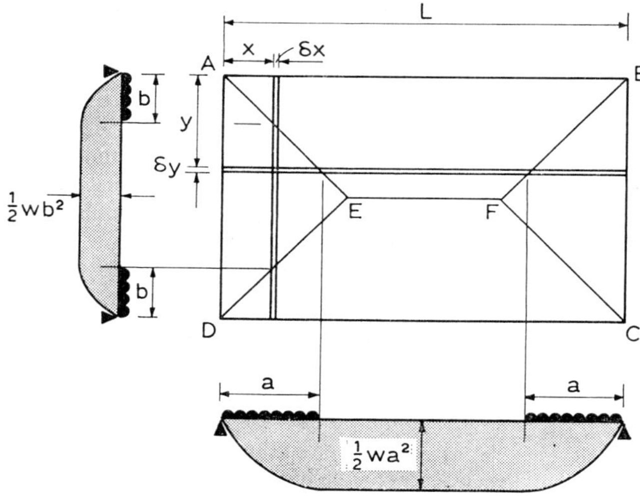

Fig 13.9.3 Example 13.9.1.

Strip δx wide

Required resistance moments near centre

$$m_y \geqslant \tfrac{1}{2}wb^2$$

where b is the value of y at intersection of AE and strip.

Required resistance moments near edge

$$m_y \geqslant wby - \tfrac{1}{2}wy^2 = wy(b - \tfrac{1}{2}y) \qquad \text{(Answer)}$$

Strip δy wide

Near centre

$$m_x \geqslant \tfrac{1}{2}wa^2$$

Near edges

$$m_x \geqslant wax - \tfrac{1}{2}wx^2 = wx(a - \tfrac{1}{2}x) \qquad \text{(Answer)}$$

where a is the value of x at intersection of AE and strip.

Hence maximum $m_x = \tfrac{1}{2}w(\alpha L)^2$ and maximum $m_y = \tfrac{1}{2}w(\tfrac{1}{2}\lambda L)^2$.

Yield line analysis

If m_x and m_y are as given above only yield line EF has a constant resistance moment per unit length. The resistance moments of the diagonal yield lines AE, BF, CF and DE all vary. At any point along

AE, for example, m_x and m_y are obtained by putting $a = x$ and $b = y$, i.e.

$$m_x = \tfrac{1}{2}wx^2 \quad \text{and} \quad m_y = \tfrac{1}{2}wy^2$$

The resistance moment of the x-direction steel at the yield line for the strip δy wide is $m_x\delta y$. Hence the total resistance moment of the x-direction steel for the slab shown in example 13.7.4 is given by the four diagonals AE, BF, CF and DE, which equals four times the resistance moment of any single diagonal. Hence

$$\sum m_x\delta y = 4\int_0^{\frac{1}{2}\lambda L} m_x dy = 2w\int_0^{\frac{1}{2}\lambda L} x^2 dy$$

after substituting $m_x = \tfrac{1}{2}wx^2$.

On yield line AE,

$$x/y = \alpha L/\tfrac{1}{2}\lambda L = 2\alpha/\lambda$$

Therefore

$$\sum m_x\delta y = 2w(2\alpha/\lambda)^2\int_0^{\frac{1}{2}\lambda L} y^2 dy$$

$$= 8w(\alpha/\lambda)^2 \times \tfrac{1}{3}(\lambda L/2)^3 = \tfrac{1}{3}w(\alpha L)^2\lambda L$$

Let m_X = maximum value of m_x, which is $\tfrac{1}{2}w(\alpha L)^2$. Hence

$$\sum m_x\delta x = \tfrac{1}{3}(2\lambda L)m_X$$

Similarly $\sum m_y\delta x$ for diagonals only equals $\tfrac{1}{3}(4\alpha L)m_Y$ where m_Y = maximum value of m_y, which is $\tfrac{1}{2}w(\tfrac{1}{2}\lambda L)^2$.

$\sum m_y\delta x$ for yield line EF is simply $(1 - 2\alpha)Lm_Y$, hence putting $\sum W\delta = \sum M\theta$ gives

$$w\lambda L^2[(1 - 2\alpha)/2 + 2\alpha/3]\delta$$

$$= 2(1 - 2\alpha)m_Y(2\delta/\lambda) + \tfrac{1}{3}(2\lambda)m_X(\delta/\alpha) + \tfrac{1}{3}(4\alpha)m_Y(2\delta/\lambda)$$

If $\alpha = \lambda/2$ then $m_X = m_Y$. Substituting in the above equation, putting $m_x = m_y = m$ and simplifying, gives

$$m = w(\lambda L)^2/8$$

which agrees with the previous value of maximum m_y (and m_x if $\alpha = \lambda/2$) given by Hillerborg's design.

Notice that a turning value is given by $dw/d\alpha = 0$ (if m is a constant) or $dm/d\alpha = 0$ (if w is a constant). However in this case m_y equals $(\lambda/2\alpha)^2m_x$. Hence if m_x is made constant, m_y varies as α and $dw/d\alpha = 0$ would give a meaningless result. When using Hillerborg's method it can be assumed that a load discontinuity at 45° is satisfactory for a 90° corner (and usually the most convenient) and that small variations from this distribution are not critical.

REFERENCES

13.1 TIMOSHENKO, S. and WAINOWSKY-KRIEGER, S. *Theory of plates and shells*, 2nd edition (McGraw-Hill, New York 1959) 580 pp.
13.2 WOOD R. H. *Plastic and Elastic Design of Slabs and Plates* (Thames and Hudson, London 1961).
13.3 SCOTT, W. F., GLANVILLE, W. and THOMAS, F. C. *Explanatory handbook on the BS code of practice for reinforced concrete, CP*114:1957 (revised 1965) 2nd edition (Concrete Publications, London 1965) 172 pp.
13.4 INGERSLEV, A. The strength of rectangular slabs *Struct Eng.* vol. 1 no. 1 Jan. 1923 pp 3–14.
13.5 JONES, L. L. *Ultimate load analysis of reinforced and prestressed concrete structures* (Chatto and Windus, 1962).
13.6 JOHANSEN, K. W. *Yield-line theory* (CACA, London 1962).
13.7 KEMP, K. O., MORLEY, C. T., NIELSON, M. P., WOOD, R. H., and JONES, L. L. Recent developments in yield line theory. *Magazine of Concrete Research* special publication (CACA, London, May 1965) 74 pp.
13.8 JONES, L. L. and WOOD, R. H. *Yield line analysis of slabs* (Thames & Hudson and Chatto & Windus, London 1967).
13.9 WOOD, R. H. and ARMER, G. S. T. The theory of the strip method for design of slabs *Proc. Inst. Civil Engineers* vol. 41 Oct. 1968 pp 285–311.
13.10 KEMP, K. O. The yield criterion for orthotropically reinforced concrete slabs. *Int. J. Mech. Sci.* vol. 7 pp 737–46.
13.11 PRINCE, M. R. and KEMP, K. O. A new approach to the yield criterion for isotropically reinforced concrete slabs *Magazine of Concrete Research* vol. 20 no. 62 March 1968, pp 13–20.
13.12 NEWMAN, K. The structure and engineering properties of concrete. *Theory of Arch Dams* ed. J. R. Rydzewski (Pergamon, 1965) pp 683–712.

Exercises

1. A rectangular, orthotropically reinforced concrete slab measures 30 ft (9·15 m) by 20 ft (6·1 m). It is continuous over the shorter sides and has ultimate resistance moments of 3000 lbf ft/ft (13·3 kN m/m) against both hogging and sagging parallel to these sides. It is simply supported on the longer sides and has an ultimate resistance moment of 6000 lbf ft/ft (26·6 kN m/m) against a sagging moment in this direction. Calculate the ultimate uniformly distributed load which can be carried by this slab.

(UB3/G 1965, 36 min)

2. The triangular slab shown in figure A is isotropically reinforced to provide a yield moment of M per unit width, and carries a uniformly distributed load of w per unit area. Determine the relationship between M and w for the failure mode shown.

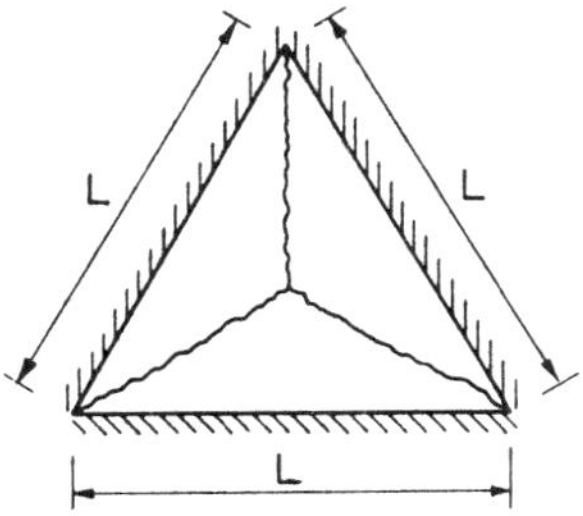

Ch. 13 Fig A Question 2.

3. The rectangular slab shown in figure B is simply supported on two sides and built in on two sides. The reinforcement is isotropic

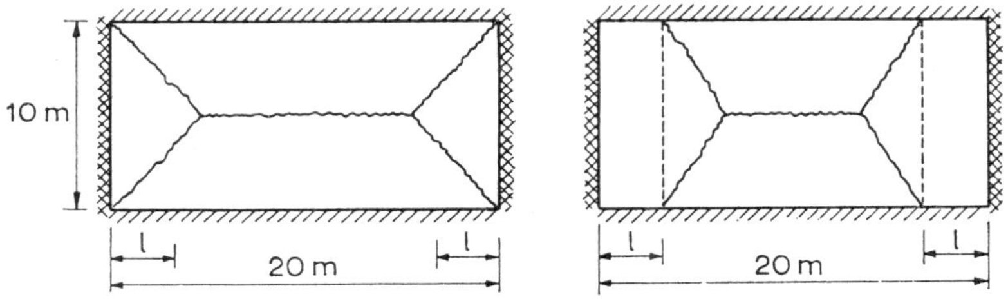

Ch. 13 Fig B Question 3.

and provides sagging and hogging yield moments of $+m$ and $-m$ per unit width respectively. The top steel is curtailed at a distance l from the built-in supports. If the slab is uniformly loaded, two possible modes of failure are as shown. Determine the extent (l) of the top steel required for these two modes to occur at the same load, and determine the value of m required for this condition if the loading is 10 kN/m².

4. A reinforced concrete slab is subjected to a uniformly distributed load per unit area of w. The layout of the slab is as shown in figure C.

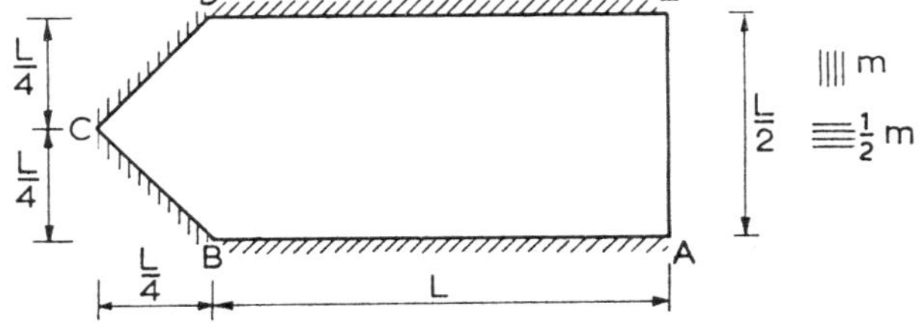

Ch. 13 Fig C Question 4.

It is simply supported along sides AB, BC, CD and DE and unsupported along AE. The reinforcement in the direction parallel to AE produces an ultimate moment per unit length for a positive yield line parallel to AB of m. The reinforcement parallel to AB produces an ultimate moment per unit length of $\frac{1}{2}m$. Determine the value of w if $L = 40$ ft (12·2 m) and $m = 6000$ lbf ft/ft (26·7 kN m/m).

(UB3/H 1968, 36 min)

5. A square reinforced concrete slab is rigidly built in at the edges and carries a central localized (point) load. It is isotropically reinforced at the bottom (m) and at the top (m'). A credible failure mechanism under overload would involve yield lines across the diagonals and around the sides. Alternatively, there might be numerous radial yield lines bounded by chords forming a multi-sided regular polygon contained within the boundaries of the slab. What is the percentage error in the failing load calculated by the less critical mode? What is the error if the load is uniformly distributed?

(UL3 1969, 36 min)

6. The reinforced concrete slab ABCD shown in figure D is to be designed for a uniformly distributed collapse load of 20 kN/m². Determine the required value of m'_x at E, point (0, 2), and of m_x and

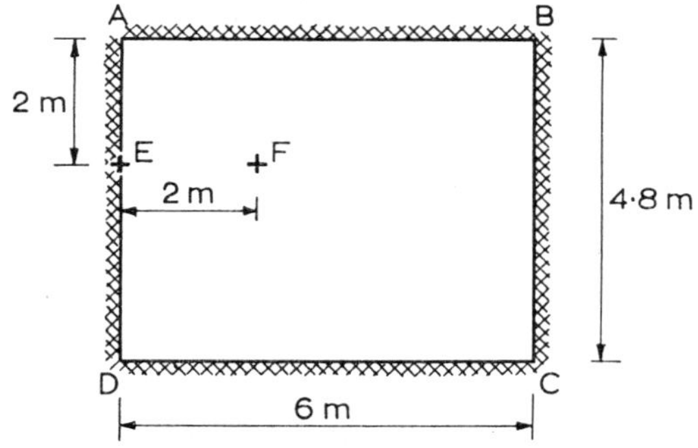

Ch. 13 Fig D Question 6.

m_y at F, point (2, 2), using Hillerborg's strip method. The co-ordinates are referred to A, point (0, 0), as origin. m and m' are the resistance moments per unit width for positive and negative bending moments respectively and suffices x and y refer to the x-direction and y-direction reinforcement respectively.

The proposed Unified Code recommends that the ratio of the support moment to the midspan moment should be between 1·5 and 2·0 for fixed slabs. Would a reduction of 15% in m'_x at E as obtained above be sufficient to satisfy this clause?

Comment, very briefly, on the essential differences in the disposition of the steel reinforcement for a design based on the strip method compared with one based solely upon the usual yield-line analysis approach. (UB3/H 1969, 45 min)

7. A rectangular orthotropically reinforced concrete slab is continuous over the longer sides and one shorter side as shown in figure E. The positive and negative resistance moments per unit

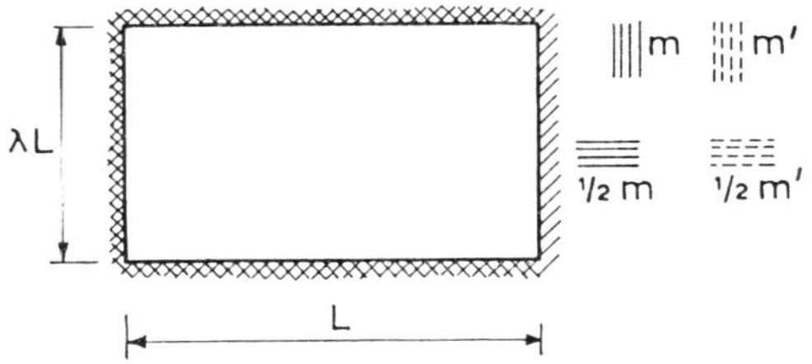

Ch. 13 Fig E Question 7.

length of the steel parallel to the longer side L are $\frac{1}{2}m$ and $\frac{1}{2}m'$ and the corresponding values for the steel parallel to the shorter side λL are m and m'.

Obtain a general expression for the ultimate uniformly distributed load w in terms of m and m'. Hence calculate w if $L = 7{\cdot}5$ m, $\lambda = 0{\cdot}6$ and $m = m' = 40$ kNm/m. (UB3/H 1970, 45 min)

Chapter 14

Design calculations for reinforced concrete

The present chapter is based upon the (draft for comment) report which was prepared for the committee of the Design and Development Division of the Concrete Society [14.1] so that standard design calculations could be presented in as clear, consistent and economical manner as possible. Computers are also considered since they are now a very useful aid to design, and it is emphasized that design calculations, even when computer-assisted, must be capable of being understood under all circumstances.

14.1 Standardized procedure for the presentation of calculations

Calculations are prepared by an engineer for three main reasons:

(i) To satisfy himself as to the suitability of his design.

(ii) For use by the detailer preparing the detailed drawings.

(iii) For submission to other authorities for checking purposes to ensure conformity with Codes of Practice and Building Regulations currently in force.

It is evident that the results of the calculations must be clearly indicated so that the detailer can easily extract the reinforcement required. They must also be presented in such a way that they can be readily followed by a checking authority or by any engineer who may have cause to refer to them subsequently. The report considered that these requirements could be achieved by utilizing:

(i) a consistent set of units and notation

(ii) a standard method of documentation

(iii) a logical sequence for the presentation of the calculations.

The units (SI) and the notation system should be consistent with those used in the Unified Code, although it is possible that an international notation system may be adopted eventually in view of the increasing collaboration between engineers throughout the world. Both the design sequence and the documentation are now considered.

Design process

The report considers two principal stages in the design process, "preliminary design considerations" and "analysis and detailed design", thus following generally the principles set out in the Institution of Structural Engineers' document *Aims of structural design*. A *typical* sequence for considering and presenting the design process and the resulting calculations is as follows:

1.0 PRELIMINARY DESIGN CONSIDERATIONS

(See separate paragraph on page 321).

1.1 USER REQUIREMENTS
1.1.1 *Description of main requirements*
e.g. nature of use—crane requirements etc.

1.2 CONDITIONS
1.2.1 *Internal*
Fire rating, heating requirements etc.
1.2.2 *External*
Wind, frost, temperature, adjoining buildings etc.
1.2.3 *Subsoil*
Soil conditions, chemical content, existing services, mining risk etc.
1.2.4 *Others*
Vibration, earthquakes, noise etc.

1.3 LOADING
Imposed, wind, inertial etc.

1.4 TYPE OF CONSTRUCTION
1.4.1 *Superstructure*
Framed, crosswall etc.
1.4.2 *Foundations*
Strip, raft etc.

1.5 STABILITY
1.5.1 *Overall stability*
1.5.2 *Progressive collapse*

1.6 MATERIALS
1.6.1 *Structural*
Concrete mix, reinforcement etc.
1.6.2 *Cladding*
Precast concrete, asbestos sheet etc.
1.6.3 *Floor finishes*
Lino tile on screed, granolithic etc.
1.6.4 *Partitions*
Breeze block, plaster-board etc.
1.6.5 *Jointing*
Type of precast joint, expansion joint, contraction joint etc.

Factors influencing the choice of the above materials include cost, appearance, construction and workmanship, location and availability.

2.0 ANALYSIS AND DETAILED DESIGN

(See pages 321–3).

A particular design could include additional items and exclude others. For example, prestressing losses (2.2.1.3) and bursting reinforcement (2.2.7.3) would only be considered in prestressed concrete design.

Documentation

The documentation based on the procedure given in the report should consist of four parts:

(i) design information sheet
(ii) structural summary sheet
(iii) index sheet
(iv) calculation sheets

Design information sheet

The design information normally includes many of the items in 1.0 above. A recommended design information sheet is used in figure 14.2.1 (see later). The purpose of the sheet is to summarize the information on which the engineer has based his design and includes those items which are of particular interest to the checking authority.

Structural summary sheet

This sheet supplements, as necessary, the information given on the design information sheet and illustrates the main structural elements, the form of the structure and the provisions for overall stability. Continuation sheets may be necessary to show all the relevant information for larger jobs. A recommended summary sheet is shown in figure 14.2.2 (see later).

Index

An index sheet should be prepared for each set of calculations to allow anyone who has to refer to them to appreciate readily the overall layout and find quickly a particular aspect they wish to examine. The form of the index should be simple, consisting of a description of the item and the page number on which it starts. It is also advantageous to include at the end of the index the names of any reference books used (e.g. reference 14.2). A special index sheet was not designed as it was considered that the standard calculation sheet or structural summary sheet could be used simply by deleting the existing headings and writing in "index" and "page no.".

Calculation sheets

Calculations should be written on a sheet of standard format which includes columns for member reference, the main calculation and the output of the calculation. 5-mm squares are used to assist in drawing sketches. A recommended calculation sheet is used in figure 14.2.3 (see later).

Preliminary design considerations

This stage commences with an appreciation of the client's requirements combined with the collection of facts on which the design will be based. Design decisions are taken and a preliminary design is carried out to determine sizes of structural elements for the information of other members of the design team. This design is usually produced with a minimum of analysis. The decisions and design data for this stage (1.0) are summarized on design information and structural summary sheets, as indicated previously.

Structural analysis and detailed design

The second stage involves a full structural analysis to support the design decisions taken at the appraisal stage and to check the scheme for soundness in service and during erection. The structural analysis and detailed design (2.0) must be clearly presented on calculation sheets and the report outlines the extent of each item as follows:

2.1 STRUCTURAL ANALYSIS

2.1.1 *Geometry of structure and approximate element sizes*
General sketch of the structural layout and choice of member sizes. Output: frame and element dimensions.

2.1.2 *Idealization of structure*
Representation of frame as an analytical model.

2.1.3 *Consideration of interaction of structural elements*
Determination of degree of fixity at joints and degree of stiffness of members. Effect on analysis of local variations in member sizes.

2.1.4 *Loading*
An assessment of the dead, imposed, and wind loadings on the structure. Output: (characteristic) loads.

2.1.5 *Overall stability*
Ensure that the restoring moment is greater than the applied overturning moment multiplied by an appropriate safety factor.

2.1.6 *Progressive collapse*
A statement of assumptions made and method adopted.

2.1.7 *Choice of critical limit state*
The frame will generally be analysed for the limit state of collapse. However, if special features exist, limit state of deflection or local damage may govern. Design loads are calculated from product of characteristic loads and partial safety factors. Output: 2.1.7.1 design loads.

2.1.8 *Frame analysis*
Choice of analysis is made and carried out for different loading combinations in order to determine the worst conditions in members, resulting in nodal bending moments, torsions, direct forces, shears, displacements and rotations. Output: bending moments etc.

2.2 ELEMENT DESIGN

2.2.1 *Design data reference*

This section will either refer to the page giving the design data (2.2.1.1, concrete; 2.2.1.2, steel; 2.2.1.3, prestressed concrete losses), or state the data if it is local to this part of the calculations.

2.2.2 *Section and element geometry*

The section size is confirmed. The effective span or height (2.2.2.1) is calculated. The section dimensions (2.2.2.2) are given. The span/breadth, span/depth ratios etc. (2.2.2.3) are calculated and reference is made to the limit state of deflection if this governs. Output: section size.

2.2.3 *Forces and moments from structural analysis*

The dead and imposed loadings are set out, together with the end forces and moments taken from the structural analysis. If there is no structural analysis, the end forces and moments are calculated in the next section. A loading diagram is very helpful.

2.2.4 *Section design—forces and moments*

This section sets out the calculations giving the maximum bending moments (2.2.4.1, bending moment envelope), shear forces (2.2.4.2, shear force envelope) axial loads (2.2.4.3) and torsions (2.2.4.4) making allowance for redistribution etc. Output: Final bending moments, shear forces, axial loads and torsions.

2.2.5 *Section design—concrete stresses*

Check that the compressive (2.2.5.1), tensile (2.2.5.2) and shear (2.2.5.3) stresses in the concrete are within the upper limits. Output: concrete stresses.

2.2.6 *Longitudinal reinforcement*

The longitudinal bars required at the critical sections are calculated. The minimum and maximum permissible quantities of reinforcement are checked. Output: Reinforcement number, type, size and pitch.

2.2.7 *Transverse reinforcement*

The quantity of stirrups, links, distribution and bursting reinforcement are calculated from the shear (2.2.7.1), torsion (2.2.7.2), bursting (2.2.7.3) and distribution (2.2.7.4) requirements. The minimum and maximum permissible quantities of reinforcement are checked. Output: reinforcement number, type, size and pitch.

2.2.8 *Bond and anchorage*

This section checks that suitable reinforcement is used to satisfy flexural bond (2.2.8.1) requirements and calculates the anchorage required for each bar, the stopping-off points for curtailed bars, and the lap lengths (2.2.8.2). Output: anchorage and cut-off lengths.

2.2.9 *Bearing stresses*

Where bearing stresses may become high (e.g. beam seatings) a check calculation should be made.

2.2.10 *Checks for other limit states*

(*a*) Limit states of deflection. This block checks that the deflection of the element is within the permissible range and will usually be a reference to the criteria for, and the use of, the four tables in the Unified Code of Practice, draft clause 6.4.

(*b*) Limit states of local damage. The nominal concrete cover is calculated. This section will usually state that the conditions comply with draft clauses 6.5.1.1 or 6.5.1.2 of the Unified Code of Practice.

2.2.11 *Specific detailing problems*
Check calculations should be made for problems specific to the job, e.g. stresses in concrete developed inside bends of reinforcement.

14.2 Typical examples of design calculations

Examples of designs of reinforced concrete members based on the recommended procedure are included in the Concrete Society Report [14.1]. Figures 14.2.1 to 4 have been reproduced from the design examples given in the report and serve to illustrate the method of presentation. They show a design information page, an index, and calculations for a typical beam and a typical column. Only one page is included in each of figures 14.2.1 to 3 and the reader should refer to the report itself for the complete calculations. Figure 14.2.4 includes the complete design example for a typical column.

It is important in the preparation of calculations that

(i) all headings are underlined
(ii) a horizontal line is drawn across the page after each stage in the calculation
(iii) the output of the calculation is clearly indicated in the right hand column.

14.3 Application of small computers

Computers are an aid to the designer. They can relieve the designer of routine calculations and can offer alternative methods of solution which are not possible or practicable using a manual approach. Small computers can usually operate most economically by carrying out the computations involved in manual methods. Larger computers, however, can usually operate most satisfactorily using techniques which need not be suitable for solution by manual methods and are therefore considered separately. The final documentation for *all* computer calculations should follow the same general principles and be intelligible to the reader.

Documentation for computer calculations

The documentation should follow the same principles that have been indicated for manual calculations. The report realized that the type of computer which is used will have a considerable effect on the format of the calculations; nevertheless the following principles were considered important:

(i) The calculation must be completely self-explanatory and must not assume that the reader of the calculations has any knowledge of computers.

A. D. Vance & Partners Consulting Engineers 123a Victoria Street London, S.W.1.	Contract: Residential Block Creekgate Sub-section: **DESIGN INFORMATION**	Job Ref A/704 Sheet No: 1 Made by: M.R.Smith Checked by: T.D.Brown Date: 15.4.70

Details	Heading
Client: County Borough of Creekgate Architect: M. J. Cranthorne Esq. F.R.I.B.A. Borough Architect.	Engineer Responsible: K.R.E. Richards
Chief Engineer County Borough of Creekgate	Building Regulation Authority or other
1. The Building Regulations, 1965 2. C.P. 114 Part 2, 1969, "The structural use of R.C. in buildings" 3. C.P.3 Chap. V Loading, Part 1, 1967, "Dead & imposed loads." 4. Draft C.P.3 Chap. V Loading, Part 2, 1969, "Wind loading"	Relevant Building Regulations and Design Codes
45 individual study/bedrooms 3 dormitories with bathrooms etc. Dining hall and kitchen Recreation room Basement garage, boiler house & tank room.	Intended use of Structure
2 hours resistance	Fire resistance requirements
Roof — 1·5 kN/m² Bedrooms & dormitories — 1·5 kN/m² Dining & recreation rooms — 2·0 kN/m² Kitchen — 3·0 kN/m² Corridors — 4·0 kN/m² Finishes & partitions — 1·5 kN/m²	General loading conditions
Speed: 44 metres per second Factors: $S_1 = 1{\cdot}0$, $S_2 = 0{\cdot}86$, $S_3 = 1{\cdot}0$	Wind loading conditions
Normal atmospheric	Exposure conditions
Sulphate resistant cement required in foundations. Buried C.E.G.B. high-tension cable across north end. Allowable bearing pressure – retaining walls = 250 kN/m² – end bearing on piles = 500 kN/m²	Subsoil conditions
Bored piles to accommodate extra loading for future extension. Retaining wall around basement area.	Foundation type
Concrete: Minimum cube crushing strengths at 28 days:— Columns — 30 MN/m² Beams & slabs — 25·5 MN/m² Foundations — 21 MN/m² Reinforcement: Maximum permissible working stresses High tensile: p_{sc} = 175 MN/m², p_{st} = 210 MN/m² (if dia. > 20 mm) & 230 MN/m² (if dia. ≯ 20 mm) Mild steel: p_{sc} = 125 MN/m², p_{st} = 140 MN/m²	Material data
Mass density of reinforced concrete = 2400 kg/m³ Equivalent value in force units = 23·6 kN/m³	Other relevant information

Fig 14.2.1 Design information page.

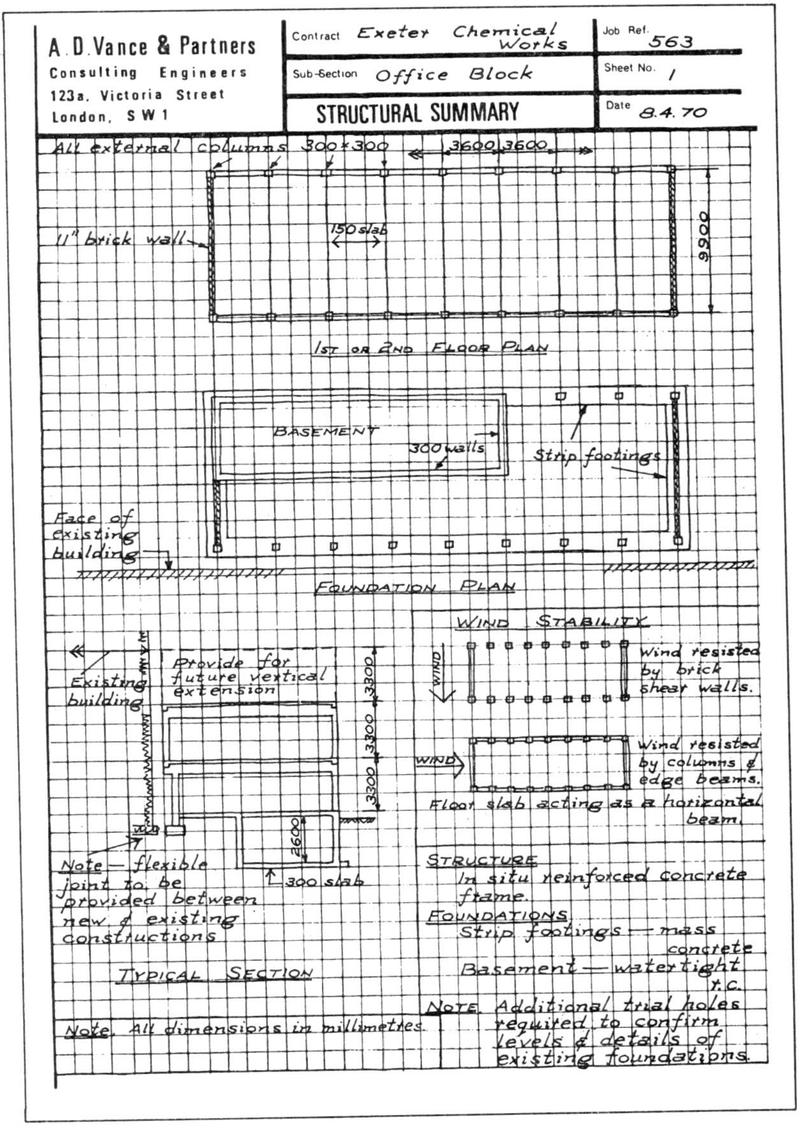

Fig 14.2.2 Structural summary page.

A.D.Vance & Partners Consulting Engineers 123a. Victoria Street London. SW1	Contract: Physics Block, Eastford	Job Ref. 668
	Part of Structure: Floor to upper laboratory — beam no.2c	Calc Sheet No. 72
	Drawing Ref. 173 · Calculations by T. D. Brown · Checked by M. R. Smith	Date 17.3.70

Member Ref	CALCULATIONS	OUTPUT
Beam 2c	DESIGN DATA	
	Min. cube crushing strength at 28 days	f_{cu} = 30 MN/m²
	Max. permissible comp. stress in concrete in bending	p_{cb} = 11 MN/m²
	Max. perm. tensile stress in high bond reinforcement: diameter ≯ 20 mm	p_{st} = 230 MN/m²
	diameter > 20 mm	p_{st} = 210 MN/m²
	Max. perm. tensile stress in mild steel reinf.	p_{st} = 140 MN/m²
	Max perm. shear stress in unreinforced concrete	p_v = 0·87 MN/m²
	Bond stresses for high-bond bars — clause 310(k)	
	Max perm. average bond stress in concrete	p_b = 1·4 MN/m²
	Max. perm. local bond stress in concrete	p_{svb} = 2·06 MN/m²
	SECTION AND ELEMENT GEOMETRY	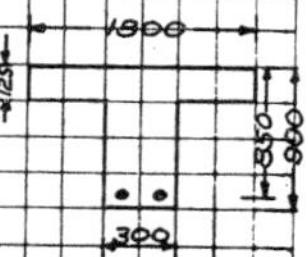
	Try final section dimensions of b_w = 300 mm	
	h = 900 mm	
	h_f = 125 mm	
	Take beam effective depth as: d = 850 mm	
	Clause 311(a), effective span is lesser of:	
	(i) Distance between centres of supports = 10·5 m	
	(ii) Clear distance between supports + d	
	= 10·3 + 0·85 = 11·15 m	l = 10·5 m
	Clause 311(e), flange breadth is lesser of:	
	(i) ⅓ of effective span = 3500 mm	
	(ii) Distance between ribs = 6000 mm	
	(iii) $12h_f + b_w$ = 1800 mm	b = 1800 mm
	From table 13, max. perm. span/depth ratio for a continuous beam = 25 × 0·85 = 21·25	
	Actual span/depth ratio = 10·5 × 10³/900	
	= 11·7 ≯ 21·25	
	FORCES AND MOMENTS FROM STRUCTURAL ANALYSIS	
	Live loading — imposed = 3·5 kN/m²	
	Dead loading — finishes = 1·9 kN/m²	
	— slab = 2·9 kN/m²	
	4·8 kN/m²	
	— Assuming density of concrete in equivalent force units = 23·6 kN/m	
	Hence total D.L. from slab = 4·8 × 6 = 28·8 kN/m	
	Add beam self wt. = 0·3 × 0·775 × 23·6 = 5·5 kN/m	
	Total dead loading = 34·3 kN/m	D.L. = 34·3 kN/m
	Total live loading from slab = 3·5 × 6 = 21·0 kN/m	L.L. = 21·0 kN/m
	NOTE: This example uses C.P.114: Part 2:1969	

Fig 14.2.3 Calculation page to CP 114 for typical beam.

A.D.Vance & Partners Consulting Engineers 123a. Victoria Street London, SW1	Contract: Physics Block Eastford	Job Ref. 668
	Part of Structure: Columns, ground-first floors — col. no. C2	Calc Sheet No. 112
	Drawing Ref. 765 · Calculations by M.R.Smith · Checked by T.D.Brown	Date 17.3.70

Member Ref	CALCULATIONS	OUTPUT
Column C2	DESIGN DATA	
	characteristic strengths: concrete	f_k = 37.5 MN/m²
	longitudinal reinf.	f_{yl} = 410 MN/m²
	transverse reinf.	f_{yv}, f_{yh} = 250 MN/m²
	SECTION AND ELEMENT GEOMETRY	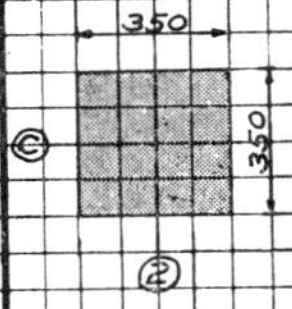
	Member under consideration is the lower half of a double storey perimeter column, whose base is assumed to be encastré.	
	Overall height of column, ℓ_c = 5.0 m	
	Section dimensions, b × h = 350 mm × 350 mm	
	Assume d/h ≏ 0.9	
	From clause 6.3.2.1 — column is unrestrained	d/h = 0.9
	From clause 6.1.5:	
	$I_e = \frac{(350)^4}{12} = 1250 \times 10^6 \text{ mm}^4$	
	$\therefore \frac{EI_e}{\ell_c} = \frac{1250 \times 10^{6-12} E}{5.0} = 250 \times 10^{-6} E$	
	(working in basic units, i.e. E in N/m²)	
	For portion of column from first floor to roof	
	ℓ_c = 3.5 m	
	$EI_e/\ell_c = 1250 \times 10^{-6} E/3.5 = 357 \times 10^{-6} E$	
	$\therefore \Sigma EI_e/\ell_c = 607 \times 10^{-6} E$	
	From beam 2c (sheet 82)	
	$EI_b/\ell_b = 2780 \times 10^{-6} E$	
	From beam 2d (sheet 85)	
	$EI_b/\ell_b = 608 \times 10^{-6} E$	
	$\therefore \Sigma EI_b/\ell_b = 3388 \times 10^{-6} E$	
	$\alpha_{c2} = 607/3388 = 0.179$	α_{c2} = 0.179
	α_{c1} = 1.0	α_{c1} = 1.0
	Clear distance between end restraints,	
	$\ell_u = 5.0 - 1.0 = 4.0$ m	
	Effective height, ℓ_e, is smaller of:	
	$\ell_u[1.0 + 0.15(\alpha_{c1} + \alpha_{c2})]$	
	and	
	$\ell_u[2.0 + 0.3\alpha_{c\,min}]$	
	$\therefore \ell_e = 4.0[1.0 + 0.15 \times 1.179] = 4 \times 1.177 = 4.710$ m	ℓ_e = 4.710 m

Fig 14.2.4 Calculation pages to draft Unified Code for typical column.

A. D. Vance & Partners
Consulting Engineers
123a. Victoria Street.
London. SW1

Contract: *Physics Block Eastford*			Job Ref. 668
Part of Structure: *Columns, ground–first floors – col. no. C2*			Calc Sheet No. 113
Drawing Ref. 765	Calculations by M.R.Smith	Checked by T.D.Brown	Date 17.3.70

Member Ref	CALCULATIONS	OUTPUT
Column C2	Clause 6.1.7.3	
	$l_e/b = 4.710/0.350 = 13.5 \not> 30$	
	$l_e h/b^2 = 4.710 \times 0.350/(0.350)^2 = 13.5 \not> 150$	∴ condition satisfied
	LIMIT STATE OF COLLAPSE	
	FORCES AND MOMENTS FROM STRUCTURAL ANALYSIS	
	From Structural Analysis, sheet 6* worst loading combination is condition one	
	(*Ed. note. Not included in this extract)	
	Axial load, $N = 1488$ kN	$N = 1488$ kN
	Maximum bending moment, $M_{ix} = 168$ kN m	$M_{ix} = 168$ kN m
	SECTION DESIGN – FORCES AND MOMENTS	
	Clause 6.3.2.8.	
	$\frac{l_{ex}}{h} = \frac{l_{ey}}{b} = \frac{4.710}{0.350} = 13.5 > 12$	∴ long column
	Clause 6.3.2.3.	
	M_{ix} is the greater of 168 kNm or $0.05Nh$	
	($\not< 0.05Nh$ is satisfied by using design chart)	
	Clause 6.3.2.9.	
	– subsection 3 – unrestrained long column	
	$Nh = 1488 \times 0.350 = 521$ kN m $> M_{ix}$	
	Since $2Nh > (Nh + M_{ix})$ use	
	$M_x = M_{ix} + \frac{Nh + M_{ix}}{2500}\left(\frac{l_{ex}}{b}\right)^2$	
	$= 168 + \frac{1488 \times 0.350 + 168}{2500}\left(\frac{4.710}{0.350}\right)^2$	
	$= 168 + 49.5 = 217.5$ kN m	$M_x = 217.5$ kNm
	SECTION DESIGN – CONCRETE STRESSES AND LONGITUDINAL REINFORCEMENT	
	A column design chart has been used and consequently the compressive and tensile stresses of the concrete and reinforcement, respectively, are assumed to be within the permissible range.	
	Minimum steel ratio – clause 6.2.1.3.	
	$\frac{A_s}{bh} \not< \frac{15 N_{max}}{f_y bh} = \frac{15 \times 1488 \times 10^3}{410 \times 350^2} = 0.45\%$	

Fig 14.2.4 Calculation pages to draft Unified Code for typical column.
(*continued*)

A.D.Vance & Partners
Consulting Engineers
123a. Victoria Street
London. S W 1

Contract: Physics Block Eastford			Job Ref. 668
Part of Structure: Columns, ground-first floors – col. no. C2			Calc Sheet No. 114
Drawing Ref. 765	Calculations by M.R.Smith	Checked by T.D.Brown	Date 17.3.70

Member Ref	CALCULATIONS	OUTPUT
Column C2	Maximum steel ratio (vert. cast column) = 6%	
	Information for use of design chart:	
	$\frac{N}{bh} = \frac{1488 \times 10^3}{350 \times 350} = 12.15\ (MN/m^2)$	
	$\frac{M}{bh^2} = \frac{217.5}{350 \times 350^2} \times 10^{3+3} = 5.07\ (MN/m^2)$	
	Referring to design chart attached to draft Unified Code (f_{cu} = 37.5 MN/m², f_y = 410 MN/m², d/h = 0.9)	
	[Ed. note. Alternatively, $\rho' f_y\ (= \rho_s f_y)$ = 12.7 from Fig E4]	
	then ρ_s = 3.1% > 0.45% and < 6%	∴ safe
	$A_s + A_s' = \frac{3.1 \times 350 \times 350}{2 \times 100} = 1900\ (mm^2)$	Provide 3no. Y25 each face (2950 mm²)
	TRANSVERSE REINFORCEMENT	
	Links – clause 6.2.2.3	
	Pitch to be the minimum of: (i) 350 mm	
	(ii) 300 mm	
	(iii) 300 mm	Provide R8
	Hence take pitch as 300 mm and dia. as 8 mm.	links at 300 mm pitch
	BOND AND ANCHORAGE	
	Lapping of reinforcement – clause 6.3.6.5.	
	Max. anchorage bond stress (table 6.7) = 3.1 MN/m²	
	Tension lap length	
	$= \frac{A_s f_y/1.15}{\Sigma b_o \times f_b} = \frac{1900 \times 410/1.15}{2(78.5 \times 3) \times 3.1} = 465\ (mm)$	Long. steel lap length = 465 mm
	CHECK REINFORCEMENT COVER (LIMIT STATE OF LOCAL DAMAGE)	
	d/h = 0.9 ∴ d = 0.9 × 350 = 315 mm	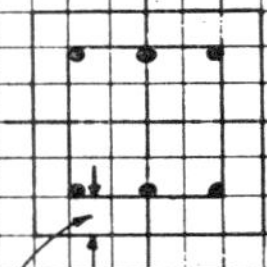
	∴ cover to longitudinal reinforcement = 35 – 12.5 = 22 mm, say 20 mm	©
	∴ cover to links = 14 mm – hence specify rendering of fire resistant plaster (see note calc. sheet no. 2.)	
	[Ed. note: alternatively, increase cover to main bars to 25 mm by using d/h = 0.85 since $A_s + A_s'$ = 2950 mm² >> 1900 mm²]	Cover ② to main bars = 20 mm
	NOTE This example uses the Unified Code of Practice, Sept. 1969, draft document no. 69/23007	

Fig 14.2.4 Calculation pages to draft Unified Code for typical column. (*continued*)

(ii) The calculations must state clearly the equations or process by which an answer is obtained (this can usually be satisfied simply by attaching preprinted calculation sheets to the computer answer sheets).

(iii) A description, and the magnitude, of all data and answers must be clearly stated.

(iv) In order to facilitate checking, intermediate answers should be given which will permit the calculations to be checked at an interim stage.

Operation of small computers

Small (desk-top) computers have very limited storage facilities and must therefore utilize familiar manual techniques, with human intervention often being required. For example, a series of programmes may be necessary in order to obtain the required final output. In addition it may be necessary to use the same reasoning with the output from a particular program as is used in a manual approach. These two points, as well as the documentation, are illustrated by reference to work which has been carried out by GKN Reinforcements Limited [14.3].

This work demonstrates that not only can a small computer readily perform tedious arithmetic, it can also greatly reduce the time spent in writing out calculations if the system is also backed, as in the GKN Reinforcements' method, with pre-printed calculation sheets and answer sheets to which the computer output is attached. Such standard texts and setting-out of equations eliminates any need for further description.

Frame and beam analysis by small computer

General framework analysis programmes covering various loading conditions for virtually any configuration of the members are readily available for large computers based on matrix methods. However, frameworks may also be analysed by the use of small computers. For example, a method of solution has been developed by GKN Reinforcements Limited for use with the Olivetti Programma 101 desk-top computer. Because of the storage limitation of this machine, the problem is solved with a number of small related programs. The complete "package" of programs determines both bending moments and shear forces at five points along each member of a frame or beam system of any size. The structural system may be subjected to any loading configuration but the stiffness of any member must be constant along its length.

The technique which is used follows the same sequence as if analysis were being carried out manually using the moment-distribution method and is shown diagrammatically in figure 14.3.1. First the loading data is specified for each member, and "program 20" used to calculate the fixed end moments. Next the designer chooses one from four available programs to carry out the moment-distribution process. The final member support moments determined by moment

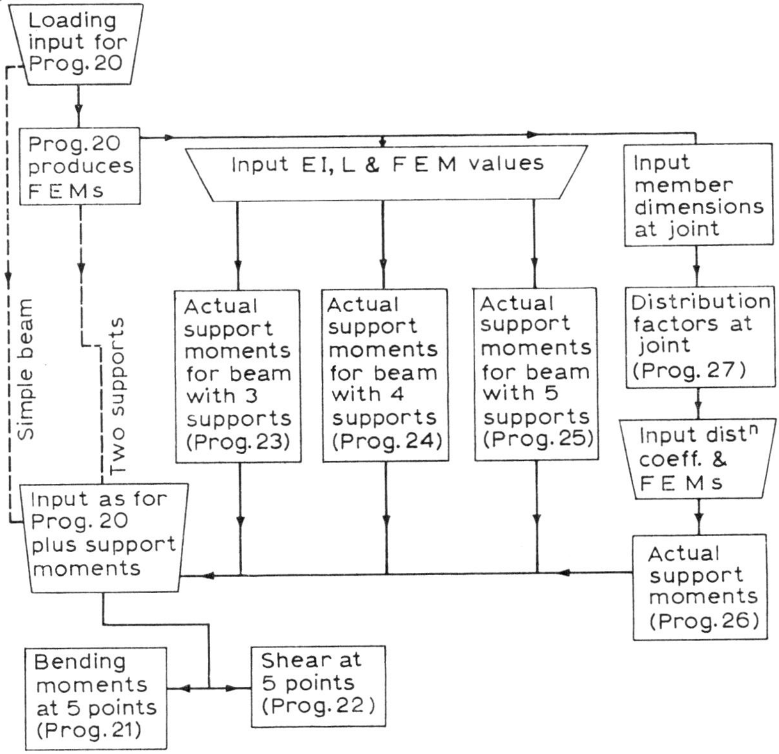

Fig 14.3.1 Example of GKN approach for small computer.

distribution are then added to the original loading data and re-input to programs 21 and 22 which calculate the bending moments and shear forces.

The choice of moment-distribution program depends on the system being analysed. For beams on three, four or five supports, there are programs 23–5, which are quick in operation and require as data the values of *EI*, *L* and the fixed end moments for each span. For more general systems, program 26 is available, which requires as data the distribution coefficients and fixed end moments for each member.

This program is slower in operation and often a subsidiary program—program 27—is used first to determine the values of the distribution coefficients.

An explanation of the data sequence and input values for the GKN programs is normally available in the program manual and is condensed for the present description. After the program number, the first data required is the effective *EI*; thus three-quarters of the actual *EI* is needed for an outside span if the member is free to rotate over an end support (i.e. a pinned or projecting cantilever).

The spans are assumed continuous over internal supports but the user must specify the type of end support. A zero marker is used for a fixed end. Although the programs are designed primarily for pinned or fixed-end outside supports, the effect of a projecting cantilever may be catered for by initially reducing the fixed end moment at the appropriate first internal support by half the value of the cantilever moment. In addition, because of built-in checks associated with a zero marker, the final outside support moment must then be adjusted to minus the value of the cantilever moment value. The use of program 24 is demonstrated in example 14.3.1.

EXAMPLE 14.3.1

Determine the final support moments for the beam on four supports shown in figure 14.3.2 for the two load cases given below. The actual

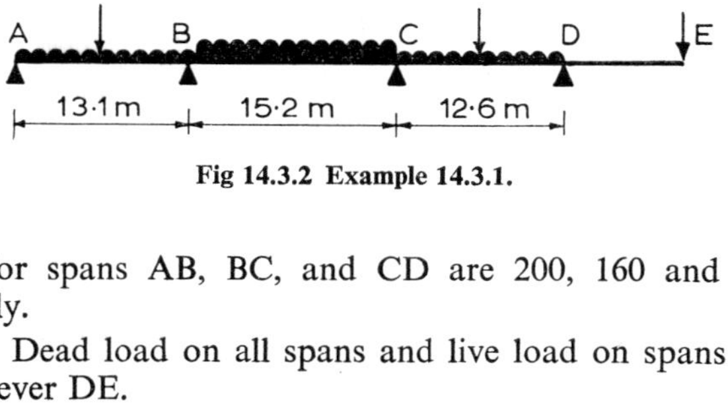

Fig 14.3.2 Example 14.3.1.

I-values for spans AB, BC, and CD are 200, 160 and 160 dm^4 respectively.

Case 1: Dead load on all spans and live load on spans AB, BC and cantilever DE.

Case 2: Dead load on all spans and live load on spans BC and DE only.

For case 1 the fixed end moments, for pinned supports at A and D, are $M_{AB} = 0$, $M_{BA} = 960$, $M_{BC} = -750$, $M_{CB} = 750$, $M_{CD} = -570$, $M_{DC} = 0$ and $M_{DE} = 220$ (all in kNm units).

For case 2 the fixed end moments, for pinned supports at A and D, are $M_{AB} = 0$, $M_{BA} = 420$, $M_{BC} = -750$, $M_{CB} = 750$, $M_{CD} = -570$, $M_{DC} = 0$ and $M_{DE} = 220$.

COMPUTER SERVICE

MOMENT DISTRIBUTION FOR A BEAM WITH FOUR SUPPORTS	PROGRAM 24 ANSWERS	Job No.	*Ex. 22.3.1*
		Sheet No.	
		Initials	
		Date	

INPUT SPECIFICATIONS

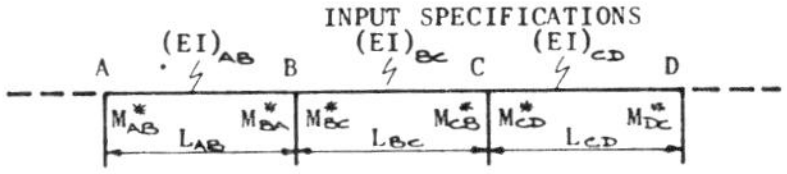

END SUPPORT MARKERS
Simple Support - 0
Fixed Support - 1

CASE 1: LIVE LOAD ON SPANS AB, BC, DE		ITEM	INPUT CHECK
		24	
24	S	$(EI)_{AB}$	
150	S	L_{AB}	
13·1	S	$(EI)_{BC}$	
160	S	L_{BC}	
15·2	S	$(EI)_{CD}$	
120	S	L_{CD}	
12·6	S	M^*_{BA}	
960	S	M^*_{BC}	
-750	S	M^*_{CB}	
750	S	M^*_{CD}	
-460	S	Mark A	
0	S	M^*_{AB}	
0	S	Mark D	
0	S	M^*_{DC}	
0	S	RW	
	RW	M^{**}_{AB}	
-0·00	A0	M^{**}_{BA}	
883·84	A0	M^{**}_{BC}	
-883·84	A0	M^{**}_{CB}	
583·24	A0	M^{**}_{CD}	
-583·24	A0	M^{**}_{DC}	
220 ~~-0·00~~	A0		

CASE 2: LIVE LOAD ON SPANS BC, DE		ITEM	INPUT CHECK
		24	
24	S	$(EI)_{AB}$	
150	S	L_{AB}	
13·1	S	$(EI)_{BC}$	
160	S	L_{BC}	
15·2	S	$(EI)_{CD}$	
120	S	L_{CD}	
12·6	S	M^*_{BA}	
420	S	M^*_{BC}	
-750	S	M^*_{CB}	
750	S	M^*_{CD}	
-460	S	Mark A	
0	S	M^*_{AB}	
0	S	Mark D	
0	S	M^*_{DC}	
0	S	RW	
	RW	M^{**}_{AB}	
0·00	A0	M^{**}_{BA}	
648·98	A0	M^{**}_{BC}	
-648·98	A0	M^{**}_{CB}	
648·02	A0	M^{**}_{CD}	
-648·02	A0	M^{**}_{DC}	
220 ~~-0·00~~	A0		

BEAM *ABCDE (FIG. 22.3.2)*	BEAM *ABCDE (FIG. 22.3.2)*
LOAD CASES *ONE*	LOAD CASES *TWO*
F.E.Ms ON PAGES —	F.E.Ms ON PAGES —
MOMENT UNITS *kN m*	MOMENT UNITS *kN m*
SIGN CONVENTION ↻ CLOCKWISE	SIGN CONVENTION ↻ CLOCKWISE

FINAL SUPPORT MOMENTS NOTATION

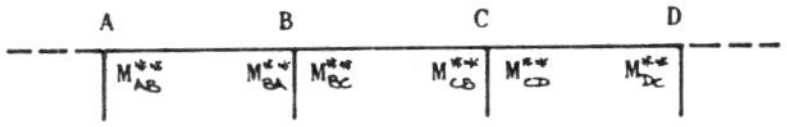

If there is a cantilever projecting over support A and/or D, the values of M^{**}_{AB} and/or M^{**}_{DC} must be adjusted for the cantilever moment value

Fig 14.3.3 Typical GKN Answer Sheet.

Solution The data for load case 1 are shown as the first 15 items listed on the left-hand side of the answer sheet in figure 14.3.3. It should be noted that three-quarters of the *I* value is in each case taken for spans AB and CD, and M_{CD} is reduced by one-half the value of the cantilever moment to give

$$-570 - \tfrac{1}{2}(-220) = -460$$

The next four items of data are the marker and fixed end moments at the end supports. The calculated support moments are the last six items listed and, as indicated by the note at the bottom of the sheet, the value of M_{DC} must be adjusted to 220. Similarly, the data and answers for case 2 are shown on the right hand side of figure 14.3.3.

Note: It cannot be over-emphasized that the mechanization of a manual method such as moment distribution should only be considered for a small computer. Manual techniques are not normally used for more sophisticated computers.

Drawings and schedules using computers

Drawings which must be to scale require an X-Y plotter. However, if drawings need not be scale it is possible to have a few standard drawings, e.g. one for a single-span beam, one for a two-span beam, one for a column, etc., and to do all the printing by relatively inexpensive automated processes, using a scheduling system for the standard structural units. Only general arrangement and non-standard members would then require to be done manually. Bar schedules lend themselves ideally to being computerized now that standard shape codes are available. The adoption of a shorthand, schematic or computer method of drawing and detailing will undoubtedly become more common in the future. Such scheduling and any other symbolism used must be clearly explained, if necessary on a special drawing prepared for the purpose, and the basic methods used should still conform with the principles of detailing given elsewhere (see appendix H).

14.4 Application of large computers

Computers are both fast and accurate and therefore have an ability to avoid the mistakes which characterize and mar hand calculations. This splendid efficiency is sufficient in itself to justify the use of machines, as evidenced by the success of small computers. There is, however, much greater justification for using large computers: first, their ability to make decisions and, second, the possibility they provide of utilizing techniques based on matrix methods. The large computer has had extensive influence on structural *analysis* and will also increasingly affect the process of *design*, which is largely concerned with decision-making. Although there is a complete range of computers with gradually increasing sophistication it is convenient in this section to consider only the largest digital machines.

Relative merits of computers

Large computers have large storage facilities and can therefore handle the basic input data and produce the final answers with a minimum of manual manipulation. They can accept comprehensive programs which can be made very general in application, again because of the large storage facilities. The advantages in structural analysis of large-computer techniques over manual methods can be briefly summarized as follows:

(i) Much larger frameworks can be contemplated for a single analysis.
(ii) Members of general orientation, in three dimensions if necessary, can be included without difficulty.
(iii) Axial, torsional and shear deformations can be taken into account as well as those due to bending.
(iv) A fast and accurate solution is readily obtainable.
(v) Several loading conditions can be analysed at once.

The disadvantage of large computers has been their extremely high initial cost and possible remoteness from the individual concerned with a particular problem. This has caused errors to occur in the input data and led to a reluctance to investigate the possible advantages of using any kind of computer.

Early development of computers

The first electronic computer was commissioned in 1946 at the University of Pennsylvania and the first automatic computer in the UK began operation at Cambridge in 1949. Early pioneers in structural computing realized that merely adapting existing techniques of manual calculation was not satisfactory for "large" computers despite the relatively small—by modern standards—working store capacity which was then available. They saw very clearly the concept of a unified computer program which would expedite the complete process of structural analysis from raw engineering data to final print-out of stresses.

By the early 1950s Livesley had produced practical computer programs for the general analysis of skeletal frames by stiffness methods [14.4,5]. Rapid developments took place in computing which also stimulated renewed interest in matrix algebra. The development of matrix methods not only provided computer programmers with practical techniques for solving problems but also helped to provide a unified approach for all structures, which could replace the traditional diverse approaches for beams, frames, trusses, arches, shells, etc. These developments produced a better understanding of fundamental principles and a general realization that methods of structural analysis can generally be classified into either force methods or displacement methods, although techniques are now available which can combine the two approaches.

Matrix algebra

Algebra is a generalized form of arithmetic in which letters are symbols which represent the numerical data which can be substituted for each individual calculation. Matrices take the generalization a stage further by providing relationships which may hold between collections of numbers without the need to specify at the outset how many numbers are involved. Argyris emphasized in 1958 that matrix methods did not mean the application of matrix algebra merely to the solution of the linear equations in the unknowns. Little, if any, progress would have been achieved if the introduction of matrix operations had been confined to this essentially trivial task [14.4]. Thus matrix methods enable a cylindrical-shell roof and a pin-jointed frame, for example, to be analysed in the same general way. Furthermore, matrix methods are eminently suited to processing by computer since computer programming is also concerned with separating the logical sequence of processes from the numerical data which are being processed. Generalized methods which enable structures with two, or two thousand, members subjected to one, or one thousand, loadings to be analysed with the same program represent an enormous economy of effort.

The first book setting out the detailed application of matrix techniques to problems in mechanics was written by Frazer, Duncan and Collar in 1937 following work mainly directed towards aircraft flutter. By 1947 Jenkins produced a book on shell structures which contained all the basic concepts that are currently used in the application of matrices to structures. By 1964 Livesley could produce a book with particularly outstanding coverage (see, e.g., references 14.4 and 14.5).

Conventional elastic analysis of beam frameworks

Many different methods of solving problems associated with the deformation of frameworks under load evolved over the years as the theory of structures developed. The classical slope-deflection method systematically provides a series of simultaneous equations and effectively describes the elastic behaviour of statically indeterminate frameworks. However, the equations for equilibrium, compatibility and slope-deflection (load deformation) cannot be solved by hand using ordinary techniques for other than simple systems.

A significant advance was made by Hardy Cross in the early 1930s when he derived the method of moment distribution. This method can be regarded as providing a numerical solution to frame problems (which implicitly incorporates satisfaction of the slope-deflection requirements) which is suitable for hand computation. It has the advantage that the beam end moments which are of primary interest are obtained directly. It was only after the advent of the digital computer that matrix methods could be implemented for the solution of large numbers of simultaneous equations and thus enable much larger and more complex structures (not only frameworks) to be solved.

Historical outline of modern structural analysis

Both of the main methods of modern elastic structural analysis—the force, flexibility or compatibility method and the displacement, stiffness or equilibrium method—had been developed by the 1860s by Maxwell (UK) and Clebsch (Germany) respectively. Both methods were later used by Mohr. Books by Muller-Breslau in 1886 and Bendixen in 1914 again aroused the interest of German-speaking engineers in the force and displacement methods respectively. Although Bendixen's book was not translated into English the displacement method as applied to beams and frames became well-known in the UK and America as the "slope deflection" method [14.5]. In 1926 Ostenfeld showed, for the first time in Germany, that the two methods were complementary, although it was not until the methods were subsequently expressed in matrix form that this was explained more completely.

The formal methods of structural analysis were little known in English-speaking countries until the late 1940s since only the "conventional" methods were in general use. The moment distribution method, for example, was a technique of successive approximations of increasing accuracy for the analysis of frames which was, in fact, based on the displacement relationships but which had not been generally recognized as such. However, by 1959, for example, Morice's book could outline the bases of the force and displacement methods and their relationship to the energy approximation methods of the 1930s, and Argyris had already given an interesting review of developments in the analysis of complex elastic aeronautical structures (see, e.g., reference 14.5).

The advent of computers and matrix methods has encouraged development within the whole field of structural analysis including special techniques for continuous-plate and 3-dimensional problems. The "finite-element" technique uses the displacement method to find displacements at discrete points having idealized the body into a number of suitably chosen elements. These elements are usually selected to suit the particular problem concerned.

Computer operation

The computer program contains the decisions and associated lists of operations and is communicated to the computer through the medium of paper tape or punched cards. The program is stored in the machine and is obeyed automatically so that a long sequence of operations can be followed, and the appropriate decisions made, without human intervention during the computation. The operations are performed on data that are fed in and stored at the start and further storage is required for the answers as they are produced. A computer with the necessary storage and facilities may not always be available and even if it is the cost in time and money of preparing the program and using the machine may outweigh its other advantages over a small desk computer. In general the problem must have more than a certain minimum degree of intricacy before it becomes economical to use a large computer for single or limited runs.

Process of computing

The process of computing involves three main stages once the problem to be solved has been clearly defined and the method or procedure for solving the problem has been decided. The identification of the problem and determination of the form of solution is sometimes known as "systems analysis" although this term can embrace flow charting as well.

(*a*) *Flow charting* A flow chart is a visual display of the precise logical procedures to be used. It is independent of any particular computer or programming language. Flow charts are considered in more detail in section 14.5.

(*b*) *Coding* Coding (or computer programming) concerns the translation of the procedures given by the flow chart into a form which is acceptable to the computer. The programmer must have a good working knowledge of the appropriate computer language in order to write the instructions to the computer.

(*c*) *Operation of computing* The actual computing sequence is as follows:

(i) *Punching and input* A card punch (or tape punch) operator punches the information from the coding sheet prepared by the programmer.

(ii) *Compilation* The cards (or tape) are fed into the computer once the computer's memory has been cleared and the standard compiler program has been read into it from magnetic tape. The computer is able to translate the information from the cards into machine language and punch further cards (or tape) corresponding to the original cards (or tape).

(iii) *Execution and output* The new cards (or tape), i.e. those in machine language, are put in the card reader of the computer, with added data if any, and it is these that are actually used in the solution of the problem.
(iv) *Development of the program* All necessary checks should be incorporated at this stage and any inconsistencies or errors eliminated (a process sometimes called "de-bugging").
(v) *Production runs* The program is now available for actual production runs.

Procedure-orientated languages

The above procedure assumes the use of a symbolic programming language such as FORTRAN or ALGOL. The idea of writing a program in normal English and mathematical symbols, and getting the computer itself to do the translation into the appropriate internal code, dates from the early 1950s or perhaps even earlier. FORTRAN became available in the USA in about 1956. It is clear that a good symbolic language leaves the programmer free to concentrate all his thoughts on the logic of his procedure and this is desirable whatever the level of his expertise. Symbolic programming languages are much easier to learn since they eliminate any need for a detailed knowledge of the operations peculiar to a particular machine. Programs written in such languages are both permanent and universal since they are then based essentially on ordinary human logic. Programs written in machine code tend to die with their machines, and many programmers have spent tedious hours rewriting programs as a result. A program in a symbolic language, however, can be run on any machine equipped with a compiler for that language. Although compiler programs are very complex they are written by teams of programmers employed by a computer manufacturer and are delivered as part of the "software" with the computer.

14.5 Flow charts for computer programming

The more difficult a problem, the more complicated it is to prepare a computer program directly from the basic specification. However, the pattern of the computation can be clarified by drawing a flow chart, which is a visual statement in the form of a network, of the routes through the computation and of the conditions that govern the choices of route. A flow chart is also the best method of documentation, since the program itself is difficult to comprehend as it inevitably includes the symbolism and grammar of a particular programming language and a particular computer. The flow chart, on the other hand, is completely general and can be made as detailed as the circumstances require.

Flow chart notation

A flow chart consists of connected boxes each of which contains the description of an operation. The connecting lines have arrows to show the direction of flow through the network. The boxes are of various shapes, chosen to identify different types of operation. This can be done in a number of ways to suit the whim of the program writer, hence the simplest of choices and the use of four shapes only are considered here: ovals, trapeziums, diamonds and rectangles. Other shapes can be considered if desired [14.6,7].

Oval

Oval shapes are used to mark the beginning and the end of a program.

Trapezium

A trapezium indicates input or output. The starting-point of the computation is defined by the data available. As far as possible the program should be of a general nature so that the values of the variables are fed in as data rather than being built into the program itself. This is of particular importance when the program is intended (as it often would be in design work) for repeated use. The form of output is important too since it is desirable to arrange in the program for the answers to appear in the final form and so avoid a further stage of processing. Output includes presentation in the form of graphs and, increasingly, of engineering drawings, as well as ordered arrangements of numbers.

Diamond

A diamond indicates a decision. In most cases the decision is made by answering "yes" or "no" to a question in the form of a numerical comparison. For example, is M_r (the moment of resistance of a section of a beam) greater than M_a (the moment applied to the section)? There are thus alternative routes away from the decision box corresponding to the "yes" and "no" answers. Computers can readily be programmed to cycle; that is, to repeat a sequence of computations a number of times with a change in the data for each cycle. This is very useful in many forms of numerical analysis (such as integration by Simpson's rule) and it depends on a decision that controls the cycle by moving on from the sequence when the required number of repetitions has been completed.

Rectangle
A rectangle indicates a computation or an operation which is not a decision, an input or an output (e.g. start, stop). This is normally an arithmetical process such as the evaluation of a formula, but it can also be the setting of a variable to one of a pre-selected range of values. In writing the program the four rules of arithmetic are shown by simple symbols; most programming languages include built-in sequences for evaluating common functions such as trigonometrical ratios and square roots, which are called up by the use of a short name. Thus quite involved expressions can be written in a simple and short manner.

EXAMPLE 14.5.1
Construct a suitable flow chart for determining the area of steel required in a number of sections for various rectangular reinforced concrete beams according to the analytical approach of Hognestad, Hanson and McHenry. The following data can vary for each section: the applied moment M_a at collapse; the beam breadth, b, and effective depth, d_1; the material strengths (u_w for concrete, f_y for steel in tension and f_{yc} for steel in compression).

The concrete stress block parameters, K_1K_3 and K_2, should be incorporated in the program, which should be able to process any chosen number, M, of beam sections in a single run.

Solution The flow chart will require two separate loops, one to consider the need for compression steel and one to enable all M sections to be processed, and not just one section. A suitable flow chart is shown in figure 14.5.1. It should be realized that example 14.3.1 referred to a practical problem, whereas example 14.5.1 is essentially an illustrative problem for teaching purposes.

Computer-orientated processes
The above example, although relatively simple, still demonstrates the use of processes which were inconceivable before the advent of computers. It is seen that an analytical technique, for example, can be used as a design method because the hand computations can be suitably mechanized on a computer. Other processes, such as optimization of an analysis, can be considered when using computers whereas with hand computation a design would only be repeated if it was clearly uneconomical. A computer, however, can give a series of solutions for the designer's consideration at very little extra cost.

Problem-orientated languages
The next stage in development now that procedure-orientated languages are becoming well established will be problem-orientated

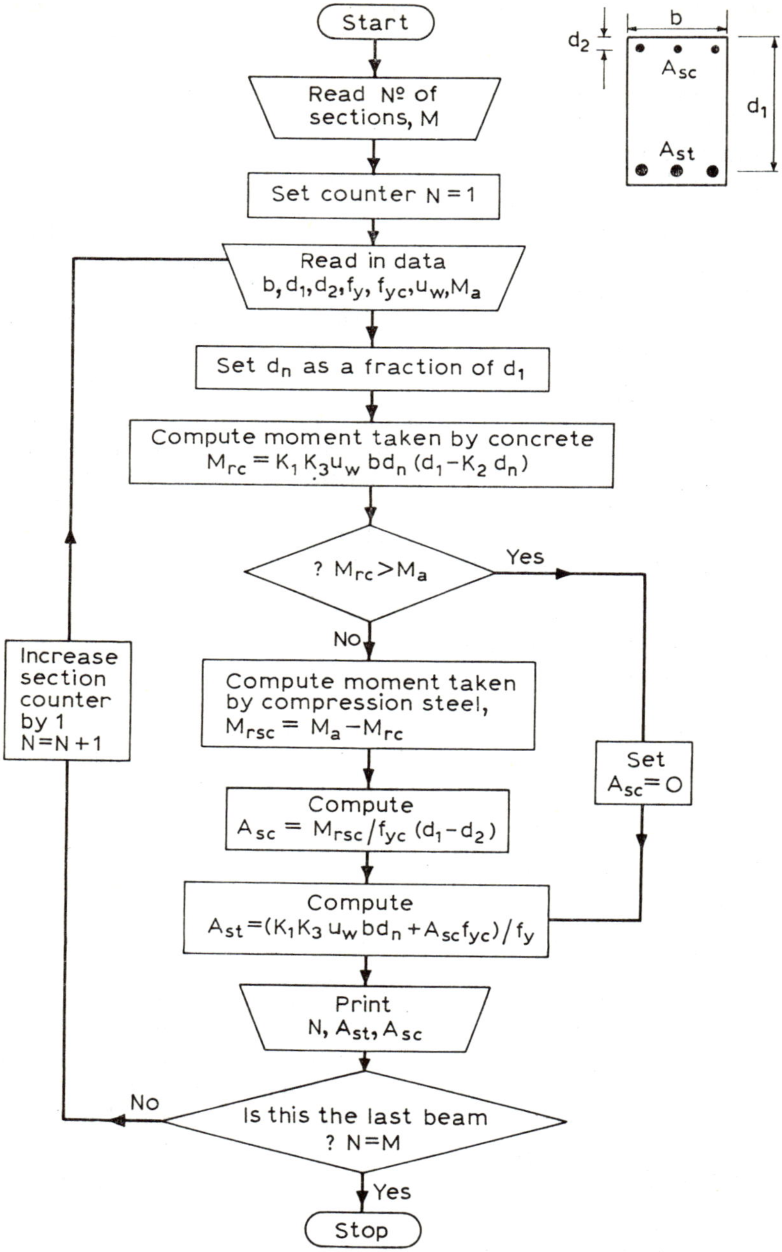

Fig 14.5.1 Example 14.5.1.

languages such as are used in systems like ASKA and GENESIS. Programs using such systems will require to be fed with even less information and, by calling up standard routines, be able to provide even more answers. For example, optimization *in design* should become a reality in the future. At present optimization is usually based on analysis *after* the overall layout, material strengths, etc., have been decided (often for quite arbitrary reasons). Optimization of a series of analyses can point towards optimization in design but should not be confused with the latter in its widest sense.

REFERENCES

14.1 CONCRETE SOCIETY *A standardized procedure for the presentation of calculations for reinforced and prestressed concrete structures* (draft for comment) Concrete Society, London, May 1969.

14.2 REYNOLDS, CHAS. E. *Reinforced concrete designer's handbook* (Concrete Publications, London).

14.3 JONES, L. L. *Structural Design Computer Weekly*, March 20th 1969.

14.4 LIVESLEY, R. K. The pattern of structural computing 1946–1966 *Struct. Eng*. vol. 46 no. 6 June 1968 pp 177–82.

14.5 MORICE, P. B. The new look in structural analysis—a historical survey *Concrete* vol. 3 no. 10 Oct. 1969 pp 415–7.

14.6 FENVES, S. J. *Computer methods in civil engineering* (Prentice-Hall, 1967).

14.7 WEAVER, W. *Computer programs for structural analysis*. (Van Nostrand, London 1967).

Exercise

1. A computer program is required for the design of shear steel, in the form of stirrups only, in a simply supported reinforced concrete beam carrying a uniformly distributed load over the whole span. Draw up a flow chart to show how this might be done, showing clearly the relations and conditions that are needed, and listing the starting data. Assume that one bar size only is available and note any other restrictions of this kind that you consider necessary.

(UL2 1969, 36 min)

Grades of concrete

Grade		Characteristic cube strength 7 days MN/m²	28 days MN/m²	(lbf/in²)	Possible application is shown against the lowest grade*
–	3 NF	2·0	3·0	(450)	No-fines concrete for plain walls
–	5 NF	3·4	5·0	(700)	
7·5	7·5 NF	5·0	7·5	(1050)	Plain concrete
10		6·7	10·0	(1450)	
15	15 SLA to	10·0	15·0	(2200)	Reinforced lightweight concrete
22·5	22·5 SLA	15·0	22·5	(3250)	Reinforced concrete
30	–	20·0	30·0	(4350)	
37·5	–	26·0	37·5	(5450)	Post-tensioned concrete
45	–	32·0	45·0	(6500)	Pre-tensioned concrete and post-tensioned pre-cast concrete
52·5	–	38·0	52·5	(7600)	
60	–	45·0	60·0	(8700)	
67·5	–	52·0	67·5	(9800)	
		Characteristic flexural strength			
	F3	2·5	3·0	(440)	
	F4	3·5	4·0	(580)	
	F5	4·6	5·0	(720)	
		Characteristic indirect tensile strength			
	IT2	1·6	2·0	(290)	
	IT2·5	2·3	2·5	(360)	
	IT3	2·7	3·0	(440)	

* e.g. 22·5 is the lowest grade for reinforced concrete.

Table A1 Grades of concrete for different applications

Appendix B

Practical rules for limit state design

(Extracts from the British Standard Code of Practice for the Structural Use of Concrete, (the Unified Code), sections 2 and 5.* Cross-references to sections in the Unified Code are shown in brackets. Conditions that are *not* as given by the Code are indicated by footnotes).

Limit states (§5)

(*a*) Collapse
(*b*) Deflection
(*c*) Local damage
(*d*) Vibration, fatigue, durability, fire resistance or other relevant limit state as necessary.

Design on basis of most critical limit state and then check that the remaining limit states are satisfied.

* It should be understood that the extracts are from the Draft Code for Comment (the latest version available to the author) and that this is subject to modification. Only the published document is authoritative.

Partial safety factors (§5.3)

Limit states and load combination	Partial safety factors γ_l Maximum	Minimum	$\gamma_m(\&1/\gamma_m)$ Steel*	Concrete
Collapse				
I Dead & imposed	$1{\cdot}4W_d+1{\cdot}6W_i$	$1{\cdot}0W_d$	1·15 (0·87)	1·5 (0·67)
II Dead & wind	$1{\cdot}4W_d+1{\cdot}4W_w$	$0{\cdot}9W_d+1{\cdot}4W_w$	,,	,,
III Dead, imposed & wind	$1{\cdot}25(W_d+W_i)+1{\cdot}2W_w$	—	,,	,,
Deflection				
I Dead & imposed	$1{\cdot}0(W_d+W_i)$	$1{\cdot}0W_d$	1·0	1·0
II Dead & wind	$1{\cdot}0(W_d+W_w)$	$1{\cdot}0(W_d+W_w)$	,,	,,
III Dead, imposed & wind	$1{\cdot}0W_d+0{\cdot}8(W_i+W_w)$	—	,,	,,
Local Damage	As for deflection	As for deflection	1·0	1·3 (0·77)

The minimum intensities apply when a reduction in load provides a more adverse condition; e.g. alternate spans in continuous systems for cases I and II.

* Steel in tension. γ_m varies for steel in compression as follows $1{\cdot}15 + f_y/2000$ (beams); 1·5 or 1·67 (axial columns).

Table B1 Specified γ_l and γ_m values for different limit states

Design loads (§2)

CHARACTERISTIC LOAD

W_k = working load given in CP3: Chapter V or other BS.

DESIGN LOAD

$_*W = W_k \times \gamma_l$

WEIGHT OF REINFORCED CONCRETE

Take as 24 kN/m^3 under standard gravity conditions.

CREEP, SHRINKAGE AND TEMPERATURE

Limit state:	Collapse	Deflection	Local damage
Creep	None	Permanent loads	Permanent loads
Shrinkage	None	Low compression-steel ratios	Low tension-steel ratios
Temperature	None	Large temperature gradients	High temperatures or temperature gradients

Table B2* Usual occurrence of significant load effects

Design strengths

CHARACTERISTIC STRENGTH

σ_k = strength for which there are 5% failures

DESIGN STRENGTH

$_*\sigma = \sigma_k/\gamma_m$

Analysis (§5.4)

ANALYSIS OF STRUCTURE

Force distributions should be determined from material properties associated with characteristic strengths. Elastic analysis may be used with relative stiffnesses based on any one of the following cross-sections throughout:

(*a*) Concrete section—given by bd

(*b*) Gross equivalent section—given by $bd + mA_s$

(*c*) Transformed section—given by $bd_n + mA_s$

For prestressed concrete gross section only should be used.

Moment redistribution is permitted for limit state of collapse only:

$\not>$ 30% if members possess adequate ductility

$\not>$ 10% for structures with not less than 4 storeys and relying on columns for stability

See text for detailed assumptions concerning moments in columns and moment redistribution in beams.

* Not as draft Code.

ANALYSIS OF SECTIONS

Material properties associated with the design strengths appropriate to the limit state being considered should be used as follows:

(*a*) Collapse—inelastic analysis based on short-term stress-strain curves.
(*b*) Instantaneous deflection—elastic analysis based on short-term modulus.
(*c*) Total deflection—either elastic analysis based on long-term modulus or using separate assessment for additional effects due to creep, shrinkage and cracking.
(*d*) Local damage—elastic analysis based on 0·5 × short-term modulus (concrete only) or inelastic analysis as appropriate.

See text for detailed assumptions for each limit state.

SUSTAINED LOADING

Short-term stress-strain curves may require modification for sustained loading as follows:

Reinforcement: none
Prestressing tendons: due allowance for relaxation
Concrete: due allowance for shrinkage and creep

MODEL ANALYSIS AND TESTING

May be used instead of above analyses.

EXPERIMENTAL DEVELOPMENT

Actual tests of prototype units and similar structures may be used to justify analytical or empirical designs based on novel methods.

Limit state requirements

COLLAPSE (§5.4.3)

Ensure adequate resistance to:

(*a*) bending, shear and axial loads
(*b*) buckling
(*c*) overturning

DEFLECTION (§5.4.4)

(*a*) Horizontal members (deflection below horizontal):

Final long-term deflection $\not>$ span/250 (generally)
and $\not>$ span/200 (special cases)

Precamber (above horizontal) $\not>$ span/250 (generally)
(Also $\not>$ span/300 for prestressed units with finishes and non-uniform camber)

Additional deflection after application of finishes $\not>$ span/350 and $\not>$ 20 mm unless provision is made for movement. Reduce for brittle finishes.

(*b*) Lateral deflection $\not>$ height/500

LOCAL DAMAGE (§5.2.4)

(*a*) Steel strain (RC beams) $\not>$ 0·0015
and $\not>$ $0{\cdot}75 f_y/E_s$

(*b*) Crack width at surface $\not>$ 0·3 mm (RC, all members)
and $\not>$ 0·2 mm (Prestressed class 3)
No cracks (class 1 and 2) and no tension (class 1)
No cracks at initial prestress for all prestressed concrete

(*c*) Crack width at surface nearest to main reinforcement for an aggressive environment $\not>$ 0·1 mm (RC and prestressed class 3)

(*d*) Compressive stresses both finally and at transfer shall be limited to prevent (i) excessive losses in prestress (prestressed concrete only) and (ii) spalling of concrete in compression zone.

(*e*)* Maximum width of thermal and shrinkage cracks extending throughout the thickness of members such as walls and slabs
$\not>$ 0·1 mm for liquid-containing members
and $\not>$ 0·2 mm for exposed external members.

(*f*)* The flexural tensile stresses, or crack widths, on the liquid-containing face, are limited for different classes of liquid-containing structures as follows:

Class A—no cracks (as class 1 above).
Class B—maximum surface crack width $\not>$ 0·1 mm.

* Not as draft Code.

Appendix C

Practical rules for RC design

(Extracts from the British Standard Code of Practice for the Structural Use of Concrete, (the Unified Code), sections 4 and 6.* Cross-references to sections in the Unified Code are shown in brackets. Conditions that are *not* as given by the Code are indicated by footnotes).

General requirements (§6.1)

DESIGN LOAD ARRANGEMENTS (§6.1.3)

Use arrangement causing most severe effect. For framed structures under uniformly distributed imposed loading assume:

(*a*) Beams and slabs: *either* alternate *or* adjacent spans loaded.

(*b*) Columns: maximum out-of-balance moment at either end.

(*c*) Wind: sideways movement of columns produces reactions on all beams or slabs.

EFFECTIVE SPAN OF BEAMS (l) (§6.1.4)

$l \ngtr$ distance between centres of bearings (all spans)

and $\ngtr$ clear span plus effective depth (simple spans only)

and $\ngtr$ length to support face plus half effective depth (built-in cantilever only)

EFFECTIVE HEIGHT OF COLUMNS (h) (§6.1.5)

(*a*) *Restrained (i.e. braced) column*

$h \ngtr H$

and $\ngtr H[0{\cdot}70 + 0{\cdot}05(K_{ru} + K_{rl})]$

and $\ngtr H[0{\cdot}85 + 0{\cdot}05(K_r)_{min}]$

(*b*) *Unrestrained column*

$h \ngtr H[1{\cdot}0 + 0{\cdot}15(K_{ru} + K_{rl})]$

and $\ngtr H[2{\cdot}0 + 0{\cdot}3(K_r)_{min}]$

where H = clear distance between restraints in plane considered.

$$K_r = \sum \frac{EI_c}{L_c} \bigg/ \sum \frac{EI_b}{L_b} \quad \text{(for plane considered)}$$

* See footnote on page 346.

Subscripts u and l refer to upper and lower ends of column

$(K_r)_{min}$ = lesser of K_{ru} and K_{rl}

L_b, L_c = length of beam and column respectively along centre-line from centre to centre of joints.

I_b, I_c = second moment of area of concrete section of beam (if $H/t < 25$) and column (all cases) respectively.

I_b = second moment of area of transformed section at mid-span if $H/t > 25$.

t = least lateral dimension of column or beam, as appropriate.

(*c*) *Column cast onto footing*

If base is designed to resist moment, $K_r = 1{\cdot}0$

If base is not designed to resist moment, $K_r = 10$

MOMENTS IN LONG COLUMNS (i.e. $h/t > 12$)—see text.

EFFECTIVE HEIGHT OF PLAIN WALLS (h) (§ 6.8)*

Given by:

$X \times$ mean height (or length) between horizontal (or vertical) supports

or $Y \times$ mean height (or length) if one edge is free.

(*a*) *Braced wall*

(i) If lateral supports resist rotation and lateral movement

$X = \frac{3}{4}$ and $Y = 2$.

(ii) If lateral supports resist lateral movement only

$X = 1$ and $Y = 2\frac{1}{2}$

(*b*) *Unbraced wall* (i.e. no vertical supports present)

$X = 1\frac{1}{2}$ and $Y = 2$.

Note: Lateral supports provide resistance to rotation if construction is monolithic *or* if bearing on wall of precast floors (along or across span) $\nless \frac{2}{3} \times$ wall thickness *or* if the supported floors (or roofs) occur at same level on both sides of the wall.

EFFECTIVE HEIGHT OF REINFORCED WALLS (h) (§6.9)

(*a*) *Braced wall*

(i) Top and bottom connections continuous with adjoining slabs:

As columns except that $K_r = \sum \frac{EI}{H} \text{(walls)} \Big/ \sum \frac{EI}{L} \text{(slabs)}$

where L = span of slabs framing into wall

I = second moment of area of concrete section in all cases.

* Not as draft Code.

(ii) Top and bottom connections hinged but having effective lateral restraint: as for plain walls.

Also

(iii) Lateral vertical restraints (if present): as for plain walls.

If height of any opening $> H/3$, treat wall as two independent part walls.

(*b*) *Unbraced wall*—as for columns.

EFFECTIVE LATERAL RESTRAINT OF WALLS (§6.8.2.2)

Local resistance to horizontal thrust at point of lateral support

$\nless$ total applied horizontal load acting on the edge
and $\nless$ (1/40) × total vertical design load for wall.

LOAD CARRIED BY PLAIN WALLS (§6.8.4.4)

$$\frac{W}{bt} = \alpha \left[1 - \frac{2(e_0 + e_{add})}{t}\right] u_w$$

where W/bt = permissible direct stress in the wall
e_0 = eccentricity due to primary moments ($\nless 0{\cdot}05\, t$)
e_{add} = eccentricity due to additional moments
α = coefficient dependent on wall dimensions, grade and type of concrete as given in table C1.

Type & grade of concrete	Clear height between supports		
	$\nless 1{\cdot}5b$	b	$\ngtr 0{\cdot}5b$
Fully compacted. $u_w \nless 22{\cdot}5$ MN/m²	0·4	0·45	0·5
Fully compacted. 10 & 15 MN/m²	0·35	0·375	0·4
No fines. Aggregate to BS882 or BS1047	0·3	0·325	0·35

Table C1 Values of coefficient α for walls*

See text for analysis of plain walls.

EFFECTIVE WIDTH OF FLANGED BEAMS (§6.1.6)

$b \ngtr$ Actual flange width
and $\ngtr$ Web width + XY

where X = 0·1 for L-beam and 0·2 for T-beam
Y = distance between points of zero moment (= 0·7 × effective span for a continuous beam)

See appendix EW of the Code for accurate method.

* Not as draft Code.

LIMITS TO SLENDERNESS (§6.1.7)

(*a*) *Simply supported and continuous beams*

$L/b \not> 50$, and $Ld_1/b^2 \not> 250$ in all cases

where L = clear distance between *lateral* restraints

b = width of *compression* flange or face midway between restraints.

See Code regarding effective restraint provided by adjoining monolithic slab.

If $L/b > 30$ provide minimum stirrups throughout span in all cases and at least one bar in each corner of compression face.*

(*b*) *Cantilevers*

$L/b \not> 20$, and $Ld_1/b^2 \not> 100$ in all cases. (If $L/b > 12$ provide shear reinforcement as for (*a*)) where L = clear distance from face of support to end of cantilever.

(*c*) *Columns*

(i) Restrained columns: $H/t \not> 50$ and $Hd/t^2 \not> 250$

(ii) Unrestrained columns: $H/t \not> 30$ and $Hd/t^2 \not> 150$

where t = least lateral dimension

d = depth perpendicular to t.

MOMENT REDISTRIBUTION IN BEAMS (§6.3.1.1)

See text.

Reinforcement

SIZES OF REINFORCEMENT (§4.1.4)

6, 8, 10, 12, 16, 20, 25, 32, 40 and 50 mm.

SPECIFIED STRENGTHS AND MINIMUM ELONGATION OF REINFORCEMENT (§4.1)

Type of reinforcement	BS	Sizes (mm)	Strength (MN/m²)		Minimum elongation on
			f_y	Ultimate	$5{\cdot}65\sqrt{(S_0)}$
Mild steel	4449	All sizes	250	$\not<1{\cdot}15f_y$	22%
Hot rolled deformed bars	4449	All sizes	410	$\not<1{\cdot}15f_y$	14%
Cold worked deformed bars	4461	>16	425*	$\not<1{\cdot}10f_y$	14%
		$\not>16$	460*		12%
Hard drawn wire	4482	$\not>12$	485	$\not<1{\cdot}10f_y$	

S_0 = original cross-sectional area of test piece

* 0·2% proof stress may be assumed to be given by 0·4% total strain.

Table C2 Specified strengths and minimum elongation of reinforcement

* Not as draft Code.

MAXIMUM AND MINIMUM STEEL AREAS (§6.2.1 to 3)

(*a*) *Beams*

(i) $A_{st} \nless 0{\cdot}0015 \times 410bd_1/f_y$ (f_y in MN/m^2) and f_y is taken as $\ngtr$ 410.*

(ii) A_{st} and $A_{sc} \ngtr 0{\cdot}04bd_1$ for calculation purposes, where b = width of beam at level of centroid of tension reinforcement for (i) and (ii).

(iii) Stirrups: see shear
Pitch of stirrups $\ngtr$ 12 × diameter of any compression bar.

(iv) Transverse reinforcement across full effective width of flange of T- or L-beam near top surface:
$\nless$ 0·3% of cross-sectional area of flange.

(*b*) *Solid slabs*

$r' \nless$ 0·12% in *each* direction.†

(*c*) *Columns*

(i) Area of longitudinal steel
$\nless 0{\cdot}15\,{}_*W/f_y$
and $\ngtr$ 6% of gross area of columns cast vertically
or $\ngtr$ 8% cast horizontally, where ${}_*W$ = maximum design load for collapse.
Area of steel at laps $\ngtr$ 10%

(ii) Number of bars $\nless$ 4 if rectangular
and $\nless$ 6 if circular

(iii) Longitudinal bar diameter $\nless$ 12 mm

(iv) Transverse bar diameter
$\nless$ 6 mm
and $\nless \frac{1}{4}$ × diameter of any longitudinal bar

(v) Pitch of transverse steel
$\ngtr$ 0·3 m
and $\ngtr$ least lateral dimension of column
and $\ngtr$ 12 × diameter of smallest longitudinal bar.

(vi) Transverse steel must change direction by $\nless$ 45° at each corner and at each alternate longitudinal bar (or bundle). Intermediate longitudinal bars $\ngtr$ 150 mm clear distance from bars laterally supported as above do not require further change in direction of stirrup.

(*d*) *Reinforced concrete walls* (i.e. walls acting as columns for which greatest lateral dimension $\nless$ 4 × least lateral dimension).

(i) $r' \nless$ 0·2% (horizontal)† and $\nless$ 0·4% (vertical)

* Not as draft Code.

† This is the condition given in the Code. However, higher values of r' are often necessary to control thermal and shrinkage cracking.

(ii) Diameter of transverse links
$\not< 6$ mm
and $\not< \frac{1}{4} \times$ vertical bar diameter

(iii) Pitch of transverse links $\not> 1{\cdot}0$ m (horizontal and vertical).

(*e*) *"Plain" concrete walls*

(i) External: $r' \not< 27\%$ (horizontal)* and $\not< 0{\cdot}13\%$ (vertical)

(ii) Internal: at junctions with floors and beams,
$r' \not< 0{\cdot}20\%$ (horizontal)* and $\not< 0{\cdot}10\%$ (vertical)

HOOKS AND BENDS (§6.2.4)—use only where necessary

(*a*) Minimum *internal* radius $\not< X_{min} \times$ bar diameter
where $X_{min} = 2$ (mild steel) or 3 (high-yield steel).

(*b*) Anchorage value of bends (*including* length forming anchorage)
$\not> 24$ bar diameters
and $\not> 2X \times \theta/45$ bar diameters
where θ = angle of bend in degrees.
E.g. standard hooks equivalent to straight bar lengths of 16 bar diameters (mild steel), and 24 bar diameters (high yield steel).

c) Length beyond end of curve
$\not< 4$ bar diameters (generally)
and $\not< 8$ bar diameters (stirrups and links)

(*d*) Stirrups (and links) are fully anchored if $\theta \not< 90°$, or $\not< 135°$ for torsion, around longitudinal bar which is $\not<$ size of its stirrup.

(*e*) Minimum distance inside support face before start of hook $\not< 4$ bar diameters.

COVER TO REINFORCEMENT (excluding finishes) (§6.2.5)

Actual cover $\not< 15$ mm and $\not<$ bar diameter
Nominal cover = Actual cover + 5 mm

Cover may need to be increased where appropriate for durability (see, e.g., reference 11.3), fire resistance, calcium chloride, lightweight and porous aggregates, surface treatment, large size aggregate. The "bar diameter" for bundled bars is the size of a single bar of the same total area.

BUNDLED BARS (§6.2.6)

3 or 4 parallel bars may be grouped in contact if:

(i) they terminate either at different points with a stagger of $\not< 40$ bar diameters, or all at a support

* These are the conditions given in the Code. However, higher values of r' are often necessary to control thermal and shrinkage cracking.

(ii) $\ngtr$ 4 bars occur even at laps, and
(iii) minimum stirrups are provided in all cases

DISTANCE BETWEEN BARS (§6.2.7)

(*a*) Single bars: horizontal: $\nless d_{max} + 5$ mm
vertical: $\nless \frac{2}{3}d_{max}$ (gaps vertically in line)

where $d_{max} \nless$ maximum aggregate size
and $\nless$ bar diameter − 5 mm

The corresponding bars in the various rows should be vertically in line.

(*b*) Pairs of bars:
Generally as for single bars using equivalent bar diameter (= diameter of single bar of equivalent area to pair)
Except that if pairs of bars are side by side horizontally then vertical distance between pairs $\nless d_{max} + 5$ mm.

(*c*) Bundled bars:
Horizontally and vertically $\nless d_{max} + 15$ mm
where $d_{max} \nless$ maximum aggregate size
and $\nless$ equivalent bar diameter − 15 mm.

(*d*) Immersion vibrators:
Leave sufficient room for use if required.

(*e*) Reduced spacings may be used if supported by suitable evidence.

CURTAILMENT OF BARS (§6.3.1.3 to 5)

(*a*) Tension reinforcement over supports for *beams*:
Distance into span $\nless 0{\cdot}1L$
and $\nless 1{\cdot}2 \times$ full anchorage length

(*b*) Tension reinforcement at midspan for *beams*:
Distance from interior supports $\ngtr 0{\cdot}2L$
Distance from exterior supports $\ngtr 0{\cdot}1L$
Percentage extending to supports (but need not be continuous through them) $\nless$ 50%

(*c*) Tension reinforcement over supports for *one-way slabs*:
Distance into span: as for beams
Percentage extending $0{\cdot}3L$ into span $\nless$ 50%

(*d*) Tension reinforcement at midspan for *one-way slabs:* as for beams
For two-way slabs, see text.

All limit states

See appendix B for assumptions in analysis. Framed structures may be idealized as frameworks with member lengths equal to the distances between the intersections of the centre lines framing into the joints. Alternatively, for dead and imposed loads only, a simplified sub-frame can be considered for each loaded member in which the ends of the beams and columns remote from the joints adjacent to the loaded span are assumed fixed.

In framed structures which rely on the columns for stability, moments due to lateral forces may be assessed assuming that there are points of contraflexure at the centres of beams and columns.

Inelastic analysis may be used, for which guidance is given in appendix RM of the Unified Code.

Limit state of collapse

MOMENT REDISTRIBUTION IN BEAMS—see text.

Shear (§6.3.4)

Nominal design shear stress calculated from

$$q = {}_*Q/bd_1$$

where ${}_*Q$ = design shear force for limit state of collapse
b = breadth of beam (or web for T-beams)
d_1 = effective depth

MINIMUM STIRRUPS

Minimum shear reinforcement in form of *stirrups* must always satisfy*

$$(0{\cdot}87 f_{yq})(A_{sq}/bs)(\sin\theta + \cos\theta) \not< 0{\cdot}4 \quad (\text{MN/m}^2 \text{ units})$$

where f_{yq} = characteristic strength of stirrups (taken as $\not> 425$ MN/m^2)
A_{sq} = cross-sectional area of stirrups (e.g. both legs for 2-leg stirrup)
θ = inclination of stirrups to longitudinal axis of member ($\not< 45°$)
s = stirrup spacing along longitudinal axis of member, $\not> \frac{3}{4}d_1$ and $\not>$ clear web depth (I-beams)

Lateral spacing of legs of stirrups $\not> \frac{3}{4}d_1$

The minimum shear reinforcement need not be provided in slabs, footings, pile caps (i.e. where $b > d_1$) or walls and members of minor importance (i.e. $q \leqslant q_c$) if $q < q_c$. If $q < \frac{1}{2}q_c$ only slender beams require minimum stirrups.

* Not as draft Code.

CRITICAL CONCRETE SHEAR STRESSES

$100A_{st}/bd_1$	Characteristic cube strength (MN/m²)				
	15	22·5	30	37·5	45 or more
0·15	0·25	0·30	0·35	0·40	0·45
0·25	0·30	0·35	0·40	0·45	0·50
0·50	0·40	0·50	0·60	0·65	0·70
1·00	0·50	0·60	0·70	0·80	0·85
1·50	0·60	0·70	0·80	0·90	0·95
2·00	0·65	0·75	0·85	1·00	1·05
2·50	0·70	0·80	0·90	1·05	1·10
3·00	0·75	0·85	0·95	1·10	1·15
3·50 or more	0·80	0·90	1·00	1·15	1·20

Table C3 Critical shear stress q_c (MN/m² units)

If $q > q_c$ provide reinforcement in form of stirrups, or bent-up bars and stirrups (minimum stirrups as above), such that*

$$(0{\cdot}87f_{yq})(A_{sq}/bs)(\sin\theta + \cos\theta) + (0{\cdot}87f'_{yq})(A'_{sq}/bs')(\sin\theta' + \cos\theta')$$
$$\nless q - 0{\cdot}8q_c$$

where f'_{yq}, A'_{sq}, θ' and s' refer to bent-up bars.

In all cases* $q \ngtr 4q_c$ and $q \ngtr u_k/5\gamma_m$

CONCENTRATED LOADS ON CORBELS, ETC. (§6.3.4.3)

If principal load is applied within $2d_1$ from support then*

$$q \ngtr (2d_1/X)q_c \quad \text{as well as}$$
$$\ngtr 4q_c \quad \text{and} \quad \ngtr u_k/5\gamma_m$$

where X = distance either from line of action of load to face of supporting member (corbels, cantilevers, continuous beams) or from edge of bearing load to line of action of support reaction (simply supported beams).

Principal loads for beams should be $\nless$ 70% of total shear. If X for a corbel $\ngtr 0{\cdot}6d_1$ consideration should be given to the anchorage of the main tension steel and provision of horizontal stirrups.

CONCENTRATED LOADS ON SLABS (PUNCHING SHEAR) (§6.3.4.4)

Punching shear, q_p, on critical perimeter distance $\frac{1}{2}d_1$ from boundary of loaded area:

$$\ngtr 3q_c \text{ in all cases.}$$

* Not as draft Code.

If $q_p \not> 2q_c$ and $\not> 1{\cdot}4$ MN/m^2 no shear steel is necessary.
(A_{st}/bd_1 is given by:

b = length of critical perimeter

and A_{st} = area of tension reinforcement passing through all sides of critical perimeter.)

Shear reinforcement (for $2q_c < q_p < 3q_c$) should satisfy:*

$$(0{\cdot}87f_{yq})(\sum A_{sq}/b_p d_1)(\sin\theta + \cos\theta) \not< 0{\cdot}4$$

$$\text{and} \not< q_p - q_c$$

on critical perimeter (see below for all perimeters progressively $\frac{1}{2}d_1$ further out), where

$\sum A_{sq}$ = area of shear reinforcement provided on perimeter being considered.

b_p = length of the perimeter being considered.

q_p = nominal shear stress around the perimeter being considered.

Spacing of reinforcement around the perimeter $\not> \frac{3}{4}d_1$

CONCENTRATED LOADS ON SLABS (SHEAR)*

Shear reinforcement (for $q_c < q < 2q_c$) on perimeters distance d, $1\frac{1}{2}d$, etc., as for punching shear (i.e. with q replacing q_p).

$q \not> q_c$ for outermost perimeter of reinforcement provided.

Torsion (§6.3.5)

ANALYSIS

Assume that torsional rigidity is

$$GJ = 0{\cdot}4EJ$$

where J = torsion constant = $\frac{1}{2}$ × polar moment of inertia of gross concrete section.

RECTANGULAR BEAMS (§6.3.6.2)

The torsional shear stress, q_t, at any cross section is given by

$$_*T = \tfrac{1}{2}q_t x^2(y - x/3)$$

where $_*T$ = design torsional moment for collapse

x = length of shorter side

y = length of longer side

TORSIONAL REINFORCEMENT

Provide torsional reinforcement in the form of closed hoops (rectangular) with *extra* longitudinal reinforcement, wherever torsion is a significant effect.

q_t and $(q + q_t) \not>$ appropriate values in table C4 in *all* cases.

* Not as draft Code.

	Characteristic cube strength				
	15	22·5	30	37·5	45
Max q_t	2·10	2·60	3·00	3·30	3·60
Max $(q+q_t)$	2·90	3·60	4·10	4·60	5·00

Table C4 Limiting *design* stresses for sections reinforced for torsion (MN/m^2 units)

Areas given by

$$_{*}T = Kx_1y_1(0{\cdot}87f_{yh}) \times A_{sh}/s \quad (\text{Put } K = 0{\cdot}60)$$

$$A_{sa} = (A_{sh}/s)(f_{yh}/f_{ya})(x_1 + y_1)$$

where A_{sh} = cross sectional area of two legs of hoop
f_{yh} = characteristic strength of hoop reinforcement (not greater than 425 MN/m^2)
A_{sa} = area of longitudinal reinforcement provided for torsion (distributed evenly inside hoops with 4 bars minimum).
f_{ya} = characteristic strength of longitudinal steel (not greater than 425 MN/m^2)
s = hoop spacing in direction of member (not greater than 0·2 m, x_1 or $\frac{1}{2}y_1$)
x_1 = aller centre line dimension of hoop
y_1 ger centre line dimension of hoop

Clear spacing of longitudinal bars $\not>$ 300 mm.*

Extend longitudinal steel a distance $x_1 + y_1$ beyond point where torsion steel is no longer required.

T- AND L-BEAMS (§6.3.5.3)

T-beams similar to rectangular beams except that

$$_{*}T_c = 0{\cdot}4q_t\sum x^2y$$

where x and y refer to component rectangles of the section, each containing all or part of a reinforcing cage consisting of rectangular hoops and longitudinal steel. The cage provided in the rib must extend into the flange(s). The torsional resistance of a given cage should be approximately in accordance with resistance of rectangle and hence approximately equal to:

$$x_n^2y_nT/\sum x_n^2y_n$$

where x_n = smaller dimension of component rectangle being considered
y_n = larger centre-line dimension of hoops in the cage being considered.

* Not as draft Code.

CLOSED BOX SECTIONS (§6.3.5.4)

As rectangular sections except that

$q_t = {}_{*}T_c/2Ab$

where A = area enclosed by median line of walls of box-section
b = minimum wall thickness

NON-RECTANGULAR CAGES

As rectangular cages except that x_1y_1 is replaced by cage area, and $2(x_1 + y_1)$ by cage perimeter.

Bond and anchorage (§6.3.6)

Bar type & stress	Characteristic cube strength				
	15	22·5	30	37·5	≮45
Plain					
tension	0·9	1·3	1·5	1·7	1·9
compression	1·2	1·7	1·9	2·1	2·4
Medium bond strength					
tension	1·3	1·9	2·2	2·4	2·7
compression	1·6	2·4	2·7	3·0	3·3
High bond strength					
tension	1·6	2·4	2·8	3·1	3·4
compression	2·0	3·0	3·5	3·9	4·2

Table C5 Maximum anchorage bond stress (MN/m^2 units)

ANCHORAGE (AVERAGE) BOND STRESS (§6.3.6.1)

Given by:

$$\frac{\text{Force in bar}}{\text{Anchorage length} \times \text{perimeter}}$$

(Use design value at collapse)

FLEXURAL (LOCAL) BOND (§6.3.6.2)

$$\text{Flexural bond stress} = \frac{{}_{*}Q \pm ({}_{*}M/d_1)\tan\theta}{0{\cdot}8d_1 \times \sum o}$$

where $\sum o$ = sum of perimeters of tension bars
${}_{*}Q$ = design shear force for collapse at section considered
${}_{*}M$ = design bending moment for collapse at section considered
θ = angle between compressive face and tensile steel (= 0 usually)

Use minus sign if M increases numerically in same direction as d_1.

Critical sections for maximum flexural bond stress are:

(*a*) Faces of supports (in all cases)

(*b*) Where tensile bars are curtailed and at points of contraflexure (only if anchorage bond stresses > 0·8 × maximum values in table C5).

Bar type	Characteristic strength of concrete 15	22·5	30	37·5	45 & over
Plain	1·6	2·4	2·8	3·1	3·4
Medium bond strength	2·0	3·0	3·5	4·0	4·4
High bond strength	2·4	3·6	4·2	4·8	5·3

Table C6 Maximum flexural bond stress (MN/m^2 units)

BUNDLED BARS (§6.3.6.3)

Effective perimeter for bond stresses ≯ 1·6 × perimeter for single bar*

CURTAILMENT OF BARS (§6.3.6.4.)

Minimum anchorage generally ≮ d_1 and ≮ 12 bar diameters from reference section (i.e. first section at which the curtailed bars are no longer required for flexure).

Minimum anchorage at simple end supports must satisfy either:

(*a*) ≮ 12 bar diameters beyond centre line of support, or

(*b*) ≮ 12 bar diameters + $\frac{1}{2}d_1$ from face of support provided that straight length ≮ $\frac{1}{2}d_1$, or

(*c*) ≮ 30 mm and ≮ $\frac{1}{3}$ × support width, if flexural bond stress ≯ $\frac{1}{2}$ maximum values.

Tensile reinforcement may be curtailed only if *one* of the following conditions is satisfied at reference section for all possible bending moment distributions at collapse:

(*a*) Anchorage length for design strength of $0{\cdot}87f_y$ is provided for cut-off bars

or (*b*) Shear capacity ≮ 2 × shear force

or (*c*) Continuing bars provide 2 × steel area for flexure

or (*d*) A simplified design procedure (e.g. table 13.2.1) is used.*

LAPS IN BARS (§6.3.6.5)

Lap length ≮ anchorage length

Also ≮ 25 diameters + 150 mm (tension)

or ≮ 20 diameters + 150 mm (compression)

* Not as draft Code.

LOCAL BEARING STRESS AT BENDS (§6.3.7)

Local bearing stress $= F_{st}/Rd$

$\not> 0{\cdot}4u_w$ (unconfined: i.e. side cover $< 2d$) or

$\not> 0{\cdot}67u_w$ (confined) if average stress $\sum F_{st}/Rb \not> 0{\cdot}4u_w$

where R = internal radius of bend (innermost layer)
d = bar diameter
b = breadth of section
$\sum F_{st}$ = sum of design tensile forces in all bars at section for limit state of collapse.

Limit state of deflection (§6.4)

Maximum deflections need not be calculated if span/effective-depth ratio $\not> 60$ and $\not> XYZ(l/d_1)$ (§6.4.2), where

l/d_1 = basic ratio of effective span to effective depth
X, Y, Z = various multipliers (see below)

Deflection:	$\not> L/200$				$\not> L/250$			
Span (metres)	All spans	<10	10	12	14	16	18	20
Cantilever	8	7	—†	—†	—†	—†	—†	—†
Simply supported	24	20	20	18	16	14	12	10
Continuous	31	26	26	23	21	18	16	13

† Values to be justified by calculation.

Table C7* Basic l/d_1 ratios for various spans and support conditions

r'(%) / (f_yMN/m^2)	$\not> 0{\cdot}5$	0·75	1·0	1·25	1·50	2·0	3·0	6·0
250	1·80	1·80	1·62	1·49	1·40	1·24	1·11	1·02
300	1·80	1·57	1·39	1·25	1·13	1·04	0·91	0·85
350	1·57	1·34	1·19	1·07	0·99	0·89	0·79	0·73
400	1·38	1·18	1·04	0·93	0·87	0·78	0·69	0·64
420	1·31	1·12	0·99	0·89	0·83	0·74	0·66	0·61
450	1·22	1·05	0·92	0·83	0·77	0·69	0·62	0·57
460	1·20	1·02	0·90	0·81	0·76	0·68	0·60	0·56
500	1·10	0·94	0·83	0·75	0·70	0·62	0·55	0·51

Table C8 *X*-multipliers for tension reinforcement

$r' = A_{st}/bd$ at midspan (beams) or support (cantilevers)

f_y = characteristic strength *required* for collapse in MN/m^2

* Not as draft Code.

Permanent load / Total load	A_{sc}/A_{st} at midspan $\not>0{\cdot}3$	$0{\cdot}5$	$\not<0{\cdot}7$
0·85–1·00	0·90	1·10	1·30
0·65–0·85	1·00	1·20	1·40
<0·65	1·15	1·35	1·55

Table C9 *Y*-multipliers for long-term effects

Total load = $1{\cdot}0W_d + 1{\cdot}0W_l$

Permanent load = $1{\cdot}0W_d$ + live load of long duration

Category	Shrinkage coefficient (ε_{sh})	Creep coefficient (C_t)	*Z* for deflection: critical	non-critical
1	$<0{\cdot}6\times10^{-3}$	$<3{\cdot}5$	1·0	1·2
2	$<0{\cdot}75\times10^{-3}$	$\not>4{\cdot}0$	0·85	1·0
3	$>0{\cdot}75\times10^{-3}$	$>4{\cdot}0$	0·7	0·85

Table C10 *Z*-multipliers for mix proportions and environment

If deflections are to be calculated, see appendix B and text.

Limit states of local damage (§6.5)

Analysis of framed structures should be based on elastic theory (see appendix B). No allowance need be made for *additional* moments in columns. For bending moments in slabs use either an elastic analysis or 70% of corresponding values for collapse. However, for all beams and slabs, compliance with limit states of collapse together with the following additional requirements also satisfies local damage:

BEAMS IN FLEXURE (§6.5.1.1)

(*a*) Moment redistribution at collapse $\not>$ 10%
(*b*) $f_y \not> 460$ MN/m²
(*c*) Clear distance between longitudinal bars in tensile face $\not>$ 150 mm
(*d*) Clear distance from corner in tension face of beam to nearest longitudinal bar $\not>$ 70 mm
(*e*) Clear distance between longitudinal bars near side faces of "deep" beams (overall depth > 0·75 m): $\not>$ 0·2 m

Effective longitudinal bars for the above are $\not<$ 12 mm diameter.

Note: the above simplified precautions are usually inadequate for an aggressive environment.

SOLID SLABS IN FLEXURE (§6.5.1.2)

(*a*) $f_y \ngtr 460$ MN/m^2

(*b*) If overall depth $> 0{\cdot}25$ m then clear distance between longitudinal bars

$\ngtr 0{\cdot}3$ m

and $\ngtr 1{\cdot}5/r'$mm (150 mm minimum if $r' < 0{\cdot}01$)

where r' = tensile steel ratio on gross concrete area.

SHEAR, TORSION, BOND, ANCHORAGE AND BEARING (§6.5.5)

No additional requirements

FLEXURAL CRACKING (§6.5.4)

Characteristic (maximum) surface crack width = $2{\cdot}3ce$ where c is distance to nearest longitudinal bar or to neutral axis, whichever is the lesser, and

$$e = \varepsilon_c - 10^{-6} \times 4(d_c - d_n)/r(d - d_n)$$

where $r = A_{st}/b_r d_1$

b_r = width of section at centroid of tensile steel

d_c = depth to point at which crack width is required

ε_c is found from an elastic analysis without allowance for stiffening effect of concrete in tensile zone.

THERMAL AND SHRINKAGE CRACKING*

Maximum width of thermal and shrinkage tension cracks in restrained, lightly-reinforced walls and slabs may be found from

$$\Delta_{max} = s_{max}(5\theta_1 + 10\theta_2) \times 10^{-6}$$

where s_{max} = estimated maximum crack spacing

θ_1 = temperature fall from hydration peak to ambient

θ_2 = temperature fall due to seasonal variations

θ_1 should be taken as $\nless$ 30°C for summer concreting (where initial temperatures can exceed 20°C) and $\nless$ 10°C for winter concreting (where initial temperatures are less than 10°C). These values may need to be increased if the cement content exceeds 400 kg/m^3, the wall thickness exceeds 300 mm, or rapid hardening cement is used.

The maximum crack spacing, s_{max}, is given by

$$s_{max} = u_t D/2r'\rho$$

where u_t/ρ = ratio of tensile strength of concrete to bond strength between concrete and reinforcement

D = diameter of reinforcing bars

r' = steel ratio (based on gross concrete section)

$u_t/\rho \simeq$ unity for plain round bars, 4/5 for medium bond strength bars and 2/3 for high bond strength bars.

* Not as draft Code (entire section).

The above formulae assume that the cracking is controlled by the reinforcement. Hence

$$r' \nless r_{crit} \simeq u_t/f_{yt}$$

where u_t = direct tensile strength of concrete at 3 days
f_{yt} = characteristic tensile strength of reinforcement.

u_t may be taken as 50% of the estimated flexural strength values at 7 days given in table 3.3 of the Unified Code.

Flat slab construction (§6.6)

WIDTH OF COLUMN STRIP

width = width of drop (if present)
or = $\frac{1}{2}L$ (if drops are not used)

where $L = \frac{1}{2}(L_1 + L_2)$ and L_1, L_2 are the long and short spans respectively (centre to centre of columns)

COLUMN HEADS

(*a*) Angle of greatest slope of head $\ngtr$ 45° from vertical.
(*b*) D $\ngtr \frac{1}{4}L$ and $\ngtr$ diameter at level of 40 mm below soffit of slab (or drop if present), where D = diameter of circle of same area as cross-section of head.

SLAB THICKNESS

Slab thickness $\nless$ 125 mm in all cases

OPENINGS

Provided that a reduced section can resist the redistributed shear and moment, the sizes of openings which can be made are as follows:

(*a*) Within area common to two intersecting middle strips: maximum dimension parallel to a centre-line of panel $\ngtr$ 0·4*L*.
(*b*) Within area common to two interesecting column strips: Aggregate length or width $\ngtr \frac{1}{10}$ × width of column strip.
(*c*) Within area common to one column strip and one middle strip: Aggregate length or width $\ngtr \frac{1}{4}$ × width of column strip.

Otherwise openings must not encroach upon a column head or drop and must be completely framed on all sides with beams to carry the loads to the columns.

CONTINUOUS FRAME METHOD OF ANALYSIS (limit state of collapse)

(*a*) Divide structure longitudinally and transversely along centre-lines of panels to provide strips of slab which form frames with the rows of columns.

(*b*) Use elastic analysis of each frame in its entirety.
Or, for vertical loads only, consider each strip of floor (or roof) separately assuming that the far ends of the columns above and below are rigidly fixed.
Or, for unrestrained columns, apportion moments due to lateral loading in proportion to column stiffnesses.

(*c*) Use gross concrete cross-section to determine relative stiffnesses of members. Ignore stiffening effect of drops and flared column heads; also solid part of recessed or coffered slabs if solid part does not extend more than $0{\cdot}15L$ into span from centre line of columns.

(*d*) Maximum bending moments in the slabs may be assumed to occur due to either (i) alternate spans loaded and all others unloaded, or (ii) any two spans loaded and all others unloaded. The negative moments may be reduced by not more than 15% provided that the sections possess adequate ductility (as for any framed structure).

(*e*) Assume maximum support moments occur at $\frac{1}{2}D$ from column centre-line. Divide moments between column and middle strips as in table C11.

Strip	Width	Apportionment of total moment	
		Support	Span
Column	$\frac{1}{2}L$	75%	55%
Column	$<\frac{1}{2}L$	May decrease*	May decrease*
Middle	$\frac{1}{2}L$	25%	45%
Middle	$>\frac{1}{2}L$	May increase*	May increase*

* Adjusted moments must ensure that total negative and total positive moments resisted by column plus middle strip are not reduced. In all cases, maximum positive moment plus average of negative moments used in design of any one span of slab, for *whole* panel width, is $\nless (L_1 - 2D/3)^2 WL_2/8$.

Table C11 Bending moments in panels of flat slabs designed as continuous frames

PANELS WITH MARGINAL BEAMS OR WALLS

If supported by marginal beam with $d > 1{\cdot}5d_s$ or by wall then:

(*a*) Beam or wall carries a uniformly distributed load of $\frac{1}{4}$ × total load on panel in addition to those loads which act directly on beam or wall.

(*b*) Bending moments on *half* column strip adjacent to beam or wall are reduced by 50% (i.e. $\frac{1}{4}$ × moments obtained by continuous frame analysis).

COLUMNS IN FLAT SLAB CONSTRUCTION

Design according to continuous frame procedures taking I_b as the second moment of area of width of slab extending over column strip.

SHEAR REINFORCEMENT IN FLAT SLABS

Reinforcement for punching shear only effective if

$d_s \nless 250$ mm

DEFLECTION OF FLAT SLABS

Additional multiplier for tables C7 to C10 equals 0·9 generally, or 1·0 if total width of drop $\nless \frac{1}{3} \times$ span for *either* direction.

LOCAL DAMAGE OF FLAT SLABS

As solid slabs.

REINFORCEMENT FOR FLAT SLABS

Satisfy following conditions, where $L = L_1$ or L_2 as appropriate:

(*a*) Each strip is reinforced over its full width, and slab is reinforced in both directions.

(*b*) Provide full area of positive steel for distance $\nless 0{\cdot}3L$ from centre line of panel (i.e. total length of span steel $\nless 0{\cdot}6L$).

(*c*) $0{\cdot}4 \times$ area of positive steel extends to within $L/8$ of centre line of columns.

(*d*) Provide full area of negative steel for distance $\nless 0{\cdot}2L$ from centre line of columns. Also, average distance for *all* negative steel $\nless L/4$.

(*e*) If $D < 0{\cdot}15L$, two-thirds negative steel in column strip should be placed centrally over column within width of half the column strip.

(*f*) At all discontinuous edges, all steel should extend to within 70 mm of the edge and be provided with right angle bends.

Ribbed and hollow block slabs (§6.7)

THICKNESS OF TOPPING

After allowance for effect of wear if necessary, topping should satisfy the following:

(*a*) Permanent blocks with clear distance between ribs of not more than 500 mm: $\nless$ 30 mm.

(*b*) As (*a*) but with blocks properly jointed with a 1:3 cement-sand mortar of 11 MN/m^2 minimum crushing strength: $\nless$ 25 mm.

(*c*) Other slabs with permanent blocks:
$\nless$ 40 mm and $\nless 0{\cdot}1 \times$ clear distance between ribs.

(*d*) Slabs without permanent blocks:
$\nless$ 50 mm and $\nless 0{\cdot}1 \times$ clear distance between ribs.

RIBS

Ribs should generally be placed along edges parallel to span of one-way slabs. Provide edge rib with width $\not<$ width of bearing along any edge which is built into a wall or rests on a beam. Consider including some reinforcement at right angles to the edge rib. In addition:

(*a*) Rib width $\not<$ 65 mm
(*b*) Rib depth, excluding any topping, $\not>$ 4 × rib width.

BLOCKS AND FORMS

These may be of any suitable material. Blocks which contribute to structural strength of slab must satisfy all the following conditions:

(*a*) Blocks should consist of concrete or burnt clay to BS3921 *Bricks and blocks of fired brickearth, clay or shale.*
(*b*) Crushing strength of blocks on net section when axially loaded in direction of flexural loading in floor slab $\not<$ 14 MN/m^2.
(*c*) *Either* properly joint blocks using 1:3 cement-sand mortar of 11 MN/m^2 minimum crushing strength *or* provide concrete topping $\not<$ 30 mm thick.

REINFORCEMENT IN RIBS

(*a*) Carry through $\not<$ 50% of maximum area of main tensile steel in bottom of rib to supports and satisfy general requirements for curtailment of bars.
(*b*) Satisfy general requirements for distance between bars.
(*c*) "Anti-crack" steel over supports in continuous slabs designed as simply supported $\not<$ $\frac{1}{3}$ × maximum area of steel in adjoining spans extending $\not<$ 0·1 × clear span into adjoining bays.
(*d*) Actual cover $\not<$ 10 mm either laterally (if permanent blocks are present), or underneath (if slip tiles are present). Otherwise cover must be increased as for general requirements.

ANALYSIS OF STRUCTURE

As for slabs generally except that ribbed and hollow tile floors, within building and enclosed against the weather, which are continuous over supports, may be designed as simply supported provided that the support steel is only reduced to nominal "anti-crack" steel.

ANALYSIS OF SECTIONS

As general assumptions except that the compressive stresses in burnt clay blocks under design collapse loads are either

$\not>$ 0·25 × strength given by axial load in appropriate direction, or
$\not>$ 0·30 × strength if 95% blocks can exceed specified strength.

Assume elastic modulus of block material equals that of surrounding concrete.

SHEAR

Allowance for shear strength of permanent hollow blocks which contribute to structural strength can be assessed as follows:

(*a*) Increase thickness of concrete rib by thickness of *one* wall of block.

(*b*) Use shear stress appropriate to the material for *both* walls of block.

LIMIT STATE OF LOCAL DAMAGE

Consider general requirements for beams except that condition for redistribution of moments (10% maximum) can be ignored.

LIMIT STATE OF DEFLECTION

Satisfy general conditions. Allowance may be made for permanent hollow blocks, if appropriate, and if deflections are calculated.

Staircases (§6.10)

Design as beams for limit state of collapse.

LOADING

Assume uniformly distributed over plan area. If two spans cross at right angles (e.g. around open wells) divide load on common area equally between two spans. If staircases or landings are built at least 110 mm into walls along part of their length, reduce loaded area by 150-mm wide strip adjacent to wall.

EFFECTIVE BREADTH OF SECTION

Effective breadth of staircase without stringer beams generally equals actual breadth. However, for any parts along span which are built into a wall, include $\frac{2}{3}\times$ embedded breadth (up to 80 mm maximum) in effective breadth.

EFFECTIVE SPAN (l)

(*a*) Simply supported and without stringer beams:

l = horizontal distance between centre lines of supports.

(*b*) Ends built monolithically into structural member spanning in direction perpendicular to span of staircase:

l = clear horizontal distance between supporting members + half the breadths of supporting members (900 mm maximum additions at both ends).

DEPTH OF SECTION

Depth of section = minimum thickness perpendicular to soffit of staircase.

LIMIT STATES OF DEFLECTION AND LOCAL DAMAGE

As general requirements for beams and slabs.

Lightweight concrete (§6.12)

Some additional restrictions are necessary for lightweight concrete due primarily to a reduced static tangent modulus of elasticity given by:

$$E_c = (850\rho^2\sqrt{u}) \times 10^{-6} \quad (\text{MN/m}^2 \text{ units})$$

where ρ = density of concrete in kg/m³ at time of test and $1440 < \rho < 2300$.

ANALYSIS FOR LIMIT STATE OF COLLAPSE

(*a*) *Short columns:* $h_a/d < 10$ and $h_t/t < 10$

(*b*) *Long columns:* Increase additional moments by factor of 1·5, e.g.

$$M_{cy} = \mu M_{ey} + \frac{3(Wt + \mu M_{ey})}{5000}\left\{\frac{h_t}{t}\right\}^2 \quad \text{etc.}$$

(*c*) *Walls:* Increase additional moments by factor of 1·5 for slender wall (i.e. if $h/t > 10$).

(*d*) *Sections:* Analysis of section generally as for dense aggregate concrete.

SHEAR FOR BEAMS AND PRINCIPAL MEMBERS

$q_c \not> \frac{1}{2} \times$ values in table C3.

Provide shear reinforcement if ${}_*q < q_c$ (minimum only) and if $q_c < {}_*q < 6q_c$. Notice that ${}_*q \not> 6q_c$ (i.e. 3 × values in table C3).

SHEAR FOR SLABS

$q_c \not> \frac{3}{4} \times$ values in table C3.

TORSION

Design torsional stresses $\not> \frac{1}{2} \times$ values for dense aggregate concrete.

BOND AND ANCHORAGE

For reinforcement held horizontally during casting (i.e. usual case) reduce design bond stresses to following values relative to dense aggregate concrete:

(i) Plain round bars $\not>$ 50%

(ii) High-bond and medium-bond bars $\not>$ 80%

For reinforcement held vertically during casting use full values.

BEARING*

Local bearing stress $\not> \frac{2}{3} \times$ values for dense concrete.

* Not as draft Code.

LIMIT STATE OF DEFLECTION

(*a*) Reduce limiting span/depth ratio for dense aggregate concrete by a further factor of 0·85 unless it can be demonstrated experimentally that a reduced depth of section is satisfactory.

(*b*) Calculations for short-term deflection should be based on the reduced modulus for lightweight concrete. The ratio of long-term deflections may be assumed to be the same as for dense aggregate concrete.

Water-containing structures*

GENERAL

See CP2007:1960 (reference 11.5) for definitions and details of water-containing structures, construction joints, etc. The basis of design, however, is superseded by the following clauses.

LIMIT STATE OF COLLAPSE

As for non-water containing structures.

LIMIT STATE OF LOCAL DAMAGE

Compliance with section 5.2.4, items (*e*) and (*f*), for all liquid-containing faces is achieved as follows:

(i) *Class A* The maximum concrete stresses calculated according to elastic theory for an uncracked section do not exceed the following permissible values based on the strength values given in table 3.3 of the Unified Code:

Flexural tension stress $\ngtr \frac{2}{3}$ flexural strength.

Direct tension stress $\ngtr \frac{2}{3}$ indirect tensile strength.

In addition, calculations based on an assumed cracked section must comply with all the requirements for a Class B structure.

(ii) *Class B* Maximum crack widths calculated as in section 6.5 not greater than 0·1 mm.

Maximum stresses in the steel reinforcement for members loaded in direct tension $\ngtr$ 83 MN/m^2.

EXPANSION AND CONTRACTION JOINTS

If movement joints are provided at spacing $\ngtr s_{max}$ then 50% of the total contraction may be assumed to occur at the joints, i.e. the permissible crack width calculated in section 6.5 as if no contraction joints were present may be increased to 0·2 mm for $s_{min}(= \frac{1}{2}s_{max})$. If expansion joints are provided at spacing $\ngtr 2s_{max}$, r' may be not less than $\frac{2}{3}r_{crit}$ (i.e. if expansion joints are separated by not more

* Not as draft Code (entire section).

than one intermediate contraction joint then r_{crit} can be reduced by one third).

LIMIT STATE OF DEFLECTION

As for non-water-containing structures.

Addendum

Anticipated changes to the draft Code or additional requirements considered necessary have been included in the text and indicated by footnotes "not as draft Code". According to the latest information available to the author it is possible that the final version may include the following amendments and additions.

LOAD CARRIED BY PLAIN WALLS

Values of e_{add} may be assumed as follows:

Short braced wall: 0

Slender braced wall: $(h^2/2\,500t) - 0{\cdot}4e_0$

Unbraced wall (i) Top of wall: 0

(ii) Bottom of wall: $h^2/2\,500t$

Use additional coefficient, β, which is unity if $b = 4t$ and 0·8 if $b = t$.

SHEAR AND BEARING FOR PLAIN WALLS

Horizontal shear force in plane of wall should be

either $\not> \frac{1}{4} \times$ vertical load

or $\not>$ 0·3 MN/m^2 (generally and $\not>$ 0·2 MN/m^2 (no-fines).

Shear strength for forces normal to wall satisfactory if

either $H/t > 6$

or they satisfy same conditions as for forces in plane or wall.

LIMITS TO SLENDERNESS

(*a*) *Beams* $L/t \not> 60$

(*b*) *Cantilevers* $L/t \not> 25$

(*c*) *Columns* $H/t \not> 60$ (all cases) and $Hd/t^2 \not> 100$ (unbraced)

(*d*) *Walls*

(i) All plain walls: $H/t \not> 30$

(ii) Reinforced walls

Unbraced: $H/t \not> 30$

Braced (vertical $r' < 1\%$) $H/t \not> 40$

Braced (vertical $r' \not< 1\%$) $H/t \not> 45$

MINIMUM STIRRUPS FOR SLABS (§6.3.5)

Unnecessary if $q < \zeta q_c$ where $\zeta = 1{\cdot}0$, $1{\cdot}1$ and $1{\cdot}2$ for overall slab depth of $\nless$ 250, 200 and $\ngtr$ 150 mm respectively.

CRITICAL CONCRETE SHEAR STRESSES (§6.3.5)

$$(0{\cdot}87 f_{yq}) \frac{A_{sq}}{bs} (\sin\theta + \cos\theta) + 0{\cdot}87 f'_{yq} \frac{A'_{sq}}{bs} (\sin\theta' + \cos\theta')$$
$$\nless q - q_c$$

where q_c has amended values as in table C3A.

$\frac{100A_{st}}{bd_1}$	Characteristic cube strength (MN/m²) 20	25	30	40 or more
0·25	0·35	0·35	0·35	0·35
0·50	0·45	0·50	0·55	0·55
1·00	0·60	0·65	0·70	0·75
2·0	0·80	0·85	0·90	0·95
3·0	0·85	0·90	0·95	1·00

Table C3A Revised q_c values (MN/m² units)

Also $q \ngtr 0{\cdot}75\sqrt{u_k}$.

TORSION (§6.3.5)

Torsion is a significant effect if $q_t >$ Min q_t

$q_t \ngtr$ Max $(q + q_t) \times (y_1/0{\cdot}55)$ for small sections with $y_1 < 0{\cdot}55$ m.

	Characteristic cube strength 15	20	25	30	35	40	45
Min q_t	0·26	0·30	0·34	0·37	0·40	0·43	0·45
Max $(q + q_t)$	3·2	3·5	3·7	4·0	4·3	4·5	4·8

Table C4A Revised q_t values (MN/m² units)

$K = 0{\cdot}80$ (instead of 0·60) in expression for $_*T$

Closed box sections with wall thickness $\nless \frac{1}{4}$ overall width (or height) can be treated as solid rectangular sections.

LOCAL BEARING STRESS AT BENDS (§6.3.7)

$$F_{st}/Rd \ngtr 1{\cdot}5u_w/(1 + 2d/c)$$

where c = clear distance between bars (or groups) *or* cover + d.

TABLE C8

The values in this table may be adjusted slightly.

Y-MULTIPLIERS FOR COMPRESSION REINFORCEMENT

Compression-steel ratio:	0·25	0·50	0·75	1·0	1·5	2·0	>3·0
Factor:	1·07	1·14	1·20	1·25	1·33	1·40	1·50

Table C9A Y-multipliers for compression reinforcement

Compression steel used for table C9A (and not considered for collapse) need not be tied in with transverse reinforcement

Z-MULTIPLIERS FOR FLANGED BEAMS AND HIGH SHRINKAGE OR CREEP

Flanged beams: $b_r < 0{\cdot}30 \times$ flange width $\quad Z_1 = 0{\cdot}8$

$b_r = 0{\cdot}65 \times$ flange width $\quad Z_1 = 0{\cdot}9$

Calculate steel ratio as A_{st}/bd_1 (i.e. use effective flange width)

$Z_2 = 0{\cdot}85$ for shrinkage coefficient $> 0{\cdot}75 \times 10^{-3}$ *or* creep coefficient $> 4{\cdot}0$.

Limit state of local damage (§6.5)

BEAMS IN FLEXURE (§6.5.1.1)

(*a*) Clear horizontal distance between longitudinal bars in tensile face $\not>$ values in table C10A.

(*b*) Clear distance from the corner of a beam or rib to surface of nearest longitudinal bar $\not> \frac{1}{2} \times$ value in table C10A.
For (*a*) and (*b*), bar diameter $\not< 0{\cdot}45 \times$ maximum bar diameter at section.

(*c*) Spacing of longitudinal bars near side faces of "deep" beams (overall depth $> 0{\cdot}75$ m) $\not> 0{\cdot}25$ m for distance from tension face of $\not< \frac{2}{3}$ overall depth.
Effective longitudinal bars for (*c*) have diameter $\not< f_y\sqrt{(d_v b_r)}$ where d_v = vertical spacing of bars.

(*d*) For a particularly aggressive environment only, value of f_y assumed for collapse $\not> 300$ MN/m^2.

f_y	Percentage redistribution to or from section considered										
	−30	−25	−20	−15	−10	0	+10	+15	+20	+25	+30
250	235	245	255	265	275	300	300	300	300	300	300
410	140	150	155	160	170	185	205	215	220	230	240
425	140	145	150	155	165	180	200	210	215	225	235
460	125	130	135	145	150	165	180	190	200	205	215
500	115	120	125	130	140	150	170	180	190	200	210

Table C10A Maximum clear distance between bars (mm)

SOLID SLABS IN FLEXURE (§6.5.1.2)

Bar spacing $\ngtr 3d_1$ (in all cases)

Satisfactory if $f_y \ngtr 425$ MN/m^2 and slab depth $\ngtr 0{\cdot}25$ m

or $f_y \ngtr 500$ MN/m^2 and slab depth $\ngtr 0{\cdot}20$ m

Clear distance between longitudinal bars $\ngtr 1/r' \times$ value in table C10A where r' = tensile steel ratio on gross concrete area and is taken as being between the limits of 1% and 0·5% only.

Long columns and biaxial bending

These expressions (see text, pages 203 and 204) may be amended.

Flat slab construction

Shear strength at each column slab connection may require to be increased slightly for unbraced columns.

Ribbed and hollow block slabs (§6.7)

Voided slabs will be specifically included.

Area of mesh in topping $\nless 0{\cdot}12\%$ of topping in either direction, with spacing of wires $\ngtr \frac{1}{2}$ rib spacing.

LIMIT STATE OF LOCAL DAMAGE

As for solid slabs except that bar spacing requirements can be ignored.

Appendix D

Design stress-strain curves for limit states of collapse

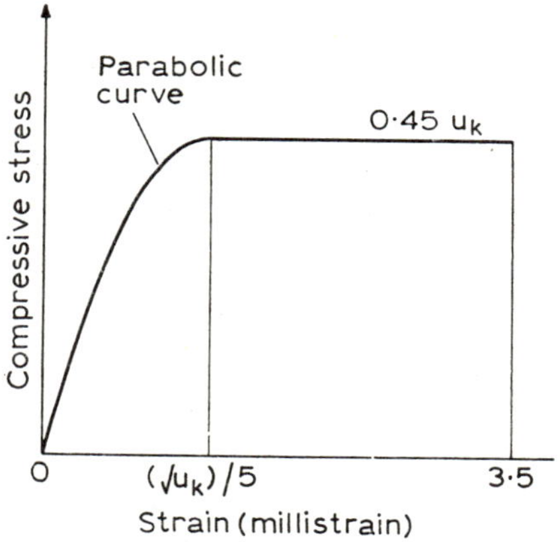

Fig D1 ***Design*** **stress-strain curve for concrete (u_k in MN/m^2 units).**

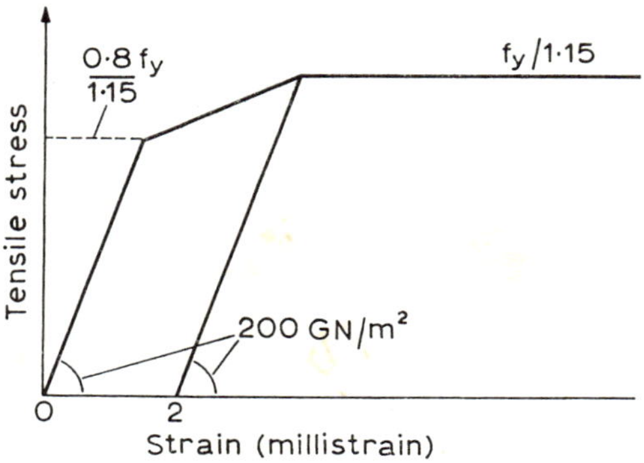

Fig D2 ***Design*** **stress-strain curve for reinforcement in tension.**

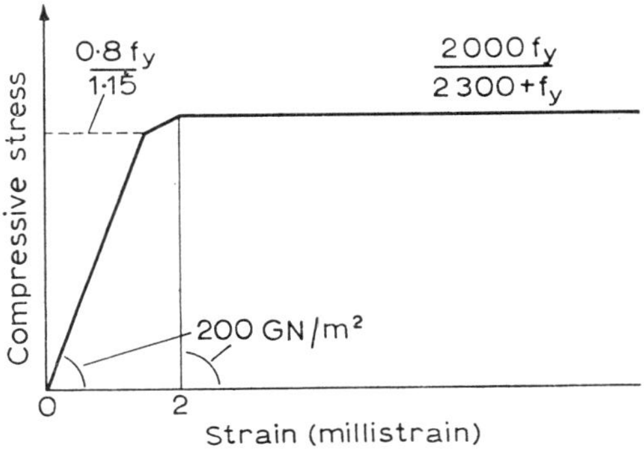

Fig D3 ***Design*** **stress-strain curve for reinforcement in compression (f_y in MN/m^2 units).**

Appendix E

Design charts for collapse

Fig E1 General design chart for singly reinforced rectangular concrete beams.

For $f_y = 250$ use upper full line for each value of u_k.
For $f_y = 410$ use chain line for each value of u_k.
For $f_y = 485$ use lower full line for each value of u_k.

Fig E2 Rectangular reinforced concrete beams. $u_k = 22{\cdot}5$ MN/m^2, $f_y = 410$–425 MN/m^2, $d_2/d_1 = 0{\cdot}10$.

For $f_y = 410$ use lower lines for values of r_c and n_1.
For $f_y = 425$ use upper lines for values or r_c and n_1.
For $f_y = 420$ use interpolation.

Fig E3 Rectangular reinforced concrete beams. $u_k = 30$ MN/m^2, $f_y = 410$–425 MN/m^2, $d_2/d_1 = 0{\cdot}10$.

For $f_y = 410$ use lower lines for values of r_c and n_1.
For $f_y = 425$ use upper lines for values of r_c and n_1.
For $f_y = 420$ use interpolation.

Fig E4 Design chart for symmetrically reinforced rectangular columns. All steel grades; $u_k = 37{\cdot}5$ MN/m^2, $d_1/d = 0{\cdot}90$.

For $f_y = 250$ use outer full lines.
For $f_y = 410$ use broken lines.
For $f_y = 485$ use inner full lines.
For other values use interpolation.

Fig E5 Design chart for symmetrically reinforced rectangular columns. All concrete grades; $f_y = 250$ MN/m^2; $d_1/d = 0{\cdot}90$.

For $u = 22{\cdot}5$ MN/m^2 use outer full lines.
For $u = 52{\cdot}5$ MN/m^2 use inner broken lines.
For other u values use interpolation.
$r'f_y/u = 0$ to $1{\cdot}2$ in $0{\cdot}1$ steps.

Fig E6 Design chart for symmetrically reinforced rectangular columns. All concrete grades; $f_y = 410$ MN/m^2; $d_1/d = 0{\cdot}90$.

For $u = 22{\cdot}5$ MN/m^2 use outer full lines.
For $u = 52{\cdot}5$ MN/m^2 use inner broken lines.
For other u values use interpolation.
$r'f_y/u = 0$ to $1{\cdot}2$ in $0{\cdot}1$ steps.

Appendix F

Design charts for local damage for walls

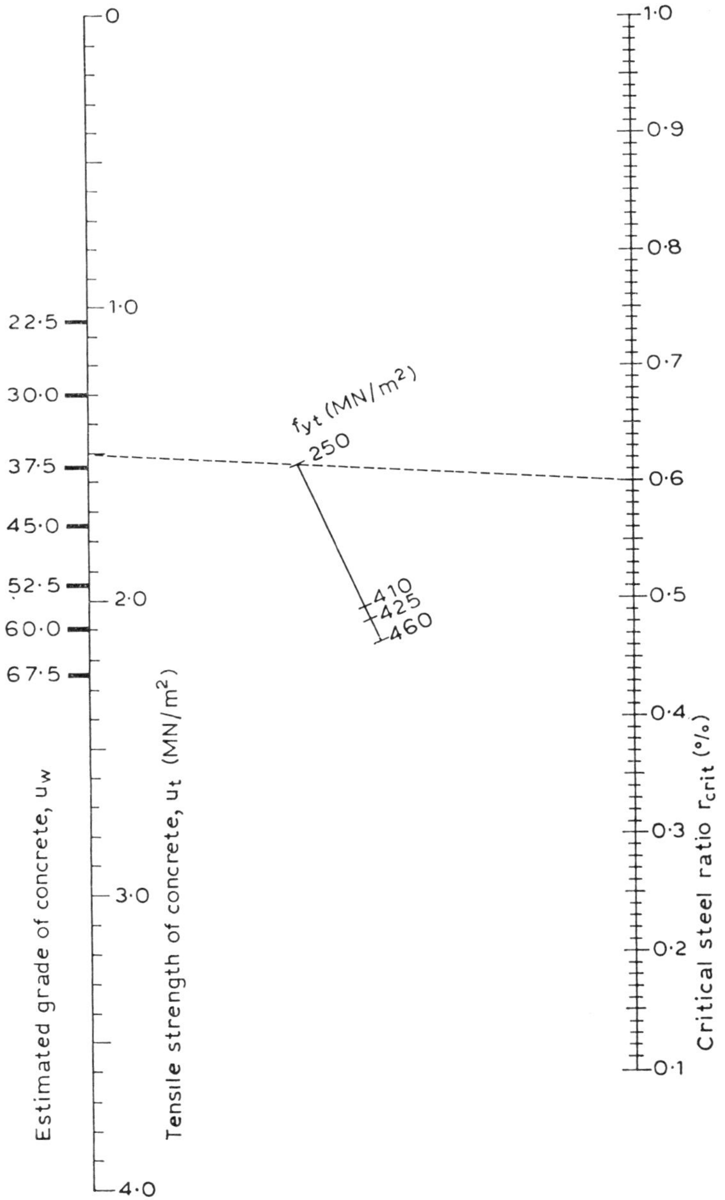

Fig F1 Critical steel ratios for various grades of concrete and steel in reinforced walls, etc.

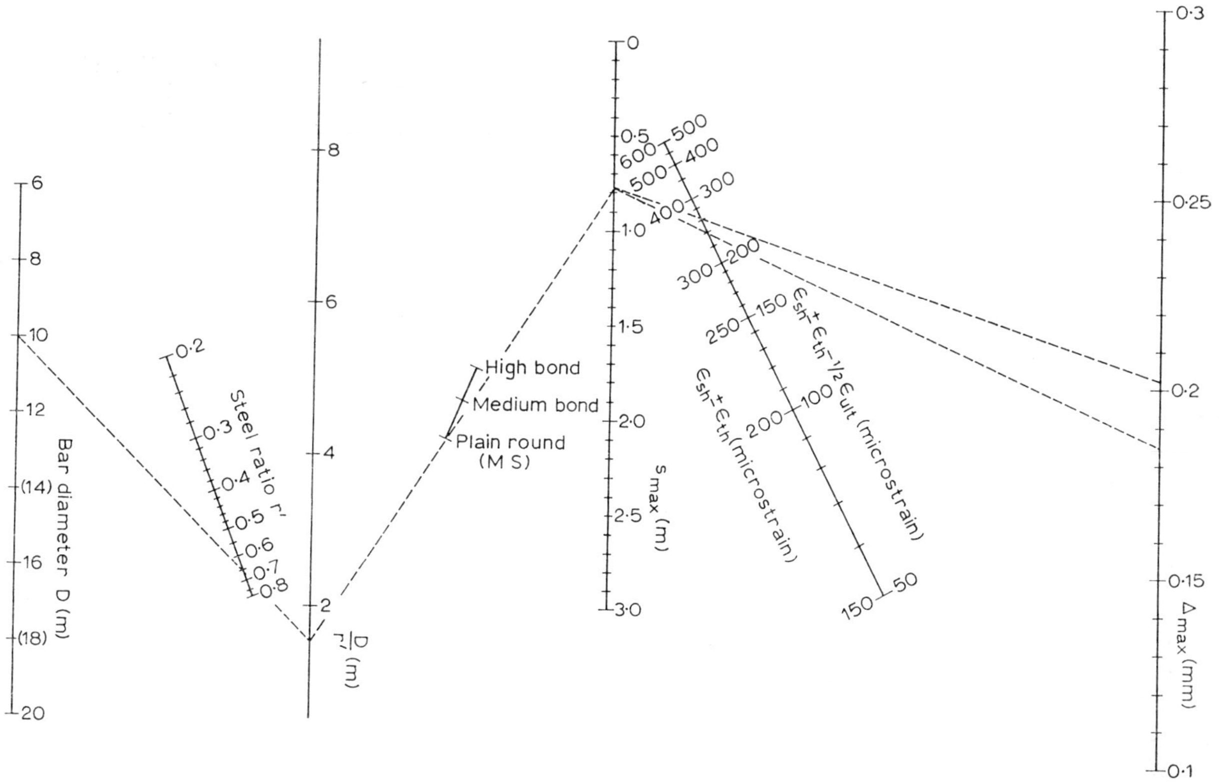

Fig F2 Maximum spacing and widths of tension cracks due to shrinkage and thermal contraction in walls.

Appendix G

Areas and weights of steel reinforcement

No. of bars	Bar diameter (mm) 6	8	10	12	16	20	25	32	40
1	28·3	50·3	78·5	113	201	314	491	804	1260
2	56·6	101	157	226	402	628	982	1610	2510
3	84·9	151	236	339	603	943	1470	2410	3770
4	113	201	314	452	804	1260	1960	3220	5030
5	142	252	393	566	1010	1570	2450	4020	6280
6	170	302	471	679	1210	1890	2950	4830	7540
7	198	352	550	792	1410	2200	3440	5630	8800
8	226	402	628	905	1610	2510	3930	6430	10100
9	255	452	707	1020	1810	2830	4420	7240	11300
10	283	503	785	1130	2010	3140	4910	8040	12600

Table G1 Cross-sectional areas of groups of bars (mm^2)

Bar spacing (mm)	Bar diameter (mm) 6	8	10	12	16	20	25	32	40
50	566	1010	1570	2260	4020	6280	9820	16100	25100
75	377	670	1050	1510	2680	4190	6550	10700	16800
100	283	503	785	1130	2010	3140	4910	8040	12600
125	226	402	628	905	1610	2510	3930	6430	10100
150	189	335	523	754	1340	2090	3270	5360	8380
175	162	287	449	646	1150	1800	2810	4600	7180
200	142	252	393	566	1010	1570	2450	4020	6280
250	113	201	314	452	804	1260	1960	3220	5030
300	94·3	168	262	377	670	1050	1640	2680	4190

Table G2 Cross-sectional areas of bars per metre width for various spacings $(mm)^2$

Bar size (mm)	6	8	10	12	16	20	25	32	40
"Weight"	0·222	0·395	0·616	0·888	1·579	2·466	3·854	6·313	9·864

Table G3 Mass per unit length of round bars in kg/m (or weight per unit length in kgf/m)

Appendix H

Guides for detailing reinforced concrete structures

(According to Technical Report TRCS2 (revised version, 1970). *Standard method of detailing reinforced concrete*, Joint Committee of the Concrete Society and the Institution of Structural Engineers; and BS 4466:1969 *Bending dimensions and scheduling of bars for the reinforcement of concrete.*)

General principles

Drawings are prepared so that the designer can communicate his requirements to the contractor's workmen in a clear and unambiguous manner. Each drawing should be complete in itself and give all the information necessary for the construction of the portion of the work shown, omitting other irrelevant detail. Lettering should be clear and not too small so that it remains legible after the drawing has been used under adverse site conditions. Written instructions should be clear and concise, kept to a minimum and include special details of cambers, sequence of construction etc. Details should be simple and bars should have as few bends as possible. It is preferable to use the largest size bars consistent with the design requirements. The spacing of bars must be considered in relation to the placing of the concrete and the possible use of immersion vibrators. Allow for double the quantity of reinforcement at positions where bars are lapped.

REINFORCEMENT SIZES

Use the following preferred sizes of round area bars: 6, 8, 10, 12, 16, 20, 25, 32, 40 mm. Deformed bars of non-circular cross section are defined by the diameter of a circular bar of equivalent cross-sectional area. Reinforcing bars are available in lengths up to 12 m but 8 or 9 m is the normal maximum and 6 m for bars less than 10 mm in diameter.

IDENTIFICATION OF BARS

The label attached to a particular group or set of bars should give *both* the bar mark number and the schedule reference number. This combination contains the essential information and ensures a unique reference.

NOTATION ON DRAWINGS

Each bar or group of identical bars is defined by the code number or "bar mark". When referring to reinforcement on drawings the sequence of description should be as follows: Number of bars, type of steel, size, mark, spacing, and comment on location within the section. Thus the full description of a group of bars might be:

20R08—63—150*T*

which means 20 round mild steel bars, 8 mm diameter, bar mark 63, spaced at 150 mm centres in the top face of the concrete. This full description should appear on a drawing *once only* either in plan or elevation. Wherever else the group of bars appears on the drawing it should be defined by the bar mark only.

H.C. POSTLETHWAITE AND PARTNERS

Bar schedule reference 0401

Site reference Wigan Public Baths

Date: 1 - 8 - 68

Member	*Bar mark*	*Type & size*	*Nº of members*	*Nº in each*	*Total Nº*	*Length of each bar*	*Shape code*	*A*	*B*	*C*	*D*	*E/R*
Beam in	6	R10	6	33	198	1760	60	230	535			
Fig. H3	9	R10	6	10	60	1525	60	125	535			
			6									
	2	Y12	6	2	12	4040	20	4040				
			6									
	1	Y16	6	2	12	1830	37	305				
			6									
	10	R20	6	4	24	280	20	280				
	5	Y20	6	4	24	2490	20	2490				
	3	Y25	6	2	12	4060	20	4060				
	4	Y25	6	2	12	3120	20	3120				
	7	Y25	6	2	12	7700	20	7700				
	8	Y25	6	2	12	5080	20	5080				

Fig H1 Typical bar schedule.

BAR BENDING SCHEDULES

A bar bending schedule lists the location, mark, type, size, number off, length and bending details of each bar or sheet of welded fabric required for a structure. The bars should be grouped together for each structural element (e.g. beam, column, etc.) as shown by figure H1. Separate schedules are normally prepared for fabric reinforcement. The form of the bar schedule and the shapes of bar used should be in accordance with BS4466 and the more commonly used shapes are given in figure H2. A reinforcement supplier often prepares cutting and bending lists from the schedules.

BAR SCHEDULE REFERENCE NUMBER

Schedules should be set out on A4 size paper and each sheet should be complete in itself. They should have simple consecutive reference numbers not exceeding six characters, and should be cross-referenced to the relevant drawing number. The first three characters can refer to the drawing number, the next two can refer to the schedule number and the last character can be used for revision letters. For example, 2908B refers to drawing 29, schedule 8, revision B. Thus each schedule has a different reference number and only refers to one drawing.

Drawings

BS1192:1969 *Drawing office practice for architects and builders* lays down standards which are generally applicable to reinforced concrete detail drawings and are summarized (with any necessary variations) in the report. The report also recommends various abbreviations.

SIZES OF DRAWINGS

The recommended normal sizes (with measurements in millimetres) are A4(210 × 297), A3(297 × 420), A2(420 × 594) and A1(594 × 841).

TYPES OF DRAWING

The following *types of drawings* are recommended.

(*a*) *General arrangement drawings* showing layout of all reinforced concrete work, setting out dimensions and sizes of all members to a small scale. Where beam sizes are given on plan the first figure should refer to the total depth, including the slab.

(*b*) *Detail drawings*

(i) *Outline or profile drawings* showing dimensions of elements of a structure (e.g. beams, columns, etc.) to a large scale,

Shape code	Method of measurement of bending dimensions	Total length of bar	Sketch and dimensions to be given in schedule
20	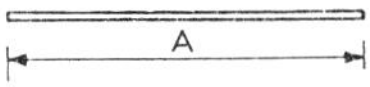	A	Straight
32	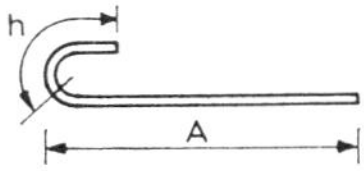	$A+h$	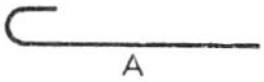
33	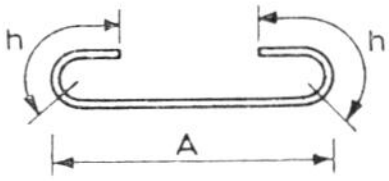	$A+2h$	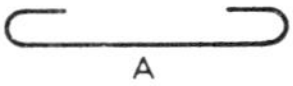
34	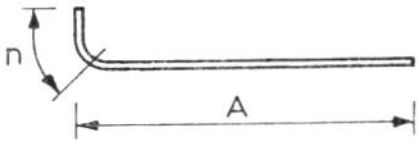	$A+n$	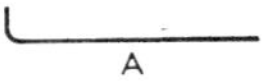
35	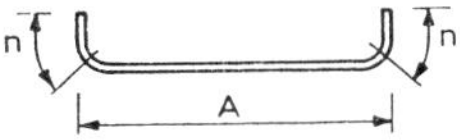	$A+2n$	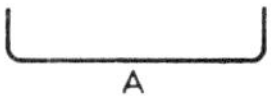
37	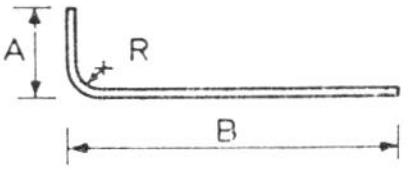	$A+B-\frac{1}{2}R-d$	
38	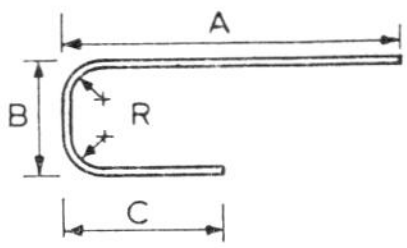	$A+B+C-R-2d$	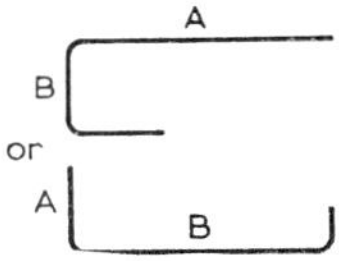
41	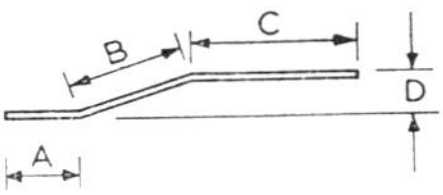	$A+B+C$	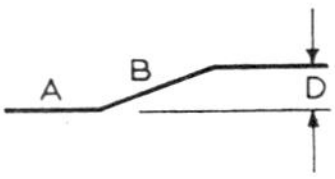

Note: all sketches on this page use outside dimensions (except *A* and *B* for shape code 41, which are along centre line).

Fig H2 Some commonly used shape codes, from BS4466:1969.

Shape code	Method of measurement of bending dimensions	Total length of bar	Sketch and dimensions to be given in schedule
42*	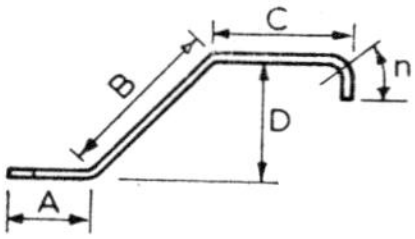	If angle with horizontal is 45° or less: $A+B+C+n$	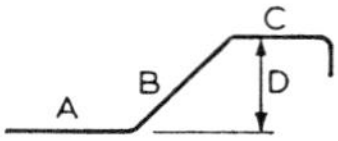
43	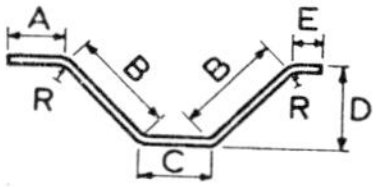	If angle with horizontal is 45° or less: A+2B+C+E	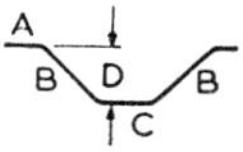
60	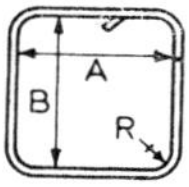	$2(A+B)+20d$	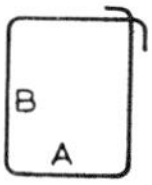
62	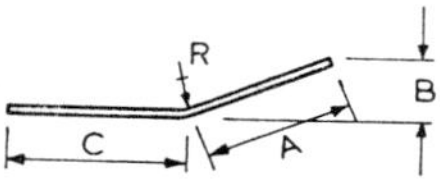	If angle with horizontal is 45° or less $A+C$	
65*	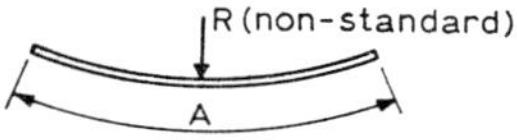	A	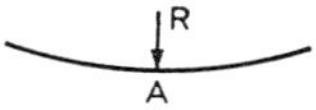
72*	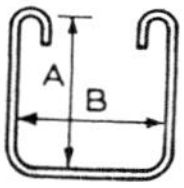	$2A+B+20d$	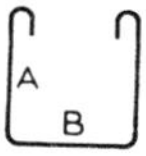
73*	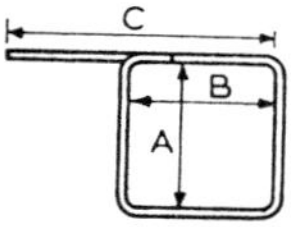	$2A+B+C+10d$	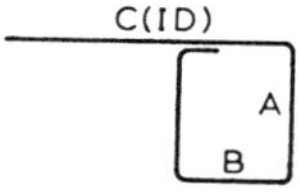
81	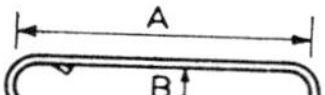	$2A+3B+22d$	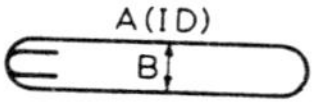
83	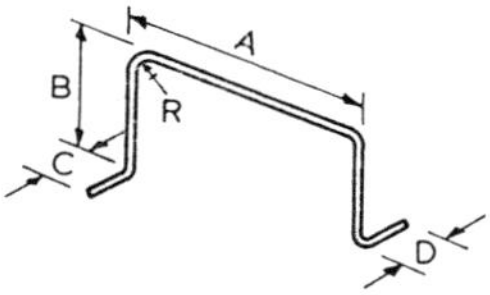	$A+2B+C+D-2R-4d$	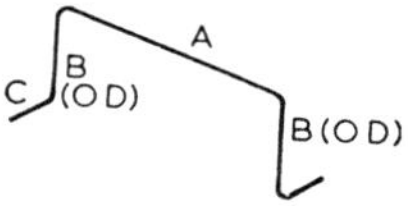

An asterisk indicates that the shape is not designated in the British Standard as a preferred shape. (I.D. = inside dimension, O.D. = outside dimension).

Fig H2 Some commonly used shape codes, from BS 4466:1969. (*continued*) sizes and positions of all holes and fixings, etc., and intended to be read in conjunction with the general arrangement drawings.

(ii) *Reinforcement drawings* showing the position of all the steel in relation to the outline of the structure. A typical example is shown in figure H3.

These two types of drawing may be combined for a relatively simple structure.

MARGINS AND PANELS

All drawings should have a 25-mm filing margin on the left hand side. A 10-mm margin should be provided on the other three sides for all sizes larger than A4. The "title panel" should be 180 mm by 60 mm and should include all the following information: job title, drawing title, job number, drawing number and name of firm. The "additional information panel" may be developed vertically or horizontally from the title panel and may include such additional information as scales, "drawn by", "checked by" date, notes and revisions.

LOCATION AND ORIENTATION

If a portion of the work has to be divided into several drawings it is convenient to have a small diagrammatic key on each drawing with the portion covered by that drawing clearly defined.

The direction of the North point should be clearly shown on all site plans. All other drawings relating to a particular building or major subdivision of a job should have consistent orientation with each other and also with the site plan orientation if possible.

THICKNESS OF LINES, LETTERING AND SPELLING

Lines should generally be about 0·4 mm thick except for main reinforcement (0·6 mm) and dimension and centre lines (0·2 mm). Reinforcement is shown approximately to scale only on cross-sections of bars and in very large scale details. Lettering should be bold and clear, and capital letters should be used for titles, lower case for notes. The spelling of words should be in accordance with the *Little Oxford Dictionary*, using the first spelling where alternatives are given.

STANDARD ABBREVIATIONS

Recommended abbreviations are as follows:

(*a*) *General*

RC	reinforced concrete	FFL	finished floor level
bwk	brickwork	SFL	structural floor level

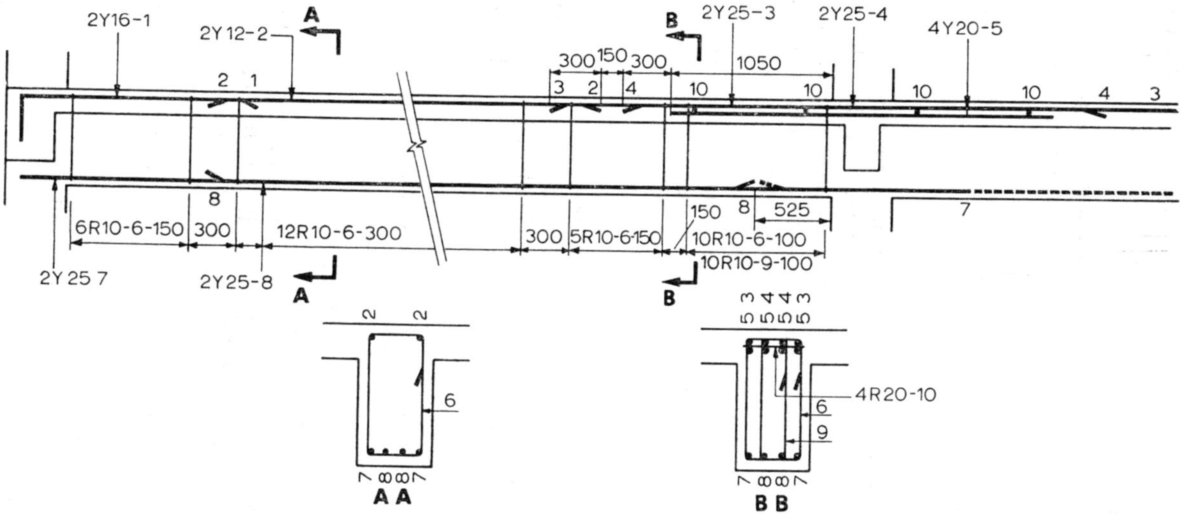

Fig H3 Typical beam detail.

drg	drawing	EL	existing level
FS	full size	hor	horizontal
NTS	not to scale	vert	vertical
		dia	diameter

(*b*) *Type of reinforcement*

R Round mild steel bars
Y High-yield, high bond-strength bars in the metric round area range
X A general abbreviation to cover any other types. X must be suitably described on drawings and schedules and in the specification.

(*c*) *Comments relating to reinforcement*

EW	each way	NF	near face
EF	each face	B	bottom
FF	far face	T	top

DIMENSIONS

The general-arrangement drawing should show all setting-out dimensions and sizes of members. Reinforcement drawings should contain only those dimensions which are necessary for the correct location of the reinforcement (see figure H3). Dimensions should be readable if viewed from either the bottom or the right-hand side of the drawing. For site layout and general-arrangement drawings the recommended unit is the metre, whilst for detail drawings the millimetre is usually more convenient. Values in millimetres should be expressed to an accuracy of 5 mm and it is not necessary to add "mm" after the figure, e.g.

5
415

Values in metres should include "m" after the figure or have 3 decimal places, e.g.

2 m or 2·000

LEVELS

A suitable fixed point should be selected as job datum such that all other levels are positive and are related to it as multiples of 5 or 10 mm as appropriate. Finished floor levels, structural floor levels and existing levels should be indicated on plans thus:

FFL [12·335] SFL [12·000] EL [11·445]

Levels on section and elevation should be projected beyond the drawing with an arrowhead indicating the appropriate line thus:

▽SFL | 12·000 |

If the FFL is referred to, the thickness of the finish should be clearly shown. The job datum must be clearly defined.

SCALES

Recommended scales are as follows:

Site layout and simple general arrangement	1:200
General arrangement	1:100 or 1:50
Simple wall and slab details	1:50
Beam and column elevations	1:20
Beam and column sections	1:20 or 1:10
Details of connections, hinges, etc	1:5, 1:2 or FS

PLANS

Plans shall be drawn as a horizontal section through all structural members immediately above the portion of structure under consideration. Dimension lines should be well clear of the structural details.

ELEVATIONS

Elevations shall normally be drawn as a vertical section immediately adjacent to the element under consideration. Structural members cut by the section shall be shown in full lines and members behind the section in broken lines.

SECTIONS

Sections shall show only the material cut by the section plane. Details behind the section are generally omitted. The reinforcement should be indicated by the bar mark only. The directions of sections shall be to the left for beams and downwards for columns. The section shall be drawn as close as possible to the plan or elevation to which it refers and should be described by letters, readable from the bottom of the drawing as shown in figure H3. Sections through concrete on drawings may be shown as *lightly* shaded areas.

GRID LINES

Grid lines provide a reference system for beams and columns. It is convenient to commence from the top left corner of the drawing using letters for one direction and numbers for the other. Grid lines *do not necessarily* have to be straight or at right angles to each other, or actually pass through beams or columns (but should be sufficiently close for the reference to be obvious). Examples:

column D4 (column at or near intersection of line D and line 4)

beam 4a (beam along or near line 4 and at or adjacent to intersection with line A)

beam a4 (beam along or near line A and at or adjacent to intersection with line 4).

Notice that beams are denoted by lower-case letters although the lines themselves are denoted by upper-ease letters. Additional grid lines can be added to provide subdivision of the main grid and a hyphened prefix can be used if location in elevation is also required, e.g. B-AA99 signifies level B on or near intersection of grid lines AA and 99.

Detailing reinforcement

All bars shall be fully described once only as indicated under *Notation on drawings* and elsewhere defined by their mark number only. Where the position of the bar is not clear the ends may be indicated by a tick and the mark number. For example, all main bars in a layer of a beam may be indicated by a single line in the elevation and ticks, suitably labelled with the mark of the bar concerned, should be used to show the ends of the individual bars (see figure H3). The position of bars should be established by dimensions to the faces of existing concrete or to the formwork. In beams all main bars are shown in full, but in columns only one bar of each type need be shown in full, the others being indicated by short lines. All bars that need to be fixed in a certain part before it can be concreted must be detailed with that part and then shown in broken lines in succeeding portions. The shape of a crank in main compressive steel should not normally exceed 1 in 10. Looped bars as in figure H4(*a*) should be avoided. Better arrangements are shown in figures H4(*b*) and H4(*c*).

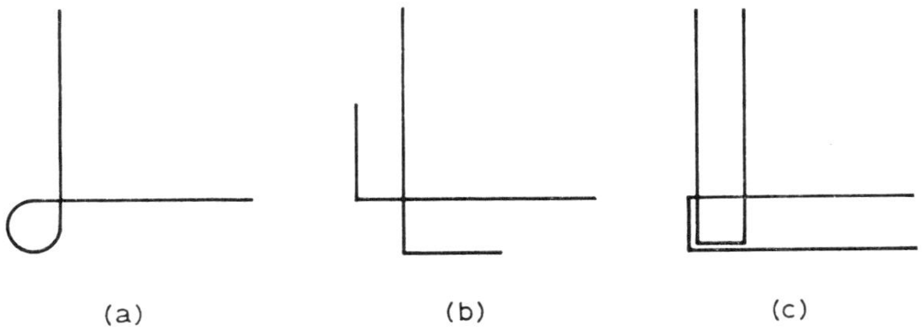

Fig H4 Angle bars.

PREFABRICATION

Detailing should be carried out with the possibility of prefabrication in mind. Reinforcement units should not be so heavy as to exceed the likely capacity of any lifting devices on the particular site and, in the case of cages, the detailing should prevent distortion during handling.

CANTILEVERS

Cantilevers are particularly susceptible to failures due to displacement of the steel during concreting. Main bars must therefore be held firmly and accurately at the support and, preferably, at the free end as well. Cantilevers are also very susceptible to thermal and shrinkage cracks: additional distribution reinforcement at the free edge, closer centres for contraction joints than in the body of the structure, and in large cantilevers the incorporation of distribution bars or fabric in the bottom of the slab, should be considered.

TOLERANCES ON REINFORCEMENT

Cover to reinforcement is liable to variation on account of small errors in the dimensions of shuttering, and in cutting, bending and fixing the steel. All reinforcement shall be fixed to the nominal cover shown on the drawings. The term "nominal cover" implies a

Overall concrete dimension (in direction of tolerance)	Deduction to determine bending dimension	Tolerance on member size
(*a*) *Position of bent stee*		
0–1000 mm	10 mm	$\ngtr$ 5 mm
1000–2000 mm	15 mm	$\ngtr$ 5 mm
over 2000 mm	20 mm	$\ngtr$ 10 mm
(*b*) *At ends of straight bars*		
Any length of bar	40 mm	$\ngtr$ 10 mm

Table H1 Deduction for tolerances in position of bent steel and at ends of straight bars

permissible negative tolerance of 5 mm, i.e. the actual cover may be up to 5 mm less than the nominal cover. When a reinforcing bar is to fit between two concrete faces, (e.g. a rectangular stirrup in a beam) the dimensions on the bar schedule should be determined as the nominal dimension of the concrete, less the nominal cover on each face, and less an allowance for all other errors as in table H1. If the tolerance on member size exceeds the values shown then either the deduction or the nominal cover should be increased.

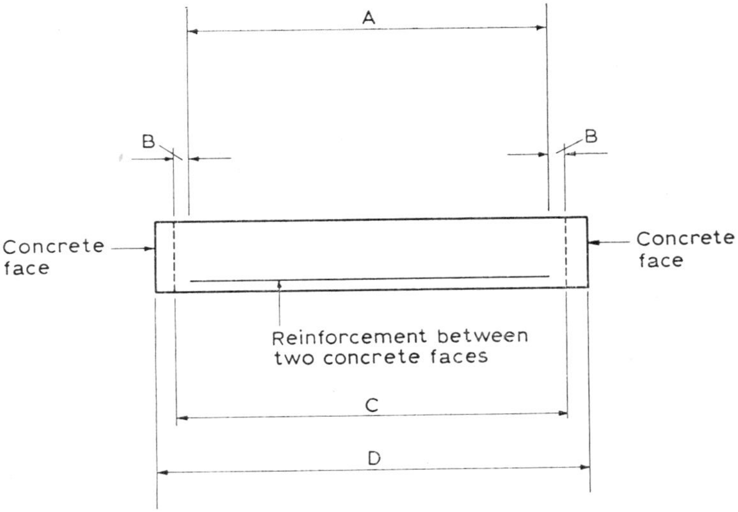

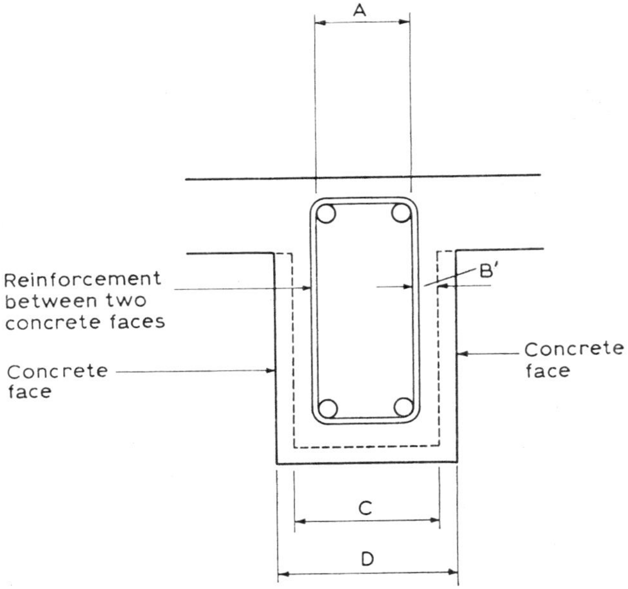

Fig H5 Tolerances.

A = bending dimension to be shown on schedule
= nominal member size less 2 × nominal cover less an allowance for other errors given in table H1.
B = nominal cover.
B′ = nominal cover to main bar (when *B′* is calculated, consideration must be given to the permissible minimum cover to stirrups).
C = nominal size less allowable tolerance.
D = nominal size of member to be shown on drawings.

BAR INTERSECTIONS

Special consideration is necessary at beam-to-column, beam-to-beam and slab-to-beam intersections to ensure that the different sets of bars avoid each other whilst still maintaining the correct cover. A system for achieving this end should be established for projects where several detailers may be working simultaneously on adjacent elements of the structure.

SLABS AND WALLS

Slabs should generally be detailed in plan. Stairs and certain slabs may be more conveniently detailed in section. Walls should be detailed in elevation. In slabs and walls one bar of each type in each group is shown in full, the extent of the group being indicated by short lines. Beam reinforcement should not be shown on plans of slabs, but it should be indicated in the sections. Full lines should be used for all bars detailed on the drawing (e.g. for both top and bottom bars in slabs): only bars detailed on another drawing are shown by broken lines. Top and bottom bars can be detailed on separate plans if necessary but the symbols T and B alone usually provide adequate clarity. Vertical bars in walls are normally in the outside layer on each face (to assist the concrete placing) unless stresses require otherwise. In walls less than 120 mm thick only one layer of horizontal and one layer of vertical bars, in the centre of the wall, need be provided if stresses permit.

Holes up to 250 mm diameter or 150 mm square need not be indicated on the reinforcement drawings, but a suitable note should be added referring to the appropriate general arrangement drawing. It is often adequate if reinforcing bars are merely adjusted on site to miss the holes. Larger holes should be indicated on the reinforcement drawing together with all trimming bars.

If a range of bars has to be detailed with progressively changing length (e.g. a wall tapering in height) it is often convenient to arrange the bars in groups and include a suitable lap so that the taper within a group can be accommodated by increasing the lap above the minimum design requirement.

STIRRUPS AND LINKS

Only one stirrup or set of stirrups in each beam, and one link or set of links in each column, need to be shown in full: short lines are used to indicate the ends of each group at the same spacing. The semi-diameter of the main bars rarely equals the internal radius of the bend at the corner of the stirrup or link containing them (and therefore do not fit snugly in position). This must be allowed for when calculating tolerances so that the main bars can be fixed in their correct positions. Compression bars in beams (as distinct from nominal "hanger" bars of, say, 12 mm diameter) may require an additional type of stirrup to provide the necessary lateral restraint to these bars. All bars in columns must be laterally restrained by links but the centre of the column should be left as free as possible from crossing steel. Transverse ties and nominal "face" bars at intermediate heights are advisable for stiffening the stirrups in very deep beams.

CHAIRS AND SPACERS

Chairs to shape code 83 (see figure H2), or welded continuous chairs, may be used to support steel in horizontal members and must be included in the bar schedules. Chairs must support the main steel and the weight of workmen during construction, e.g. for slabs the diameter must not be less than 10 mm and the spacing not more than 1·25 m in either direction at right angles. U-type spacers should be used to separate two layers of vertical steel in walls having at least one horizontal line of spacers of not less than 6 mm diameter for each lift of concrete. If straight lateral spacer bars are shown between layers of main bars in beams, they should be scheduled.

WELDED FABRIC DETAILING

Steel fabric in flat sheets is sometimes used in slabs and walls and is detailed by drawing the outline of each sheet as a rectangle, with a diagonal line to indicate its extent. The direction of the diagonal indicates whether the sheet is in the top or bottom of the slab. The reference mark of the sheet is written along the diagonal. The amount of side or end lap should be shown on the drawing. Dimensions to show the positions of the sheets should be given on sections where possible, and the fabric in sections is indicated by heavy dashed lines —for oblong mesh, long or short dashes according to whether the section is parallel or at right angles to the main wires, respectively.

Bending moment coefficients for continuous beams

M = Coefficient × total load × span × 10^{-3} (All spans loaded)

Uniformly distributed load

Support	Span	Support	Span	Support	Span	Support	Span	Support	Span	Support
		(107) **113** *125*								
△	*70* **75** (78)	△	*70* **75** (78)	△						
		(85) **90** *100*		(85) **90** *100*						
△	*80* **84** (86)	△	*25* 35 (40)	△	*80* **84** (86)	△				
		(91) **96** *107*		(61) **64** *71*		(91) **96** *107*				
△	*77* **81** (84)	△	*36* **45** (50)	△	*36* **45** (50)	△	*77* **81** (84)	△		
		(90) **95** *105*		(67) **71** *79*		(67) **71** *79*		(90) **95** *105*		
△	*78* **82** (84)	△	*34* **43** (47)	△	*46* **54** (58)	△	*34* **43** (47)	△	*78* **82** (84)	△

Load at centre point

Support	Span	Support	Span	Support	Span	Support
		(160) **169** *188*				
△	*156* **166** (171)	△	*156* **166** (171)	△		
		(128) **135** *150*		(128) **135** *150*		
△	*175* **183** (187)	△	*100* **115** (123)	△	*175* **183** (187)	△

		(137) **145** *161*		(91) **96** *107*		(137) **145** *161*				
△	*170* **178**(182)	△	*116* **129**(136)	△	*116* **129**(136)	△	*170* **178**(182)	△		
		(134) **142** *158*		(101) **107** *119*		(101) **107** *119*		(134) **142** *158*		
△	*171* **179**(183)	△	*112* **126**(133)	△	*131* **143**(149)	△	*112* **126**(133)	△	*171* **179**(183)	△

Load at third points

		(142) **159** *167*								
△	*111* **117**(120)	△	*111* **117**(120)	△						
		(114) **121** *134*		(114) **121** *134*						
△	*122* **127**(129)	△	*33* **46**(53)	△	*122* **127**(129)	△				
		(122) **129** *143*		(81) **86** *96*		(122) **129** *143*				
△	*119* **124**(126)	△	*56* **67**(73)	△	*56* **67**(73)	△	*119* **124**(126)	△		
		(120) **127** *141*		(91) **96** *106*		(91) **96** *106*		(120) **127** *141*		
△	*120* **125**(127)	△	*50* **61**(67)	△	*61* **72**(77)	△	*50* **61**(67)	△	*120* **125**(127)	△

Values in heavy type and in brackets equal coefficients for moment redistributions of 10% and 15% respectively.

Table I1 Bending moment coefficients for dead loads

M = Coefficient × total load × span × 10^{-3}

Uniformly distributed load

Support	Span	Support	Span	Support	Span	Support	Span	Support	Span	Support
△	*96* **93**(92)	(107)**113** *125* △	*96* **93**(92)	△						
△	*101* **99**(98)	(100)**106** *117* △	*75* **70**(68)	(100)**106** *117* △	*101* **99**(98)	△				
△	*100* **97**(96)	(99)**105** *116* △	*81* **76**(74)	(91)**96** *107* △	*81* **76**(74)	(99)**105** *116* △	*100* **97**(96)	△		
△	*100* **98**(97)	(99)**105** *116* △	*79* **74**(72)	(91)**96** *107* △	*86* **82**(80)	(91)**96** *107* △	*79* **74**(72)	(99)**105** *116* △	*100* **98**(97)	△

Load at centre point

Support	Span	Support	Span	Support	Span	Support
△	*203* **198**(196)	(160)**169** *188* △	*203* **198**(196)	△		
△	*213* **209**(207)	(149)**158** *175* △	*175* **168**(164)	(149)**158** *175* △	*213* **209**(207)	△

△	Span 1	△	Span 2	△	Span 3	△	Span 4	△	Span 5	△
		(148) **157** *174*		(137) **145** *161*		(148) **157** *174*				
△	*210* **206**(204)	△	*183* **176**(173)	△	*183* **176**(173)	△	*210* **206**(204)	△		
		(148) **157** *174*		(136) **144** *160*		(136) **144** *160*		(148) **157** *174*		
△	*211* **207**(205)	△	*181* **174**(171)	△	*191* **185**(182)	△	*181* **174**(171)	△	*211* **207**(205)	△

Load at third points

△	Span 1	△	Span 2	△	Span 3	△	Span 4	△	Span 5	△
		(142) **151** *169*								
△	*139* **136**(135)	△	*139* **136**(135)	△						
		(133) **141** *156*		(133) **141** *156*						
△	*144* **142**(141)	△	*100* **99**(98)	△	*144* **142**(141)	△				
		(132) **136** *155*		(122) **129** *143*		(132) **136** *155*				
△	*143* **140**(139)	△	*111* **106**(103)	△	*111* **106**(103)	△	*143* **140**(139)	△		
		(132) **140** *155*		(121) **128** *142*		(121) **128** *142*		(132) **140** *155*		
△	*143* **141**(140)	△	*108* **103**(100)	△	*114* **109**(106)	△	*108* **103**(100)	△	*143* **141**(140)	△

Values in heavy type and in brackets equal coefficients for moment redistribution of 10% and 15% respectively.

Table I2 Bending moment coefficients for variable loads

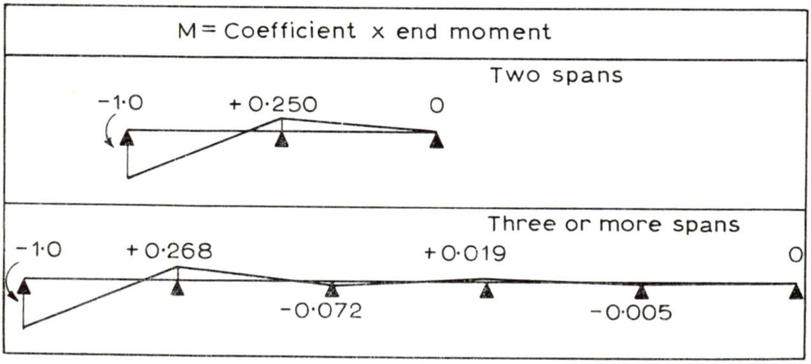

Table I3 Bending moment coefficients for fixed end moments

Curtailment of tensile bars

Sections at which bottom bars are no longer required, for a uniformly distributed load, are indicated in figure J1 and table J1 in terms of the maximum distance from centre of support to point of bending up. If bar is not bent up, sufficient bond distance must be provided from point of maximum stress in bar to point of curtailment. That is, suitable end anchorage must be provided beyond point where bar is no longer required (theoretically), from bending moment diagrams, as indicated in figure J1.

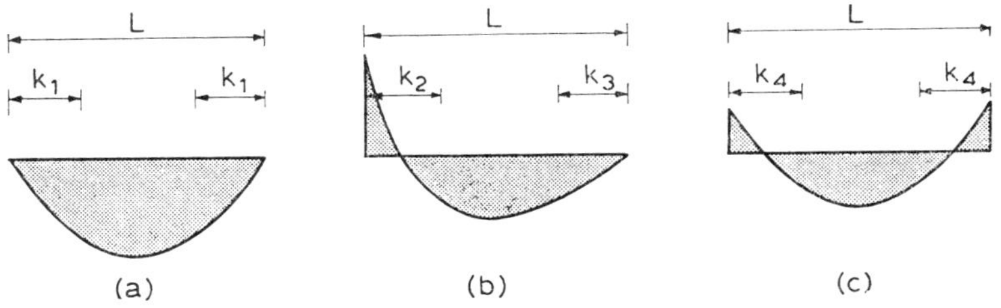

Fig J1 Theoretical positions for "bending up" bars. (a) Simple supports. (b) End span. (c) Interior span.

No. of bars at midspan	k_1 1st	2nd	3rd	4th	5th	6th	k_4 1st	2nd	3rd	4th	5th	6th
1	0						·09					
2	·15	0					·21	·09				
3	·21	·09	0				·27	·16	·09			
4	·25	·15	·07	0			·30	·21	·15	·09		
5	·27	·19	·12	·05	0		·31	·24	·18	·13	·09	
6	·30	·21	·15	·09	·04	0	·33	·27	·21	·16	·12	·09
7	·31	·23	·17	·12	·08	·04	·34	·29	·23	·19	·15	·12
8	·32	·25	·19	·15	·10	·07	·35	·30	·25	·21	·18	·15

Table J1 Order of "bending up" bars

No. of bars at midspan	k_2						k_3					
	1st	2nd	3rd	4th	5th	6th	1st	2nd	3rd	4th	5th	6th
1	·11						0					
2	·24	·11					·13	0				
3	·30	·19	·11				·18	·08	0			
4	·33	·24	·17	·11			·22	·13	·05	0		
5	·35	·28	·21	·16	·11		·24	·15	·09	·04	0	
6	·36	·30	·24	·19	·14	·11	·25	·18	·13	·08	·03	0
7	·39	·32	·26	·22	·18	·14	·29	·20	·15	·11	·07	·03
8	·40	·33	·27	·24	·20	·17	·30	·22	·17	·13	·09	·05

Table J1 Order of "bending up" bars
(*continued*)

Appendix K

Central deflections of beams

Loading	BMD	Maximum moment	Value of k
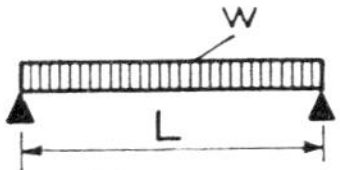	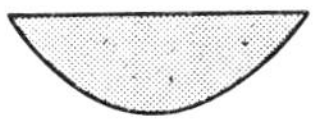	$\dfrac{wL^2}{8}$	$\dfrac{5}{48}$
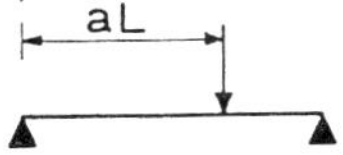	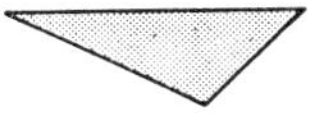	$Wa(1-a)L$	$a \nless \frac{1}{2};\ \dfrac{4a^2-8a+1}{-48a}$
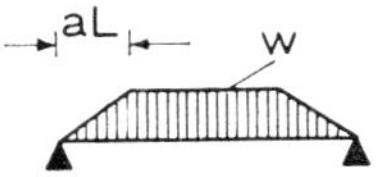	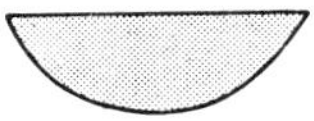	$\dfrac{(3-4a^2)\,wL^2}{24}$	$\dfrac{(5-4a^2)^2}{80(3-4a)^2}$
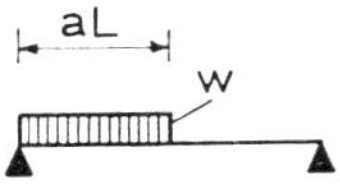		$\dfrac{a^2(2-a)^2wL^2}{8}$	$a \ngtr \frac{1}{2};\ \dfrac{3-2a}{12(2-a)^2}$ $a > \frac{1}{2};$ $\dfrac{4a^4-10a^3+9a^2-2a+\frac{1}{4}}{12a^2(a-2)^2}$
		$-M$	$-\dfrac{1}{8}$
		$-M$	$-\dfrac{1}{16}$
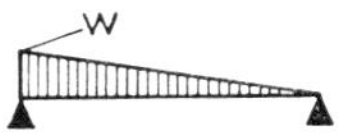	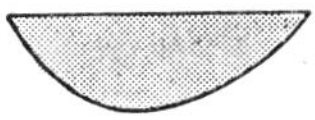	$\dfrac{wL^2}{7{\cdot}81}$	$\dfrac{5}{48}$

Table K1 Maximum bending moments and *k*-values for central deflections of beams

Appendix L

Fixed end moments

Type of loading	One end fixed $-M_A$	Both ends fixed $-M_A$	Both ends fixed $-M_B$
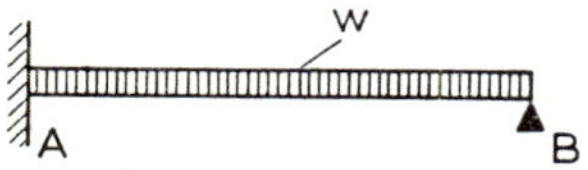	$\frac{wL^2}{8}$	$\frac{wL^2}{12}$	$\frac{wL^2}{12}$
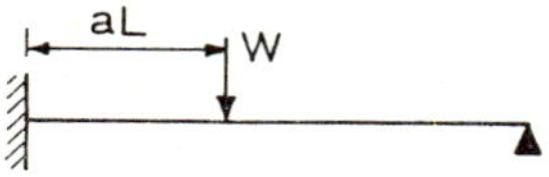	$a(1-a)(1-\frac{1}{2}a)WL$	$a(1-a)^2WL$	$a^2(1-a)WL$
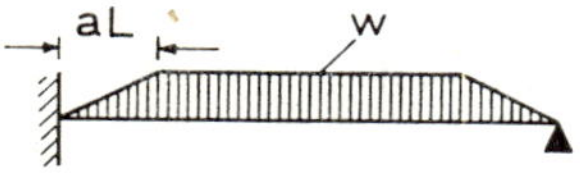	$\frac{(1-2a^2+a^3)wL^2}{8}$	$\frac{(1-2a^2+a^3)wL^2}{12(1-a)}$	$\frac{(1-2a^2+a^3)wL^2}{12(1-a)}$
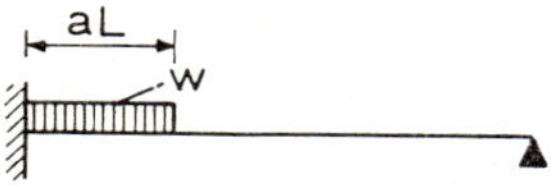	$\frac{a(4-4a+a^2)wL^2}{8}$	$\frac{a(6-8a+3a^2)wL^2}{12}$	$\frac{a^2(4-3a)wL^2}{12}$
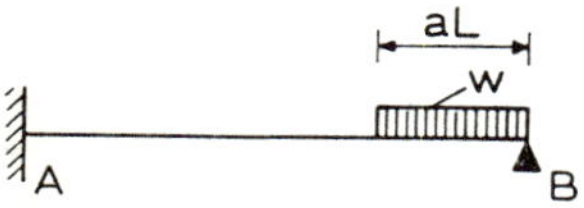	$\frac{a(2-a^2)wL^2}{8}$	$\frac{a^2(4-3a)wL^2}{12}$	$\frac{a(6-8a+3a^2)wL^2}{12}$
	$\frac{wL^2}{15}$	$\frac{wL^2}{20}$	$\frac{wL^2}{30}$
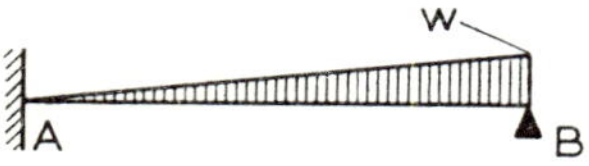	$\frac{7wL^2}{120}$	$\frac{wL^2}{30}$	$\frac{wL^2}{20}$

Note: with one end fixed stiffness should be taken as $\frac{3}{4}I/L$.

Table L1 End moments for beams fixed at one or both ends

Appendix W

Dead loadings

The dead load is the load assumed to be produced by the static weight of all walls, partitions, floors, roofs, and finishes, including all other permanent construction. The dead load allowance for partitions and tanks as given in CP3 chapter V *Dead and imposed loads*, are included in table W1 and the unit weights of some common materials are included in table W2.

Partitions shown on plans	Actual weight
Provision for partitions	Uniformly distributed load per metre: $\nless \frac{1}{3} \times$ weight per metre run of finished portion and $\nless 1$ kN/m^2 (offices only)
Tanks and other receptacles	Actual weights, full and empty

Table W1 Dead load allowances for partitions and tanks

Density (kg/m^3)	Material	Weight (kN/m^3)
800–960	Aerated concrete (typical)	7·85–9·42
2550	Aluminium (cast)	25·0
640	Ashes (dry)	6·28
2080	Asphalt—paving	20·4
1760	Ballast—gravel and sand mixed	17·3
1920	—undisturbed	18·8
1350	Bitumen (prepared)	13·2
	Bricks (common burnt clay)	
1600–1920	—stacked	15·7–18·8
1200	—uncrushed ballast	11·8
1800	—crushed hardcore	17·7
1920	Brickwork—common burnt clay	18·8
2240	—heavy pressed brick	22·0
1200–1360	Cement (Portland)—loose	11·8–13·3
1120–1280	—bags	11·0–12·6
1280–1440	—bulk	12·6–14·1
1040	Clay—(dry fill, lumps)	10·2
2080	—(solid i.e. wet, compact)	20·4

Table W2 Typical unit weights of materials

Density (kg/m³)	Material	Weight (kN/m³)
800	Clinker, or ash (loose)	7·85
	Concrete (plain OPC)	
1840	—brick aggregate (typical)	18·1
1440	—clinker aggregate (typical)	14·1
2240	—stone aggregate (typical)	22·0
2370	Concrete (reinforced)—1% steel	23·2
2420	—2% steel	23·7
2580	—5% steel	25·3
8760	Copper (cast)	85·9
1280	Earth (loamy or dry, loose)	12·6
1760–1840	Earth (moist, compact)	17·3–18·1
1600	Gravel—loose	15·7
1920–2160	—undisturbed	18·8–21·2
1920	Hardcore (well consolidated)	18·8
910	Ice	8·93
7200	Iron (cast)	70·6
2100	Macadam	20·6
1920–2080	Mortar (set)—cement	18·8–20·4
2300	—cement screeds	22·6
1600–1760	—lime	15·7–17·3
1280	Sand—fill, damp when filled	12·6
1600	—dry	15·7
2000	—undisturbed, saturated	19·6
1010–1040	Sea water	9·91–10·20
100	Snow—fresh and dry	0·98
320	—wet compact	3·14
7830	Steel (cast or mild)	76·8
2080	Stone (solid)—Bath	20·4
2640	—granite	25·9
2560	—limestone	25·1
2240	—Portland	22·0
	Stone (crushed uncompacted)	
1600	—granite	15·7
800–880	Timber (seasoned)—English oak	7·85–8·63
650	—Pitch pine	6·38
480	—Softwoods	4·71

Table W2 Typical unit weights of materials
(*continued*)

Appendix X

Imposed floor loadings

(Extracts from the British Standard Code of Practice CP3 *Code of Basic Data for the Design of Buildings*, chapter V part 1:1967 *Dead and imposed loads*).

The imposed load is the load assumed to be produced by the intended occupancy or use, including distributed, concentrated, impact, inertia and snow loads, but excluding wind loads. CP3 chapter V part 1:1967 requires that all beams shall be designed to carry the distributed load which is appropriate for the intended use.

Intended occupancy	Load intensity
Domestic and college bedrooms. Houses. Flats. Dormitories. Stairs (3 storeys or less).	1·5
Hotel bedrooms. Lounges and dining rooms. Hospital wards. Toilet rooms.* Billiard rooms.	2·0
Book stores per metre of storage height. (6·5 kN/m^2 minimum)	2·4
Reading rooms without book storage. Light workrooms. Offices. Car parking for light cars and vans only.	2·5
Chapels and churches. Class rooms. Banking halls.* Minimum for stairs (over 3 storeys). Laundries. Kitchens. Laboratories.	3·0
Offices with computing equipment.	3·5
Public halls, clubs, colleges and theatres with fixed seating.* Shop floors. Libraries. Museums (minimum). Corridors subject to crowd loading. Stationery stores per metre of storage height.	4·0
Public and drill halls, clubs, colleges and theatres without fixed seating. Hotel bars and vestibules.* Gymnasia. Maximum for stairs (over 3 storeys). Cold storage per metre of storage height (15·0 kN/m^2 minimum). Office filing and storage spaces. Minimum for garages generally.	5·0
Boiler rooms. Theatre stages.	7·5
Minimum for foundries.*	20·0
Balconies: same as rooms to which they give access.	

* No concentrated load need be considered for slabs.

Table X1 Distributed loads (kN/m^2)

All floor slabs, or beams, ribs and joists spaced at centres not exceeding 1 metre, shall be designed to carry the appropriate distributed or concentrated imposed load, whichever produces the greater stresses in the part of the floor slab under consideration. The concentrated loads shall be considered to be applied in the positions which produce the maximum stresses and, where deflection is the design criteria, in the positions which produce maximum deflections. The concentrated imposed load need not be considered where the floor slabs are capable of effective lateral distribution of this load.

CONCENTRATED LOADS

Load	*Use*
1·4	Domestic bedrooms. Houses. Flats.
1·5	Balconies per metre run concentrated at the edge.
1·8	Dormitories. Hotel and college bedrooms. Hospital wards. Stairs (3 storeys or less).
2·7	Lounges and dining rooms. Class rooms. Billiard rooms. Chapels and churches.
3·6	Public halls, clubs, colleges and theatres without fixed seating. Gymnasia. Shop floors.
4·5	Reading rooms with or without book storage. Corridors subject to crowd loading. Laundries. Minimum for kitchens and laboratories.
9·0	Minimum for drill halls, vehicle ramps and garages.

Table X2 Concentrated loads (kN), applied over an area 300 mm square

CEILINGS AND SKYLIGHTS

The supports of ceilings, ribs of skylights, etc., which may need to support the weight of a man, shall be designed for a concentrated load of 0·9 kN over an area 125 mm square.

ALLOWABLE REDUCTIONS IN IMPOSED FLOOR LOADS

Reductions in the total imposed floor loads assumed for the beams may be taken as shown in table X3 in designing columns, piers, walls, their supports and foundations. A roof may be regarded as a floor.

No. of floors	1	2	3	4	5–10	over 10
Reduction (%)	0	10	20	30	40	50

Table X3 Reduction in total distributed floor loads

Where a single span of a beam or girder supports not less than 46 m^2 of floor at one general level the imposed load *for the beam* may be reduced by 5% for each 46 m^2 supported subject to a maximum reduction of 25%. This reduction, or that given in table X3, whichever is greater, may be assumed for the columns or other members supporting such a beam. The reductions given by table X3 for factories and workshops designed for 5 kN/m² or more are subject to the limitation that the mean value assumed for all floors should not become less than 5 kN/m².

IMPOSED ROOF LOADS

	Distributed load (kN/m²)	Concentrated load (kN)*
Flat roofs (slope ≯10°) with ready access	1·5	1·8*
Flat or sloping roofs (slope ≯30°) without ready access	0·75	0·9*
Sloping roofs, slope ⩾75°	0·0	0·0
Sloping roofs, 30°< slope <75°	Use linear interpolation	

* applied over an area 300 mm square.

Table X4 Imposed roof loads other than wind loads

DYNAMIC LOADING

	Electric overhead crane	Hand-operated crane
Vertical loads	25%†	10%†
Horizontal force transverse to rails‡	10%*	5%*
Horizontal force along the rails‡	5%†	5%†

‡ at rail level.
† % of maximum static wheel load.
* % of combined weight of crab and load lifted.

Table X5 Dynamic loading, for crane gantry construction for single-crane operations

PARAPETS AND BALUSTRADES

Light access stairs ≯0·6 m wide	220
Light access stairs >0·6 m wide; Domestic landings and balconies	360
All other stairways and parapets	740
Panic barriers	3000

Table X6 Minimum horizontal forces (N/m run) **acting at handrail or coping level**

Appendix Y

Sign conventions

CONVENTION (1): used for reinforced or prestressed concrete when considering the direct forces in the steel and the concrete separately.

(*a*) Steel. Tensile stresses and forces are positive.

(*b*) Concrete. Compressive stresses and forces are positive.

This enables a rigid sign convention to be followed which conforms with the natural and logical use of the materials and therefore avoids any anomalies. For example, the prestress force in prestressed concrete is consistently positive both for the steel (which is always tensile) and the concrete (which is always compressive) since the object of prestressing is to produce compression in the concrete (which is weak in tension).

CONVENTION (2): used for all sections *when considered in their entirety*.

Tensile stresses and forces are positive.

Angles are measured clockwise positive from x-direction.

y is always positive when in direction at 90° clockwise from positive x-direction (i.e. y-axis positive downwards for beams).

A shear stress τ_{xy} is positive if it acts in the positive y-direction on the x-plane whose outward drawn normal is in the positive x-direction (hence $\tau_{xy} = +\tau_{yx}$).

(Notation for plates and slabs: first suffix indicates plane, second suffix indicates direction; for example, τ_{xy} is shear stress on plane normal to x-axis, acting in y-direction).

Appendix Z

SI units

The Système International d'Unités (SI) is essentially an extension and refinement of the traditional metric system. It embodies features which make it logically superior to any other system as well as practically more convenient: it is rational, coherent and comprehensive.

The metric system spread to several countries following the aftermath of the French Revolution and began to be adopted in scientific work in English-speaking countries towards the end of the nineteenth century. Over the years much thought was given to its extension and improvement until finally, in 1960, the Conférence Générale des Poids et Mesures, the body responsible for maintaining standards of measurements (of which the UK is an active participant), formally approved SI. The metrication of the construction industry in the UK is programmed to be effectively complete by December 1972. Nearly thirty countries have already decided to make SI the only legally accepted system and it is clearly destined to become the universal currency of science and commerce.

MAIN FEATURES OF SI

The main features of SI can be summarized as follows:

1. There are six basic units, the metre and kilogramme taking the place of the centimetre and gramme of the old metric system.
2. The unit of force, the newton ($kg\,m/s^2$), is independent of the Earth's gravitation and the often confusing introduction, in some branches of science and technology, of g into equations is no longer necessary.
3. The unit of energy in all forms is the joule (newton $\times$ metre), and of power, the joule per second (watt); thus the variously defined calories, together with the kilowatt hour, the Btu and the horsepower, are all superseded.
4. "Electrostatic" and "electromagnetic" units are replaced by SI electrical units.
5. Multiples of units are normally to be restricted to steps of a thousand and, similarly, fractions to steps of a thousandth. A prefix should normally be used in the numerator only.

BASIC SI UNITS

There are in all six basic SI units: metre, kilogramme, second, ampere, degree Kelvin and candela. The abbreviations are m, kg, s, A, °K and cd respectively. Derived units with special names include newton, joule, watt, volt, ohm, degree Celsius. (N, J, W, V, Ω, °C). Derived units without special names include stress (N/m^2), moment (Nm), etc. Selected conversion factors are given on the attached list.

PREFIXES

The names and symbols of the prefixes representing numerical factors are as follows:

k kilo	$= 10^3$	m milli	$= 10^{-3}$
M mega	$= 10^6$	μ micro	$= 10^{-6}$
G giga	$= 10^9$	n nano	$= 10^{-9}$
T tera	$= 10^{12}$	p pico	$= 10^{-12}$

These prefixes are both convenient in obviating the need to write large numbers of zeros (or high powers of ten) and also helpful in establishing familiarity with the numerical framework of modern science. Notice that the recommended prefixes are limited to $10^{\pm 3n}$. Notice also that the use of a prefix constitutes a new unit: thus mm^2, for example, $= (10^{-3}m)^2 = 10^{-6}m^2$ and *not* $10^{-3}(m)^2$.

DUPLICATED PREFIXES

SI recommends that wherever possible any numerical prefix should appear in the numerator of an expression and that two prefixes should not be used. For example, the units of stress and pressure should be expressed as N/m^2, kN/m^2, MN/m^2 etc. There are two reasons for this. First, there are fewer chances of making errors and, second, there are fewer demands on the memory. For example, consider a typical value of E for steel of 200 GN/m^2. This value can be remembered by two items only: the number 200 and the prefix giga, since one is referring them to a constant reference datum consisting of basic units. Compare this with the three chances of error when writing down 200 kN/mm^2 in terms of basic SI units as $200 \times 10^3/(10^{-3})^2$.

ADDITIONAL PREFIXES

SI allows the use of the following additional prefixes for units involving cubes or higher powers of the basic unit:

d deci	$= 10^{-1}$	da deka	$= 10$
c centi	$= 10^{-2}$	h hecto	$= 10^2$

There is a strongly felt need in such cases since the intervals between units involving cubes, for example, become as great as $10^{\pm 9}$, and with higher powers the intervals between powers of the basic units are even greater. Thus second moments of area may be expressed in terms of dm^4 or cm^4 as well as m^4 and mm^4.

UNITS OF TIME

The common units of time will persist. It is clear that many human activities are related to daily and yearly cycles and that it would be irrational to express all units in terms of ks, Ms, etc. Years, days and possibly hours should be retained.

SI UNITS—Selected conversion factors

	British	*SI*	*Reciprocal*
Length	1 in	$25{\cdot}4 \times 10^{-3}$m	39·37
	1 ft	0·3048 m	3·281
Area	1 in^2	$645{\cdot}16 \times 10^{-6}m^2$ (*1*)	1550
	1 ft^2	$92{\cdot}90 \times 10^{-3}m^2$	10·76
Volume	1 ft^3	$28{\cdot}32 \times 10^{-3}m^3$ (*2*)	35·31
	1 yd^3	0·7646 m^3	1·308
	1 gal	$4{\cdot}546 \times 10^{-3}m^3$	220·0
2nd Moment of area	1 in^4	41·62 cm^4	$2{\cdot}403 \times 10^{-2}$
Mass	1 lb	0·4536 kg	2·205
	1 slug	14·59 kg	$68{\cdot}52 \times 10^{-3}$
Density	1 lb/ft^3	16·02 kg/m^3	$62{\cdot}43 \times 10^{-3}$
Force	1 lbf	4·448 N	0·2248
	1 tonf	9964 N	$100{\cdot}4 \times 10^{-6}$
Force/unit length	1 lbf/ft	14·59 N/m	$68{\cdot}52 \times 10^{-3}$
Pressure, stress	1 lbf/ft^2	47·88 N/m^2 (*3*)	$20{\cdot}88 \times 10^{-3}$
	1 lbf/in^2	6895 N/m^2	$145{\cdot}0 \times 10^{-6}$
	1 tonf/in^2	$15{\cdot}444 \times 10^6$N/m^2	$64{\cdot}75 \times 10^{-9}$
Moment, torque	1 lbfin	0·1130 Nm	8·850
	1 tonfft	3037 Nm	$0{\cdot}3293 \times 10^{-3}$
Work, Energy	1 ftlbf	1·356 J (*4*)	0·7376
Heat	1 Btu	1055 J	$947{\cdot}9 \times 10^{-6}$
Power	1 HP	745·7 *W* (*5*)	0·001341
Temperature	1 degF	5/9 degC (Celsius)	9/5

	Old metric	*SI*	*Reciprocal*
Area	1 cm^2	$10^{-4}m^2$	10^4
Force	1 kgf	9·807 N	0·1020
Pressure, stress	1 kgf/cm^2	98070 N/m^2	$10{\cdot}20 \times 10^{-6}$
	1 bar	10^5 N/m^2	10^{-5}
Heat	1 cal	4·187 J	0·2388
Dynamic viscosity	1 poise	10^{-1}Ns/m^2	10

Notes

(*1*) $10^{-6}m^2 = 1\ mm^2$
(*2*) $10^{-3}m^3 = 1$ litre
(*3*) 1 MN/m^2 = 1 N/mm^2
(*4*) 1 joule = 1 mN. Note that dimensionally a moment of 1 Nm equals 1J
(*5*) 1 watt = 1 J/s

Answers to problems

Chapter 1

1 68·6 (exact value, 68·54); 4·74

Chapter 3

1 116 kNm

2 (*a*) 102·5 kNm ($d_n = 0{\cdot}242$ and $M_{rc} = 159$ kNm)
(*b*) 118 kNm ($d_n = 0{\cdot}114$, M_r for collapse = 189 kNm)

Chapter 4

1 150·3 kNm; 1·43%

2 434 mm^2 compression; 1930 + 360 mm^2 tension

Chapter 5

1 (*a*) ($Q = 36{\cdot}6$ kN); 975 kN/m^2; 205 mm ($0{\cdot}4 > q - 0{\cdot}8q_c$);
No, since $\frac{3}{4}d_1 = 188$ mm
(*b*) 5 bars $\sum o = 314$ mm, $f_b = 0{\cdot}58$ MN/m^2

2 ($Q = 380$ kN); 3·38 MN/m^2; 185 mm; 38 mm; 405 mm

3 (*a*) 224 mm. (*b*) 158 mm

Chapter 6

1 (*a*) −277 (support) and +204 kNm (3·66 m from A).
(*b*) $d_1 = 0{\cdot}49$; $A_{st} = 2055$ mm^2; $A_{sc} = 266$ mm^2
or $d_1 = 0{\cdot}52$; $A_{st} = 1980$ mm^2; $A_{sc} = 0$, etc.
($d_1 = 0{\cdot}49$ or thereabouts slightly preferable, economically)

Chapter 7

1 $1{\cdot}6 \times 10^6$ lbfin (181 kNm); under-reinforced beam

2 (*a*) 522 kNm (*b*) 432 kNm
Differences due primarily to partial safety factors for material strength but also to differences in stress-strain curves.

Chapter 8

1 (*a*) Yes, $M_{rt} = 39\,500$ lbfft ($> M = 39\,050$ lbfft)
(*b*) No, $M_{rc} = 44\,500$ lbfft ($\ngtr M = 46\,100$ lbfft)

2 See chapter 3, 2(*b*)

3 ($M_0 = 102{\cdot}9$ kN m; singly reinforced $M_{rc} = 63{\cdot}3$ kN m; $A_{st} = 922$ mm^2; excess moment, $M' = 39{\cdot}6$ kN m; $A_{st}' = 547$ mm^2) $A_{sc} = 648$ mm^2; $A_{st} + A_{st}' = 1469$ mm^2.

Chapter 9

1 (*b*) (i) 970 kN (ii) 1 020 kN
2 $b = H/3$; $h = 2H/3$
3 2·11 MN ($d_n = 0{\cdot}245$); 6%
4 1·515 MN ($d_n = 0{\cdot}26$); 11%
5 ($r = 5{\cdot}6\%$ for direct load and $< 5\%$ for moment) 3 500 mm^2 (e.g. 4 No. 40 mm dia bars)
6 324 kips (1·44 MN)

Chapter 10

1 0·030 in (0·77 mm); 925 lbf/ft run

Chapter 11

1 (ii) 0·245 mm
2 (i) 0·327 millistrain; (ii) 31·6 MN/m^2, 0·88 m
3 (*a*) 3·18, 2·77 (f_{ct}), 15·2 MN/m^2
(*b*) 5·05, 69·2 MN/m^2; 0·11 mm
Neither criterion is satisfactory but (*b*) could be made so by using smaller diameter bars at closer spacing.

Chapter 13

1 254 lbf/ft^2 (12·2 kN/m^2)
2 $M = wl^2/72$
3 $l = 2{\cdot}95$ m; $m = 35{\cdot}3$ kN m/m
4 143 lbf/ft^2 (6·85 kN/m^2)
5 +27·3%; 0%
6 $m_x' = 31{\cdot}1$ kN m/m at E; $m_x = 89$ and $m_y = 11{\cdot}1$ kN m/m at F; Yes, ratio = 1·95 for 15% reduction in m_x' at E
7 $\alpha = A\beta$; $\beta = \dfrac{-B \pm \sqrt{[B^2 + 24(m + m')m/\lambda^2]}}{8(m + m')/\lambda^2}$

where $A = \sqrt{(1 + m'/m)}$; $B = (m + m')/2A + m(1 + \frac{1}{2}A)$
$\beta = 0{\cdot}211$; $\alpha = 0{\cdot}299$; $w = 47{\cdot}9$ kN/m^2

Index